China and Charles Darwin

HARVARD EAST ASIAN MONOGRAPHS
100

CHINA AND CHARLES DARWIN

JAMES REEVE PUSEY

Published by COUNCIL ON EAST ASIAN STUDIES, HARVARD UNIVERSITY, *and distributed by* HARVARD UNIVERSITY PRESS, *Cambridge (Massachusetts) and London* 1983

The Council on East Asian Studies at Harvard University publishes a monograph series and, through the Fairbank Center for East Asian Research and the Japan Institute, administers research projects designed to further scholarly understanding of China, Japan, Korea, Vietnam, Inner Asia, and adjacent areas. Publication of this volume has been assisted by a grant from the Shell Companies Foundation.

Library of Congress Cataloging in Publication Data

Pusey, James Reeve, 1940–
China and Charles Darwin.

(Harvard East Asian monographs ; 100)
Bibliography: p.
Includes index.
1. China–Civilization–Occidental influences.
2. Social Darwinism. I. Title. II. Series.
DS721.P84 1983 951 82-23264
ISBN 0-674-11735-2

To my parents
Anne W. Pusey and Nathan M. Pusey,
to my mother-in-law
Wang Ch'en Wan,
and to the memory of my father-in-law
Wang Teh-chao

Contents

Acknowledgments

The thesis from which this book has come took seven years—lean or fat, a long time. But I owe my thesis advisors, Professors Benjamin I. Schwartz and John K. Fairbank, thanks for many more than seven years' good advice, inspiration, help, and friendship. Professor Schwartz's own work more than that of any other helped excite me towards my topic. Professor Fairbank's staggering efficiency at reading manuscripts and sending missives of encouragement to keep going kept me going.

Somehow six more years have now slipped by, and the list of people who have helped me with what is at last a book has lengthened. But still at the head of that list must surely come my now, alas, late father-in-law, Professor Wang Teh-chao, who helped me tremendously in Hong Kong, where I did the bulk of my research, and who continued to help thereafter, sending materials, answering queries, and solving infuriating footnote puzzles. He read the thesis; would that I had worked fast enough to give him a copy of this book.

I am also most grateful to my mother-in-law for that splendid year in Hong Kong and for much help since, sending books (and Chinese Care packages). My uncle-in-law, Mr. Wang Tzu-liang, also sent me very useful books, from Taipei.

I owe most special thanks to Professors Nicholas R. Clifford and Yü Ying-shih for more things than I can here enumerate. I know I did not manage to follow all their suggestions, so I want to state most sincerely that neither they, nor my father-in-law, nor anyone else but myself is to blame for any sins of commission or omission in this work. Most especially no one else is to blame for my conclusions.

I am grateful to two directors of the Middlebury Chinese School, Professors T. T. Ch'en and Helen Lin, who let me work on thesis or book while I taught there, to Professors Hans Beilenstein and Kuo Cheng-chao, to Mr. Wang Tzu-ch'un, and to Miss Hu Chia-yang.

I am also forever indebted to four intrepid typists, Elinor Bryant, Ethel Hartman, Margaret Marx, and my wife; to three *courriers extraordinaires,* Jerome A. Cohen, Stephen A. Orleans, and Jamie Horsley; and to one calligrapher, Diana Wang.

The reader should join me in thanking Katherine Frost Bruner for a splendid index. For, after one day's attempt, I so despaired of preparing an index myself that I was ready to offer the reader as my only bit of help the advice of the King of Hearts to the White Rabbit: "Begin at the beginning and go on till you come to the end; then stop." Actually, that is still my "best advice," but nonetheless I am thankful to Mrs. Bruner for, with such decisive skill, making it possible to look things up in this book.

There is only one other producer of this book to whom I owe greater thanks, and that is my editor, for whose erudition, patience, good taste, good cheer, and good eyesight I have endless admiration. When this struggle began, I was pathologically editor-shy, afraid that any editor would mince my words. But I have been cured. I now believe that no author could have greater good fortune than to have an editor—like Florence Trefethen.

Institutions have also helped me: New Asia College, of the Chinese University of Hong Kong; Academia Sinica, in Nankang; the Harvard-Yenching Library, and, of course, the Fairbank Center for East Asian Research at Harvard. The United States Government (and the American people) helped me learn Chinese, get through graduate school, and do research with a Fulbright grant for two years and National Defense Foreign Language Fellowships for four. And Bucknell University, through Deans Leon Pacala and Robert H. Chambers, Provost Wendell I. Smith, and Vice-President Frances

D. Fergusson, generously lightened my teaching load for two terms and rescued me from quite staggering costs of typing, xeroxing, and the like.

My final thanks are offered with apologies to all the colleagues, friends, and relatives who worried and waited and wished me, vainly though it seemed, Godspeed through all those years, for they know, alas, that I found it easier to share the sorrows of writing than the joys. So I thank my brother, sister-in-law, sister, brother-in-law, *ko-ko, chie-chie,* and *chie-fu* in law, two oldest nieces (I hope younger nieces and nephews were oblivious to all this), and good friends for all the restorative good cheer of those reunions *not* postponed by my writing.

For the same and for infinitely more besides than I will ever be able to say, I thank my parents.

And finally I thank Ch'i-fang, who helped me most of all, day after day in a thousand ways, surely suffering because of this book but always celebrating its every step of geologically slow progress—or evolution.

Together we thank Hsiao Soong and Hsiao Lan.

Thirteen years, but we have made it.

Hsieh t'ien hsieh ti.

Part One

Fair Warning—Darwin's Entry into China

Prologue: The Beagle *in the China Sea*

The *Beagle* never reached the China Sea, and Charles Darwin never got to China. But part of Darwin's "ism" sailed the China coast and inhabited her harbors before "Darwinism" was a term, and indeed before *The Origin of Species* was ever written. It did so in the minds of men like Rutherford Alcock, British Consul. Born the same year as Darwin, 1809, Rutherford Alcock, after ten exciting if exasperating years of service half inside the Gates of China, wrote in 1855 from Shanghai of "a natural and moral law which governs the life, and growth, and decay of nations, as clearly as it does the life of man. . . . Man's efforts at civilization," he announced, "invariably—when the race to be benefited is inferior and weaker, intellectually and physically, than the nation civilizing—have had but one result: the weaker has gone down before the stronger."[1]

This grim law, which seemingly gave Alcock great hope, was written four years before *The Origin of Species,* and forty years before *The Origin of Species* was really introduced to China, but it was precisely the law that the Chinese first thought Darwin's book proclaimed: The weaker go down before the stronger, the weaker nation, the weaker race. And it was precisely the presence, by then far too far within the gates, of men like Alcock, backed by "barbarian" hordes of merchants, missionaries, and men at arms, that made the law so foreboding.

"The weaker go down before the stronger"—After 1895, the Japanese-Chinese translation of the famous Spencerian slogan, "the survival of the fittest," *yu sheng lieh pai* (the superior win, the inferior lose),[2] which, however elegant in classical Chinese, was as simple in sentiment as Alcock's phrase, was to force its way into a thousand essays and dominate for a time the Chinese editorial mind as *the* argument for almost any course of action. But why did it take forty years for such a phrase, so simple without its biological pedigree, to strike fire in the Chinese imagination, and why *did* it strike fire? Did China need Darwin to tell her the strong, as a rule, beat the weak or to tell her that Europeans threatened her existence?

The forty-year difference between China's age of Darwin and that of Europe and America was partly due to the accidental timing of translation. Very few in China had heard of Darwin before Yen Fu introduced him. Of course, if Rutherford Alcock could sound like a Social Darwinist before there really were any, we can well imagine that many bona fide Social Darwinists must have reached China from England and America during the four decades that formed a veritable age of Social Darwinism in the West. But somehow those who did spread neither Darwin's theory nor his name.

We should not have expected Alcock to do so, for surely as good a diplomat as he would not have confessed to his Chinese counterparts his belief that, although "China may linger . . . in her agonies," she was doomed by natural law to perish in the end.[3] Nor should we have expected the missionaries to introduce Darwin, for he must still have seemed to some of them the greatest modern anti-Christ, and it is doubtful that many, even of those who had already managed to reconcile Darwin's discoveries with their faith, would have risked confusing their sorely won converts with an inordinate number of exhortations not to take their Bibles literally. And yet missionaries *were* the first to mention Darwin.

As early as 1873, a missionary translation of the sixth edition of Sir Charles Lyell's *Elements of Geology* at least gave Darwin's name, and in other missionary publications there were other references, in 1877, 1884, and 1891. But all were very brief. None attacked Darwin, but none mentioned the religious havoc his theory was wreaking in the West, and none mentioned the most vital concept of his theory—the struggle for existence. None stirred up intellectual ferment.[4]

Reformer missionaries like Timothy Richard, Young J. Allen, and W.A.P. Martin helped prepare the way for evolution by preaching science and progress as they tried to preach Christianity, but none of them gave enough news of evolution for Chinese to grasp it. Not even Allen's famous periodical *Wan kuo kung pao* (The globe magazine) can claim to have widely publicized Darwin's theory before Yen Fu. Darwin's name, before Yen Fu, made only odd appearances, some very odd indeed. A foreign military adviser in the 1880s commented that a band of Chinese troops drilled with such spirit "they must have heard of some of Darwin's theories about the survival of the fittest." But that was unlikely.[5]

Perhaps Darwin's name spread through the uncertain networks of private conversation. Yen Fu himself had fifteen years to talk before he wrote. But, when at last he did write, he wrote of Darwin as of an unknown. In the end, the essential fact is that, even if there were other introducers of Darwin, they have been universally forgotten. Yen Fu's own generation, and that immediately following, both acknowledged Yen Fu as China's first Darwinian.[6] Only Sun Yat-sen, educated as he was in a Church of England school in Hawaii, could claim to have learned of Darwin first from any source closer to the horse's mouth.[7]

There is another reason, however, why Darwinism did not strike fire in China until 1895. That year saw a new national mood, of unparalleled apprehension, frustration, and anger, caused by what finally caused Yen Fu to write—China's shattering defeat in the Sino-Japanese War.

It did not, of course, take that defeat to make China feel threatened. Fifty years after the Opium War, it took no new defeat and certainly no new biological theory or cosmic vision to see that national strength was of the essence and that the stakes were high. The great essayist Feng Kuei-fen, one of the first in modern times to use the phrase *tzu-ch'iang* (make ourselves strong), which has been China's aim since that day, had already written in 1860 just after French and English troops had fought their way to Peking and burned the Emperor's summer palace, that "the most unparalleled anger which has ever existed since the creation of heaven and earth is exciting all who are conscious in their minds and have spirit in their blood; their hair is bristling with rage, because the largest country on the globe today with a vast area of 10,000 *li* is yet controlled by small barbarians."[8] And Feng Kuei-fen could see without Western "natural laws" that, as "China is the largest country on earth, with ample fertile plains and marshes, numerous people and abundant resources, naturally the mouths of all nations are watering with desire."[9] If any doubted it, they could look at India. India was the fearsome proof of Westerners' intentions, proof that they played for keeps.

China knew this without Darwin's theories or Alcock's law. That was why she started the whole self-strengthening movement in the 1860s. But it took time for the full extent of China's danger to sink in. And, despite a constant series of encroachments, incidents, and insults at the hands of foreigners, there were successes in the early ventures of self-strengthening that argued against China's prophets of doom. The treaties, however unfair and humiliating, did seem to help control the barbarians, or at least slow them up. The great rebellions had been put down. Arsenals had been built and railroads begun. A navy had been created and modern troops trained. Students had been sent abroad—Yen Fu was proof of that. But none of it really worked. When the test of China's new strength finally came in 1894–1895, when upstart Japan forced China to fight over Korea, China lost, miserably, not to wild-eyed Westerners, but to "puny Japan."[10]

The shock of rage, humiliation and finally fear that consequently shook the country far surpassed even Feng Kuei-fen's "unparalleled anger" of 1860. Even those who had resisted the self-strengthening innovators were now furious that the innovations had not worked. The country demanded an explanation. Droves of memorialists demanded punishment for whoever was responsible.

Others refused to admit defeat and clamored for a continuation of the war. And, for the first time in centuries, educated men with no legitimate right to memorialize the Throne banded together and did so as a body, and created all over the country *hsueh-hui* (study societies) and journals in which to argue about what should be done.

It was this furor that started the intellectual ferment that was to doom imperial China and brought into prominence the three figures most important in the first part of our story of Darwin in China, Yen Fu, K'ang Yu-wei, and Liang Ch'i-ch'ao.

Yen Fu was the great introducer of Darwin and perhaps, in the end, China's greatest Social Darwinist. K'ang Yu-wei was both modern China's first prophet of progress and her first great Anti-Darwinist. Liang Ch'i-ch'ao was the great apostle, almost a Chinese Huxley, and also, all aspersions as to the depth of his own thought to the contrary, the great complexifier.

The concern that first made all three men write of Darwin and evolution, however, was not at all complex. It was simply China's weakness, China's peril. The burning question before them was the same that Feng Kuei-fen had asked thirty years before, "How can we be strong?" Only now it was phrased in more desperate terms, "How can we survive?" That is why the rallying cry of the Pao kuo hui (Preserve the Nation Society), which had succeeded the Ch'iang hsueh hui (Society for the Study of Strength), was *Pao chung, pao kuo, pao chiao* (Preserve the race, preserve the nation, preserve the faith). What was different, however, was the answer to Feng Kuei-fen's question. He had argued that "we have only to learn from the barbarians one thing, solid ships and effective guns,"[11] but China's guns had proved effective only in the hands of the wily Japanese, who had used them in Weihaiwei harbor to sink most of China's solid ships.

Today many Chinese have again turned to weapons to explain away this debacle. The Japanese, they say, had better ships. It is to the credit of men like K'ang Yu-wei and Yen Fu that they took this defeat to mean something quite different. They saw in it proof that the real source of national strength was not in weapons, but in attitudes and institutions. In short they saw in it proof that something was radically wrong with China and that something would have to be radically changed. Thus it was that the new patriotic cry after 1895 was *pien fa,* "Change our laws," "Change our institutions," "Change our ways."

It was in this context that Yen Fu brought in Darwin. He did so not to introduce an interesting discovery of modern science, but to usher in a Western witness to the vital necessity of change. Evolution, progress, the struggle for survival—it all spoke first for reform.

ONE

Progress

In China as in Europe, belief in progress preceded knowledge of evolution, but in China the one only barely preceded the other. In the West, if one were forced to find historical roots for faith in progress, not that such faith necessarily needs such roots, one could find them, I think, in early Christianity and the Judaic tradition before it. The Bible accustomed people to look ahead. Old Testament progress might have been a three-steps-forward, two-steps-back affair, but the struggle of the people of Israel was very much an up-from-Eden struggle, and up from Eden did not mean back to Eden. Deliverance and the Deliverer were always in the future, and getting back meant only getting back to the right path forward. Granted, once men thought the Messiah had appeared, it was possible for them in their longing for the promised coming of the Kingdom of God, to "overlook" the world, to look forward, indeed, to its end, to Doomsday and the Day of Judgment. But it was also possible, with only a subtle shifting of sights, to see in the hope that God's will "be done on earth as it is in Heaven," a vision of Heaven on earth—and from that vision, or from secularized reactions to it, came many Western utopias.

"PROOF" IN THE WEST

The greatest century for Western visions of Heaven on earth began in 1750, when the young Abbé Turgot, Baron de l'Aulne, delivered before the Sorbonne two lectures in Latin on "the progress of the human spirit." His first purpose was to glorify Christianity as the greatest force of progress in all history, and, after eloquently exorcising those who thought Christianity useful only as a gate to the hereafter, he sought by a "comparison of the Christian world with the world of idolatry" to prove "the advantages that the universe had received from Christianity" and hence demonstrate the earthly "utility of religion."[1] But obviously what really excited him and what undoubtedly most excited his later readers was the tableau he saw in history, both before and after Christ, of "the successive progress of the human race . . . marching ever towards its perfection."[2] History was the story of mankind's upward struggle from savagery to civilization and, as he told it, literally the story of man's struggle up from Eden. Turgot was not at all disturbed by the Book of Genesis. Savagery was exactly what we should have expected to have occurred after Adam and Eve were driven into the wilderness, alone and without provisions: "It was in punishment for his disobedience that man, born for a state more happy, was reduced to an ignorance and misery that he could only alleviate by the force of time and travail."[3] Savagery was the depths of this ignorance and misery, but perfection was man's destiny.

The perfectibility of man and his society became the inspiration of an age, and formed a common bond in the thought of men whose philosophies were otherwise wildly apart. Immanuel Kant saw man "as a species of rational beings who are steadily progressing from the evil to the good,"[4] and, although in a more sober moment he doubted whether one could ever "fashion something absolutely straight from wood which is as crooked as that of which man is made,"[5] he held that man, nonetheless, "should eventually struggle up from the greatest backwardness to the greatest skills, to inner perfection of mind and (as far as it is possible on earth) to blessed happiness."[6]

Turgot's disciple, the Marquis de Condorcet, pushed "the greatest backwardness" back in time beyond the Baron's "savagery" (seemingly forgetting Eden), but only to make man's progress the more remarkable. "We pass," he announced, "by imperceptible gradations from the brute to the savage and from the savage to . . . Newton."[7] And after Newton there was still all the time in the world to pursue with rapture the "limitless perfectibility of the human faculties and the social order."[8]

The not-too-modest St. Simon sought to call men to such pursuit with the voice of authority, first scientific ("Gentlemen, I have lived much in the company of scientists and artists; I have observed them intimately")[9] and later divine ("Princes, hearken to the voice of God which speaks through me").[10] "Poetic imagination," he declared, "has put the Golden Age in the cradle of the human race, amid the ignorance and brutishness of primitive times; it is rather the Iron Age which should be put there. The Golden Age of the human race is not behind us but before us; it lies in the perfection of the social order. Our ancestors never saw it; our children will one day arrive there; it is for us to clear the way."[11] For, as Jeremy Bentham added from England, "Shall [men] stop? Shall they turn back? The rivers shall as soon make a wall or roll up the mountains to their sources."[12]

Even John Stuart Mill, who gloomily noted somewhat in contradiction that "a people, it appears, may be progressive for a certain length of time and then stop"[13] (he feared that "Europe, notwithstanding its noble antecedents and its professed Christianity, [might] tend to become another China"),[14] believed enough in progress to see that "even the men and women who at present inhabit the more civilized parts of the world . . . assuredly are but starved specimens of what nature can and will produce."[15] And Alexis de Tocqueville, reporting from America while Darwin was still on the *Beagle,* stated that Americans, "all have a lively faith in the perfectibility of man," although he himself had his doubts.[16]

Progress, then, was a faith for many minds, and many great minds, long before Darwin ever wrote of evolution. Evolution itself was a concept held by many before Darwin, for, as Loren Eiseley has shown so splendidly in *Darwin's Century,* almost all the pieces of Darwin's finished theory were at hand before the *Beagle* put to sea. Still it was Darwin who convinced people (or at least most people, more or less and rightly or wrongly). It was Darwin who "proved" evolution and, to many, progress. That is probably why Karl Marx sought, though unsuccessfully, to dedicate *Das Kapital* to Darwin, seemingly in grateful recognition of his having discovered "a basis in natural science for the class struggle in history,"[17] the latter of which, of course, Marx himself claimed to have discovered long before.

One may well ask just how Darwin was supposed to have proven progress. The answer is that he had not, although people thought he had and that proved more important. Actually, one of the first effects of his work was to strike an immediate, though far from fatal, blow at the "progressionists," those proto-evolutionists who saw in the seeming relatedness of creatures proof of God's hand working progressively with steady purpose to make man.

When Kant had spoken of "a hidden plan of nature,"[18] ("or perhaps one should say of providence"),[19] he was talking of a plan in human history, as was de Tocqueville, when he said all men "have been blind instruments in the hands of God."[20] The progressionists, however, believed that "providence" had been working on man's behalf long before man himself appeared, and working longer than five days. Thus, as the earth itself was slowly prepared, so were the animals, who were "merely foetal stages of man,"[21] all taking their turn to a greater purpose. Man, as Louis Agassiz valiantly maintained long after his colleagues had almost all deserted to the camp of Darwin, was "the last term of a series, beyond which there is no material progress possible, in accordance with the plan upon which the whole animal kingdom is constructed."[22]

But Darwin's system of evolution, although he himself was long reluctant to admit it, left little place for such a plan. It discovered no purpose and no aim. The descent of man (not the ascent of him) was the story of a ground creeper, living and ever changing, to be sure, but not necessarily getting anywhere. Man existed. He had somehow come out this way, through millennia of "descent with modification." There were his cousins and his uncles and his aunts. There indeed was all life distantly related. But what a family! What weird relations! Compared with his earliest ancestors, man might well embrace the common primate ancestor he shared with the orang-utan as a sympathetic friend. And what a monstrous multitude of minuscule changes it had taken to bring *him* about, and through what eons of time! How, men wondered, and wonder still, could it all be by design?

It is ironic that originally both those who believed and disbelieved in design, those who sought and those who denied the workings of God's hand in nature, were looking for natural order. Kant believed in "providence" rather than accepting an "aimlessly playing nature,"[23] while others, to protect their systems of unbreakable laws, rejected any "providence" that could interfere in such a system even if it was his own. But Darwin, also seeking order, found it in natural laws that, however unbreakable, did seem aimlessly playing.

In his usual kindhearted manner, he seemed to be absolving God of any blame for having created mosquitoes, when he contended that "it is derogatory [to say] that the Creator of countless systems of worlds should have created each of the myriads of creeping parasites and slimy worms which have swarmed each day of life . . . on this one globe,"[24] but, by demonstrating that man himself was a product of the same forces, chance mutation and natural selection, that produced this endless variation of slimy worms, he made it hard, for many, to believe that man was not an accident.

In so doing, he should also have made it hard to believe in progress, for if all that is has come into being and survived solely thanks to chance mutation and fortuitous suitability to environment, plus, of course, a (quite mysterious) will to live, then how was one such accident better than another? Progress is a value judgment, and it is hard to find a criterion of value even in an "orderly" world of chance. If being was its own excuse for being, then what else could "success" be measured in than in the terms of the *Origin's* subtitle: *The preservation of favored races in the struggle for life?* But, if self-preservation was the contest, laurels belonged not to man, but to the amoeba, or the shark, or else to whatever genes survive in one species from another.

Most people, however, without fully realizing the metaphysical minefield into which they were stepping, blithely asserted, quite anthropocentrically, that, however erratic, nature's journey from amoeba to man was a grand advance. Things were looking up. Nature was herself. Things were getting better. That was the way. That was progress. Whose way? Well, nature's.

Thus the very people who most gleefully threw out the old designer, unwittingly ushered in a new one. They kicked God out at the front door and let him in again at the back. To be fair, one must admit that Darwin himself was partly responsible for fostering such thinking. He wrote in the last pages of *The Origin of Species,* without, alas, giving any evidence, scientific or philosophical, as to how what he said could be true or indeed what it could mean, that, "as natural selection works solely by and for the good of each being, all corporeal and mental endowments will tend to progress towards perfection."[25]

But what in heaven's name is perfection? Is there a perfect man as there is a perfect circle, perfect without function? What would be a perfect hand—one that can play Bach or one that can climb trees? If perfection is perfection *for* something, then we are back in the realm of purpose and will. But whose will? Our own? A creator's? Most gibbons climb trees without complaint, but a musician may want to be a fullback. Is it, then, will that is imperfect? Is will in the way of perfection, or is it the way? Is perfection perfect love, perfect beauty, or perfect happiness? Is it to be moral or immortal? Are things to be bigger and better, bigger or better, infinitely complex, or perfectly simple? Whatever perfection is, if it is for anything, how can it be for *our* purposes, for we, so says the theory, are but midway by-products of "the process"?

Darwin did not explain perfection. He did not prove that we are progressing towards perfection. When he left them, both terms, *progress* and *perfection,* were still perfectly mysterious. But people could fall in love with them nonetheless. Westerners could, and so could Chinese.

THE CONFUCIAN PROGRESS OF K'ANG YU-WEI

K'ang Yu-wei and Yen Fu had, by 1895, each come to an almost mystical faith in progress. They had come to their faiths, however, by very different paths and, not surprisingly, were often led by them in different directions. Yen Fu's faith was admittedly Western, tempered, to be sure, by his Chinese education and his own predispositions, but proudly acknowledged as a universal truth discovered and brought back from the West. His authorities, almost—but not quite—masters, were Western wise men Darwin and Spencer. K'ang Yu-wei's faith, on the other hand, was even more proudly proclaimed Chinese. It was the long lost "true doctrine" of Confucius, a universal truth, rediscovered in the East, waiting to be propagated in the West.

Of course, as far as Darwinian or Darwinesque concepts of progress are concerned, K'ang Yu-wei surely was indebted to Yen Fu, although he never seems to have acknowledged such a debt. His indebtedness has admittedly been confused by the fact that many of his works were begun before Yen Fu wrote anything. But the "Darwinian" passages that do appear in some of these works almost certainly were post-Yen Fu embellishments. They all speak of Darwin and evolution and progress in Yen Fu's vocabulary, a vocabulary that does not appear in those of K'ang Yu-wei's writings that can definitely be dated before Yen Fu's. They were added, I am convinced, in the revisions K'ang Yu-wei made in exile after the failure of his famed Reform Movement of 1898.[26] K'ang Yu-wei himself never created a good term for progress, but he definitely did have such a concept, and an understanding of the way he came to it is important not only for an understanding of his own thought, but for an appreciation of those factors in the Chinese tradition that helped others accept the concept of progress that Yen Fu brought back from Europe.

VISIONS. In 1878, Yen Fu, age twenty-five, was in England. He was dutifully studying naval science, as he was supposed to, but at heart he was thirsting after knowledge of Western political philosophy. He was hearing for the first time of men like Spencer, Darwin, Mill, and Montesquieu, and he was conversing, sometimes late into the night, with China's first ambassador to England, Kuo Sung-t'ao, a man who, however much his elder, shared his excitement for comparing Western and Chinese thought and institutions, in the hope of discovering the secret of Western wealth and power.[27]

In the same year, K'ang Yu-wei, age twenty, was at home in Kwangtung province having visions. Worn out from intense, impatient study of China's Classics and great histories, he had "suddenly stopped [his] lessons, given up [his] books, closed [his] door to [his] friends, and taken to sitting in meditation to nurture [his] mind." Apparently it worked. "While sitting in meditation," he recalled much later, "I suddenly saw that Heaven and earth and the ten thousand things were all of one body with me. How clear it was. I thought I was a Sage and laughed for joy. But as suddenly I thought of men's suffering and wept in despair. Then, with a start, I realized I was not being filial to my grandfather. Had I not learned anything? So I packed up my things and went to dwell near his tomb. My fellow students, seeing me singing one moment and weeping the next, thought I had gone mad."[28]

K'ang Yu-wei was not, however, a nut. He was an intelligent, sensitive, and serious young man of great feeling and fantastic imagination, who as a young boy was already so "grown up" in his concerns that he "avoided all frivolous talk and jesting." He delved into Confucian literature with such earnestness, and talked so often about "the Sage," that the village wits mockingly dubbed him "the Sage Yu-wei."[29] It is small wonder if he later came to believe that he might be one.

Certainly he came to feel that he was called upon to do great things. A year after his Confucian conscience had led him to finish proper mourning for his grandfather, he went on another retreat, this time to a place called the Cave of White Clouds, on Hsi Ch'iao mountain near his home, where, "because of the outstanding beauty of the scenery, one could practice quietude" and fittingly study Taoist and Buddhist texts.[30] But significantly, although he experienced great initial excitement and gained much lasting inspiration from such study, and although Buddhism was to color his thought throughout his later life, he again came to feel restless, and discontented with philosophies of withdrawal.

It was his uncle who virtually forced him out of the mountains and into the official examinations by threatening to cut off his allowance,[31] but K'ang Yu-wei already seemed to have made a sincere choice by himself to "enter the world" rather than escape it.[32] At the time, he wrote later, when "I concentrated on nurturing my mind, I kept thinking of how hard people's lives were, and of how Heaven had given me the intelligence, talent, and energy to save them. Feeling deep sympathy for all things and lamenting the times, I thus dedicated myself to the ordering of the world."[33]

What obviously struck him most in Buddhism was the ideal of the Bodhisattva, the saint who reaches enlightenment only to turn back voluntarily from the very threshold of nirvana to help enlighten his fellow men. With no apology, K'ang Yu-wei mixed Buddhist phrases with his own in stating his resolve: "I would not abide in Heaven but would willfully enter hell. I would not seek refuge in the Pure Land, but purposely commit myself to the muddy world. . . . I would not settle for purifying myself alone. I would not settle for private joy. I wanted not to exalt myself; but simply saw in my relatedness to all creatures a chance to save them. Everyday I thus set my heart on saving the world, and every moment took saving the world as my vocation."[34]

One may well feel that K'ang Yu-wei exhibits in this speech an extraordinarily exalted view of himself, almost a "Messiah complex," or at least a "Sage complex." But K'ang Yu-wei was no fanatic. His visions, however fantastic, were not those of a Hung Hsiu-ch'üan, the Taiping rebel leader, who thought that as Christ's younger brother and "Heavenly King" he could personally lead the world to a "Heavenly Kingdom of Great Peace,"—to be won, of course, by fire and sword. K'ang Yu-wei did not want to save the world that way. Nor did he think of himself as its only savior. He was not, it is true, above referring to himself as "K'ang Tzu,"[35] appropriating the suffix reserved for only the greatest philosophical teachers, of whom the greatest was "K'ung Tzu," Confucius; but he never tried to supersede Confucius, and he certainly never tried to call himself anything like Comte's title for himself, "Le Fondateur de la religion universelle, Grand Prête de l'Humanité"[36] —not even after Liang Ch'i-ch'ao later called him "Confucianism's Martin Luther."[37] The salvation of the world would come, but it was something one and many had to work for.

K'ang Yu-wei was not, then, a fanatic, but he had great religious zeal. He truly felt compassion for the world and a sense of mission to help it. It was no coincidence, therefore, that he was later so impressed by the spirit of Western missionaries, whom he saw in China and read about in history. He was of the same temperament.

The most significant thing about K'ang Yu-wei's missionary spirit, however, was that he was not out to save men's souls but their society. When he declared that he would give his all to help "bring all creatures to the world of Infinite Happiness,"[38] he was using a Buddhist name for a Paradise completely above and beyond the illusory world of man; but he had come to make the same subtle shifting of his sights that so many Christians or Christian renegades had made in Europe before him. His Paradise was to be a Heaven on earth. It was to be the truly great society, the *Ta t'ung,* the Great Unity, One World, with happiness for all. Thus, just as nineteenth-century Western utopias were often secularizations of Christian visions, K'ang Yu-wei's was a secularization of a Buddhist one. The salvation of the world was the salvation of *this* world. It was for humanity in the flesh.

By 1885, when he was only twenty-seven, K'ang Yu-wei had written a good enough draft of his utopian vision, his *Ta t'ung shu* (The Great Unity), which he then called *Jen-lei kung-li* (The natural laws of humanity), to allow himself to take cheer even in the face of an attack of excruciating headaches, which he fully thought would kill him. Echoing Confucius, whose true Way he thought he had resurrected, he could sigh resignedly that, "having heard the truth, having fixed the idea of Grand Unity, it is all right to die."[39]

Fortunately K'ang Yu-wei was still many decades away from death, whatever that might signify as to his distance away from truth, but the work that gave him such a comforting sense of accomplishment during his illness, the work in which he truly thought he had forever grasped the Way, was, in its final form, a fantastic document. It was a utopian plan that could and can hold its own with any of those produced in the Western world, and it unquestionably admitted K'ang Yu-wei into the questionable company of the great Western visionaries, St. Simon, Fourrier, Comte, and Marx. His work, like theirs, can even to the same individual be both inspiring and repellent, moving and terrifying. Like theirs, his work contains ideas that seem very possible and desirable—some of which have indeed already come into being—and others that one can only fervently hope will forever prove impossible, although some of these also seem close to fulfillment. But, whatever one thinks of the *Ta t'ung shu,* it is definitely the work of a remarkably unfettered mind.

The *Ta t'ung shu* was a plan for world unity and an end to war, but more than that it was a plan for an ideal world society through which men could virtually conquer evil. Evil for K'ang Yu-wei was suffering, and the root of all suffering was a kind of alienation so vast in scope that it makes that of Marx seem positively childish. K'ang Yu-wei's is a Buddhist alienation, although the deliverance he seeks is far from Buddhist. It is the alienation of the individual from the universe, the alienation of all parts from the whole, of man from man, and of man from nature, of all men from each other and from all things.

Whether or not this alienation is artificial or innate is (understandably) unclear. Alienation slipped into K'ang Yu-wei's universe by way of an "and yet" as brazenly unexplained as any in all history, an "and yet" that tried to bridge a chasm of mystery as deep as that behind any "fall" in Eden. In the story of Genesis, according to K'ang Yu-wei, all was also well in the beginning: "Men were all born of Heaven. Together they are Heaven's sons. All share the same human form. All are of one race. All are perfectly equal. And yet, in far antiquity, men set themselves up in selfishness."[40]

Once again it happened. Man, the primeval Humpty Dumpty, somehow had a great fall, although why, as always, Heaven only knew. Man fell to pieces, and selfishly started drawing lines about those pieces, lines that were boundaries, that became barriers, that ruled off one man from another and so caused men's suffering. But K'ang Yu-wei was somehow convinced that Man could put himself together again, although he might in the process have to ask most of the king's horses and men, and eventually the king himself, to step aside. For, as the barriers that K'ang Yu-wei saw were largely artificial, they were unnecessary and could be torn down.

After thirty-nine pages of painful description of thirty-eight types of human suffering, K'ang Yu-wei concluded that, "when we gather together the sources of these afflictions, we find they are caused by just nine barriers"—of nation, class, race, sex, family, occupation, species, disorder, and pain.[41] If these barriers were abolished, then men would be able to live in complete peace, harmony, and happiness. K'ang Yu-wei obviously believed such barriers could be abolished and indeed would be. The rest of his book is a detailed description of what the world would be like without them, and of how it will come to be without them. There was one redeeming factor in the present alienated world that would make the Great Unity possible, and that was stated in the first subtitle of the first chapter: *Men have compassionate hearts.* This, of course, was the famous position of Mencius, expressed negatively in Chinese as, "Men have hearts that cannot bear [that others should suffer]."[42] The positive virtue that this negative reaction proves is *jen* (compassion, love, or benevolence), and for K'ang Yu-wei it was a force innate in all men, which they would one day learn to extend to all men, and in yet a later day to all creatures.

Obviously, having already admitted that the barriers responsible for man's suffering were the result of man's own selfishness, the introduction of man's "compassionate heart" set the stage for a civil war in man himself which was unconsciously the war of which K'ang Yu-wei wrote, but which was one he never quite declared. For the moment, however, we need only be concerned with the fact of his faith that compassion would win. His 1885 draft proves that his earliest faith in the *Ta t'ung* already incorporated a firm faith in inevitable progress. Somehow K'ang Yu-wei had come to believe that the universe was not aimless, nor static, nor cyclical, nor simply out of joint. The cosmos was going somewhere, and the world had a destiny. It was destined for better things than it had ever known. The ideal world was not yet at hand. It was not waiting on the morrow, and K'ang Yu-wei would have been the first to admit that the present situation looked bleak, but he held nonetheless that, "if we speak in terms of underlying natural laws and consider the nature of the human heart, then surely we can see in the overall direction in which the world is moving that the goal of the future can be nothing less than Great Unity. It will simply take time. And the way will not be straight, or easy."[43]

AID FROM THE WEST. But how was K'ang Yu-wei so sure? How had he come to have such faith in progress? Had he really dreamt up his utopian ideas off the wild top of his head, or had he had help?

In the totality of his thought he was "an original." But, surely, along the way to his finished works he had more help than he acknowledged. Clearly he borrowed a good part of his Three-Age theory from Liao P'ing. Probably he borrowed details for his final *Ta t'ung* blueprint from Edward Bellamy, thanks to Timothy Richard's Chinese paraphrase of *Looking Backward,* serialized in *Wang kuo kung pao* from December 1889 to April 1890. And yet, even if K'ang Yu-wei plagiarized Bellamy in 1902, it was only to add embellishments to his vision, for he had written his first utopian work in 1885, three years *before* Bellamy wrote *Looking Backward,* and also before Bellamy's famous European predecessors, St. Simon, Comte, Fourier, and Marx, were ever introduced to China.[44] But perhaps K'ang Yu-wei had subtler help.

He liked to boast that he had drunk in the very best of both

Eastern and Western philosophy,[45] but Liang Ch'i-ch'ao, who eventually learned far more of Western thought than his former teacher, suggested in his reflections on K'ang Yu-wei's education that his boast, at least with regard to Western philosophy, could not possibly have been true. He stated that in 1882 K'ang Yu-wei "bought copies of all the books that had been translated by the Kiangnan Arsenal and Western missionaries, and read them all. But the books that had been translated by that time were all on common, elementary subjects, if not on technology, military science, or medicine, then on Christian scripture and exposition. There was absolutely nothing on politics or philosphy."[46]

A glance at the list of translations made by John Fryer and others at the Kiangnan Arsenal at that time largely substantiates such a view. Strictly speaking, of those books published by 1882 there were none on politics or philosophy. There was, however, one book on politics published in 1885 after K'ang Yu-wei had started writing his *Ta t'ung shu* but well in time to influence at least the first major revisions he made in 1887. This was the *Tso chih ch'u yen,* an oral translation by Fryer (transcribed into literary Chinese by Ying Tsu-hsi) of an unpretentious treatise entitled "Homely Words to Aid Governance," originally published in the mid-nineteenth century as part of *Chambers Educational Course,* edited by Messrs. William and Robert Chambers of Edinburgh. As K'ang Yu-wei most probably did read this work, and as Liang Ch'i-ch'ao himself still considered it in 1896 to be the best book on politics then available, its contents are worth studying.[47]

Tso chih ch'u yen was not a work of utopian vision, nor a work that specifically introduced the names and ideas of Western philosophers or social thinkers. It did, however, exude a firm but gentle faith in progress, which reflected an equally gentle satisfaction with nineteenth-century England, a satisfaction that great progress had been made, was being made, and would continue to be made in human life. *Tso chih ch'u yen* did not, however, present progress as the inalterable way of the universe, as some independent, transcendent power, or mystical force, that would force us to perfection despite ourselves. Rather it simply stated that it was God's will that we progress, and that since we could, we should. At the same time, however, it gave hints of a better world that could come, and in so doing it may well have helped inspire K'ang Yu-wei to see his perfect world that would come.

Progress as envisioned by the Chamberses was quite un-Darwinian. Progress was civilization, and civilization was the gradual extension of reason, knowledge, and good will. The book announced in its first paragraph (in the Chinese version) that "the Creator's original intention in giving birth to men was that all people on earth should be able to eat their fill, be well clothed, and share in health and happiness. Heaven's mind shall not find peace until all men can live from youth to old age without want and without regret."[48] The burden of progress, however, was to be on us: "Man is different from the beasts by virtue of his intelligence alone. . . . Animals . . . look only to their present needs. . . . [But man] is never satisfied in his striving to know and to be able. As there is no limit to man's talents and intelligence, however, then what today we call ingenious will in a few year's time be surpassed by the even more ingenious. Since the Creator has given us intelligence, he has clearly given us the right of invention. If we throw it away and seek our own pleasure, if we take no thought of building new things to help mankind, but instead use our intelligence to amaze the stupid and cheat the world, then do we not go against the will of creation and revert to something lower than the beasts?"[49] Man seemed to owe it to his creator to civilize himself, to work out his own salvation here on earth.

"Man," moreover, meant all men. "When we look at the histories of different nations," the Chamberses went on, (still in Chinese) "we see that their ancestors were all savages. . . . but, as the basic nature of both civilized men and savages is endowed from heaven, a people, though they begin as savages, may by the gradual accumulation of good influences become civilized."[50] England itself was "still only half civilized, as there are many of her people who can neither read nor write, and whose wisdom and intelligence are not necessarily any higher than those of savages." Still, with the recent interest in universal education, the Chamberses saw hope that real civilization could be reached.[51]

Civilized society was to be a further refinement of British society. Thus the bulk of *Tso chih ch'u yen* turned out to be a justification for law, democracy, free trade, private property, capitalism, and competition. But competition was not to be ruthless. The law of the jungle was outlawed.[52] The ideal of competition was an idealized version of British free enterprise, without any social Darwinian justification for its harsher aspects, and indeed without any acknowledgment that such harsher aspects need occur. Competition was definitely a good thing: "Within a country not only should the people not be forbidden to struggle to get ahead and strive after fame and wealth; they should be urged and encouraged to do so, so that all may be their own masters and seek wealth and fame with their own talents and energies. But there is a limit to all this, and there is no choice but to repress those who seek profit without any sense of this limit. . . . [those who] seek only to benefit themselves, without even asking whether or not their actions injure others."[53] It was true that "on earth men have similarities and dissimilarities. They have greater and lesser natural endowments . , , but a civilized country must see to it that all types of men gain basic equality. There can be no strife between strong and weak."[54]

Equality, however, did not mean economic equality. The right of private property was a natural right, perfectly understood even by uncivilized savages: "Even animals have property." What was more, private property was for the good of all, "for as the poor are largely supported by the rich, if the rich could not accumulate property, the poor would have even less on which to rely."[55] True, certain Frenchmen, "fooled by their own abstract terms, claimed that to equally distribute property was to be equally humane to all. They gained by this a certain notoriety, to be sure, but they ignored the question of the feasibility of such a plan. They wrote frequent tracts, well designed, indeed, to strike the reader's fancy, but once their theory was out, the stupid and harebrained in the land refused to be conscientious in their work and put their hope instead in the divvying up of others' property for their own use. Thus was the door opened to all kinds of brigandage and thievery. . . . No wonder that the British have never been willing to squander a word of praise for the equal distribution of property. [For] if England were to carry out [such an idea,] then all the many advantages that the county now enjoys, and the advantages that each individual enjoys in his independence, would all be obliterated, and uncountable future evils would perforce ensue."[56] No, "In all things one must be practical to be of any aid to the people. The bandying about of empty words is no help."[57]

Perhaps K'ang Yu-wei took the Chamberses' warning to heart. He certainly tried to be practical when he wrote the *Ta t'ung shu,* and he quite obviously thought he was being so, but his practical plan for a better world led, in his mind, to something far better than a better England. The Chamberses' "homely words" were hardly a blueprint for the Great Unity, and they cannot, therefore, destroy K'ang Yu-wei's claims to originality. Nonetheless, many of the Chamberses' ideas were ideas in embryo of the world K'ang Yu-wei longed for. Where the Chamberses argued that the society had the right to insist on education for all, despite family objections,[58] K'ang Yu-wei decided that society should have the right to "rescue" each citizen from the family altogether, and raise and educate him from birth to occupation. Where the Chamberses welcomed voluntary federations and looked forward to a gradual coming together of nations under international law,[59] K'ang Yu-wei looked further ahead to the abolition of nations. Where the Chamberses saw in the natural scattering of the necessities of life over the whole globe more proof of a marvelous creation, purposefully designed to bring men together in friendly interdependence through free trade,[60] K'ang Yu-wei saw the means and the necessity for One World.

K'ang Yu-wei must have liked *Tso chih ch'u yen* very much, but the book probably did more to confirm his beliefs than to inspire them. The Chamberses' Christian conviction that Heaven would not be happy until all men were was in harmony with K'ang Yu-wei's

Buddhistic belief that all creatures must one day be brought to "The World of Infinite Happiness." K'ang Yu-wei shared the Chamberses' faith in science and technology, in man's seemingly limitless ingenuity, and "applicable" intelligence. Above all, he shared their belief in a progress that could be gentle, orderly, gentlemanly, as befitted the product of intelligence and good will. But surely K'ang Yu-wei had begun to believe in such progress even before he read Fryer's translation of the Chamberses' book.

One of the Kiangnan Arsenal translations, however, definitely did have an influence on K'ang Yu-wei's progressive thinking, and that was the 1873 translation of Lyell's *Elements of Geology,* which, as we have seen, mentioned Darwin but only mentioned him. Knowledge of geological ages and of the discoveries of fossils and skeletons proving the earlier existence of monstrous species now extinct swept through China well before any detailed knowledge of Darwinian evolution, and such knowledge, which had so affected Darwin himself, definitely impressed K'ang Yu-wei.

But even so, it did not take Lyell's dizzying revelation of geological time to make K'ang Yu-wei believe in progress. Given his original religious revelation that somehow Paradise was to be built on earth, then every one of the Arsenal translations, even the driest and most mechanical, could speak to him of progress, for each spoke somehow of the astounding technological progress of the nineteenth century that had made it so easy for Westerners themselves to believe in progress philosophically. The fossil record, after all, had for years served only to draw out the time man thought it had taken God to make him. It was the technological record since the Industrial Revolution that led man to believe (at least for a while) that he himself could "invent" a perfect world. Before China had time to think about the matter philosophically, she was indoctrinated into at least a grudging belief in progress by the technological show the West put on in her harbors.

The Opium War had been a simple show of strength, Western technology versus Eastern, and Western proved the stronger. Thereafter, however, the amazing thing China was forced to discover was that barbarian strength was not just stronger; it kept getting stronger. The change from sail to steam, which, however sad aesthetically, must certainly be considered one of the greatest technical advances in history, took place at China's very doorstep, while China was still looking in envy at the clipper ships. Western ships and Western guns, and Western medicine to patch up the gun's handiwork—everything seemed to be getting better and better. Trains and telegraphs—when the foreigners first ruffled China's calm, they themselves had not yet dreamt of such things. It was discouraging to watch the West charge ahead, but what a world it was, and why could not China move too? Thus, K'ang Yu-wei must have been angered and excited. With this demonstration going on before him, and with his sensitivity to human suffering, his long-felt will to save the world, and his passionate desire to save China and insure that she be part of that world, how could he not believe in progress? He wanted to believe in progress.

PROGRESS IN THE SPRING AND AUTUMN

If there was at least some degree of wishful thinking in K'ang Yu-wei's own vision of progress, there definitely was wishful thinking in his discovery of a long-lost doctrine of progress in the Confucian canon. But wishful thinking, of course, need not be insincere (nor false). K'ang Yu-wei's conviction that Confucius himself was a prophet of progress was a true conviction, however much desired, and however much indebted he was to the writings of Liao P'ing. He did not fabricate his Confucian theory of progress in order to hoodwink conservative China into an official policy of reform. Either in discovering Liao P'ing's work or in rereading the Kung-yang commentary to the *Ch'un ch'iu* (Spring and autumn annals) and the "Li yun" (Evolution of ritual) passage in the *Li chi* (Book of rites) on his own, he must have experienced a genuine feeling of enlightment. Everything must have seemed to fit, not just as a way out of China's troubles, but as the Way out of humanity's troubles. The world was on the move, all would be well, and China's own Confucius had realized it over two thousand years before!

One must hasten to add, however, that, whenever K'ang Yu-wei felt inspired, he did prove extraordinarily adept at overlooking any minor realities that did not fit his inspiration. As his wavering disciple, Liang Chi-ch'ao, once put it, "Our teacher is a man endowed with exceptional self-confidence. Once he holds a belief, absolutely no one can shake him from it. This is true of his scholarship; it is true of his action. He refuses to alter his beliefs. His attitude is usually [that of Lu Hsiang-shan,] 'The Six Classics are but footnotes to my thought'. . . ."[62] K'ang Yu-wei's interpretations of the *Spring and Autumn Annals* and the *Evolution of Ritual* are perfect examples of his ability to mold facts and reconstrue the Classics, but they are also perfect examples of the sincerity of his sense of enlightenment.

What so excited K'ang Yu-wei in the *Spring and Autumn Annals* was the idea of the Three Ages (*san shih*): the Age of Disorder (*chü-luan*), the Age of Relative or Increasing Peace (*sheng-p'ing*), and the Age of Complete or Great Peace (*t'ai-p'ing*). To K'ang Yu-wei these phrases leapt off the page as a spine-tingling vision of man's destiny, as a picture of both the path and the purpose of human history. They gave him proof that Confucius, the greatest Sage that ever walked the globe, had realized long ago that all the world was not a stage, but three stages, through which man was destined to progress to perfect happiness. There was just one trouble with this idea, however. K'ang Yu-wei's blinding revelation blinded him to the fact that the doctrine which for him, his followers, opponents, and commentators became inalterably fixed as "The Three-Ages Doctrine of the *Spring and Autumn Annals*" was not in the *Annals*. Nor was it in the Kung-yang commentary to the *Annals*. It was only in the commentary to the commentary, written by one Ho Hsiu of the Later Han dynasty, and it did not really mean what K'ang Yu-wei said it did.

It is small wonder, of course, that the Three Ages were not to be found in the *Spring and Autumn Annals* themselves. Almost nothing is. Indeed, had Confucius not made his cryptic remark that "the goal to which I aspire is in the *Ch'un ch'iu*," Ch'in Shih

Huang would not even have found the book worth burning. It is an almost inconceivably terse record of selected happenings in the state of Lu from 722 to 481 B.C.; even if one can accept the traditional belief that Confucius hid in it moral judgments of those happenings, through subtle variants of vocabulary, it must surely be considered one of the least pregnant inspirational tracts in all literature.

In the present instance the work positively outdoes itself, and the commentary, and the commentary to the commentary are corresponding tours de force, to which K'ang Yu-wei's is only one more in the series. All that "Confucius" actually said was, "Winter. The twelfth month. Chi Po came." The commentary, however, quickly enquires, with wily astuteness, "Why did he not specify the day?" and then answers its own question, "It was too long ago. He (Confucius) uses different words for what he witnessed, for what he heard tell of, and for what others heard tell of." It was from this enticing beginning that Ho Hsiu was able to delineate three ages (As Tung Chung-shu had before him),[63] but his Three Ages were simply a division of Confucius's chronicle into three parts. What Confucius had seen himself, of course, were the events of his own time. What he had heard of were the events of his father's time. What others had heard of and passed on to him were the events of his grandfather's and great-grandfather's time. This was a helpful clarification but hardly breathtaking. The excitement, however, came when Ho Hsiu took it upon himself to go on and describe those times:

> In the age others had heard described, rule first rose out of the midst of disorder. Men's thinking was still unrefined and each kingdom thus took itself as the center and looked upon all other Chinese kingdoms as outsiders. . . . In the age Confucius himself heard described, rule became relatively peaceful. All Chinese kingdoms were considered together as the inside, and only barbarians were thought to be outsiders. . . . In the age Confucius actually witnessed, rule reached complete peace. Barbarians came into the fold and were ennobled. All under Heaven, far and near, great and small, were as one. Men thought deeply and carefully and came to value righteousness and goodwill.[64]

Here was a vision. Here was progress. But it was progress in the past, not a prophecy for the future. And did K'ang Yu-wei not see what it really was? Ho Hsiu's Three Ages were a far better description of the rise of the Han than of the decline of the Chou, which is all that Confucius and his forefathers had really witnessed. Consciously or unconsciously, Ho Hsiu's Three Ages were a description of the Warring States, of Han unity, and of the great Han Empire; and K'ang Yu-wei, consciously or unconsciously, was thus inspired by a Han imperialist vision, not by any vision of Confucius's, except, of course, insofar as the Han vision was itself Confucian. This, to be sure, it partly was. The world was to be united by righteousness and love, but Ho Hsiu's model for "the Great Peace" was the "Pax Sinica" of his own dynasty, which had hardly come into being through love and righteousness alone.

Ho Hsiu drew a picture of progress as a nineteenth century Englishman or American would have, out of pride and a sense of his country's accomplishment and superiority. He projected his picture of progress into the past, of course, to see progress in the time of Confucius, but his belief in progress was founded on the glory of the Han. K'ang Yu-wei, on the other hand, caught as he was in what seemed to be China's hour of greatest weakness, believed in progress "with a vengeance" and thus projected his Three Ages into the future. But there how neatly the old Han dream still fit! Again there were Three Ages: China divided against itself by petty selfishness, China united in resistance to the new barbarians, and China triumphant, as the inspiring genius of One World, to which all barbarians would voluntarily flock in peace.

Those who would see in K'ang Yu-wei's scholarship only the clever machinations of a master politician (which he was not) will object that he could not possibly have really believed that Ho Hsiu's Three Ages were actually Confucius's own hidden plan for future salvation. Indeed anyone may well object that, had Confucius only had a better date, there never would have been Three Ages. But, although the latter objection is perfectly justifiable, the former does not follow from it. K'ang Yu-wei's revelation was quite blinding enough for him to believe what he said he did. The tradition of accepting one's favorite commentator to the *Ch'un ch'iu* as a true revealer of Confucius's wisdom was very strong. Moreover, K'ang Yu-wei held that "the true meaning of the *Ch'un ch'iu* is not in the classical text but in what has been passed down about it by word of mouth." He saw in the similarities between the commentaries of Tung Chung-shu and Ho Hsiu proof of a legitimate oral tradition, handed down in almost secret "apostolic succession" from Confucius himself, and he could thus accept their commentaries as faithful transmissions of Confucius's true teachings.[65] What it was in his youth that so convinced him that the truth must lie with Confucius we do not know but, armed with that faith, it was "in good faith" that K'ang Yu-wei chose as the truth of Confucius that which rang most true to him.

Ritual Evolution. K'ang Yu-wei was not, however, completely satisfied with Ho Hsiu's meagerly sketched Three Ages in themselves. His own great vision of a heaven on earth required further inspiration, and the Three Ages did not assume their full significance in his thinking until he found it. When he had first come upon the *Ch'un ch'iu's* doctrine of Three Ages, he had been immediately excited by it but also disturbed: "I said to myself, in this you can almost glimpse the outline of Confucius's Great Way, but you can still not see it completely. I regretted that it was so vague and obscure, and that I could not grasp its essence in a few words. So I gave up on commentaries and searched in the classical texts themselves. It was when I got to the "Li yun" (Evolution of ritual) that I cried out in exultation, 'Here is the meaning of Confucius's successive Three Ages. Here is his true Way!' "[66]

For there was the *Ta t'ung*, the Great Unity, the name and description of a better world that was to become the professed goal of both Reformers and Revolutionaries, of Republicans, Anarchists, Socialists, Communists and Constitutional Monarchists alike, the one and only classical concept that was to gain and maintain an almost universally esteemed position in the ideological rhetoric of late Ch'ing and Republican China. From the last years of the nineteenth century to the present, it was to work its way not only into the essays of all factions, but into the names of countless schools, restaurants, banks, and shops, and into the brand names of products of every description. Even today, millions of Chinese in Taiwan and Hong Kong, and probably half those in America, give thanks for their daily rice to *Ta t'ung* electric rice-cookers, though few if any still do so with utopian dreams of the future.

But what does or did *Ta t'ung* really mean? Like so many great slogans, it is beautifully vague and defiant of translation. The *Great Unity* is fairly literal and certainly vague enough, but it takes something like *One World* to convey the attractive ring of the original. The only trouble with that is that *Ta t'ung* was used as often by nationalists as by true internationalists or universalists. Indeed almost all who used it thought first of China, and most were convinced that *Ta t'ung* society could be realized in China before it was realized in the world at large. What counts, of course, is what *Ta t'ung* meant for those who believed in it, and we must let them define it in their different ways. The only unity in their definitions was that the *Ta t'ung* was to be a way out of China's troubles. In that sense it was a Promised Land—but of course there was all the room in the world for people to disagree, and violently disagree, on what it would or should be, where it was, and how to get there.

It is not surprising, therefore, that although the great "reformer," K'ang Yu-wei, rediscovered the "Li yun" chapter for his generation and was hence ultimately responsible for bringing the *Ta t'ung* to his generation's mind, it was the revolutionary, Sun Yat-sen, who turned the *Ta t'ung* into a political slogan shouted throughout China. He too picked up the "Li yun's" paragraph-long description of the *Ta t'ung* as his favorite classical text, but he made it the goal of his revolution, the socialist goal to which the Three People's Principles were supposed to lead. As the Republic of China's "national" anthem, originally a Kuomintang party song, still puts it: "The Three People's Principles shall be our guiding faith, to build the Republic and advance to the *Ta t'ung*." Indeed the whole "Li yun" description of the *Ta t'ung* was made a song, still sung by the school children of Taiwan on Confucius's birthday, and Sun Yat-sen's inscription of the text has become the most famous specimen of his calligraphy, aside from his inscription of his favorite line therein, *T'ien-hsia wei kung* (The earth is for all).[67]

T'ien-hsia wei kung is another beautifully vague sentiment that could mean "The earth is everyman's, " "The earth is public property," "The world belongs to all mankind," or "The Empire (or the nation) is the people's." It has been happily used as a slogan for democracy, socialism, anarchism, and communism, and the whole "Li yun" description of the *Ta t'ung* also could as happily be used for all four. If Mao Tse-tung had wanted classical precedent for his policies, or Old Testament prophets to herald the coming of his state, he too could have inscribed the *Ta t'ung* text in his own calligraphy and had it hung in every household and from the T'ien An Men. He did not, of course, want classical precedent, and Confucius was the last man he would have wanted as a prophet, but if he had been honest with himself, he would have had to admit that *Ta t'ung* thinking helped pave the way to the "People's Republic," for it helped prepare the way for the thought of Karl Marx, and so finally for his own.

But we are still in the first act and must try to keep the Chairman in the wings. For now, it is enough to see that this ancient "Li yun" text was so flexible that it could both bring together the future arch-rivals, K'ang Yu-wei and Sun Yat-sen, and keep them apart. Both could seize the *Ta t'ung* as their goal, but Sun Yat-sen could find in it justification for revolution, while K'ang Yu-wei found a rationale for gradual reform, a rationale, of course, that would seem to bring him close to Yen Fu, who had found a Darwinian rationale for gradual reform. Yet, oddly enough, it was over this same "Li yun" text that K'ang Yu-wei and Yen Fu would reveal their own first great disagreement. For, although Yen Fu would be pleased to find classical support for "progress by stages," he would be incensed at the notion of "progress by Sages."

The "Li yun" very quickly, therefore, was to become embroiled in the whole Darwinian excitement, lending its ideas and vocabulary to Darwinians, non-Darwinians, utopians, and progressionists of all sorts. The *Ta t'ung* itself was for many to become an integral part of Darwinism. And so the "Li yun" chapter is well worth careful scrutiny.

The "Li yun" chapter of the *Li chi* is a story of Confucius, written at the latest in the early Han, but by whom no one knows. Although the story may be pure fiction, the Confucius of the "Li yun" is not logically incompatible with the Confucius of the *Analects,* for the *Ta t'ung* is not an un-Confucian ideal. It is only "wildly" utopian when taken to the extremes to which K'ang Yu-wei took it.

The whole "Li yun" text is only 744 characters long, of which 107 describe the *Ta t'ung.* Besides the *Ta t'ung* passage, there are two other important passages, which, to K'ang Yu-wei's delight, did make up a description of three ages, although they are not so called. Together they give a picture of an ideal world (the world as it should be), of an imperfect world (the world as it is, or was for Confucius), and a primitive world (the world as it had been in earliest times). Of these, the ideal world is described first and, taken out of context, its description can indeed be read as a prophecy of a utopian society for the future. This, of course, is how both K'ang Yu-wei and Sun Yat-sen interpreted it. In context, however, the passage is much more confusing, for it appears that, in "Confucius's" mind at least, the *Ta t'ung* was a Golden Age of the past, and there is no real promise given that it will be attained again. Most of China's utopians have in their optimism cheerfully overlooked the context, but in so doing they have also overlooked the sobering fact that the "Li yun" opens with a sigh:

> Long ago, Confucius once assisted at the great December sacrifice, and, when it was over, he went out and walked upon the tower over the temple gate, and there he sighed. He sighed for the state of Lu. Yen Yen, who stood beside him, asked, "Why, Sir, do you sigh?" and Confucius answered, "I have not attained unto the Great Way as it was practiced, nor unto the glory of the Three Dynasties, although I have longed to."
>
> When the Great Way was practiced, our land belonged to all men together. The good and the able were chosen to govern, and men valued honesty and strove to live as brothers. Thus people did not treat only their parents as parents or their own sons as sons. They made sure that

> all of the old were cared for till death, that those in their prime were given useful work, that the young were carefully brought up, and that all widows and widowers, orphans and the childless, cripples and invalids were helped in their afflictions. Each man had his work and each woman her home. People hated to see things thrown away and wasted, but felt no need to hoard things for themselves. When there was work to be done, people loved to be the first to do it, but did so not for their own rewards. Hence there were no plotters or crooks, no thieves or brigands, and people never locked their gates. Such was the *Ta t'ung*, the Great Unity.[68]

It is this last paragraph that was set to music, and what a song it made. If translated into the future tense, as it perfectly well can be, out of context, could not Turgot and Condorcet, St. Simon and Comte, Fourier and Robert Owen, Karl Marx and the architect of the Chinese communes all have joined in chorus with K'ang Yu-wei and Sun Yat-sen, or, more likely, sung it each as his own solo? And yet the song sings actually not of any of their ideal worlds, for their ideal worlds have all inspired some and frightened others, and this world could frighten no one. It is the ideal ideal, for it spells out no extremes and mentions no price. It tells no one that his cake must be given up on the eating. As long as one could be unrestrictive in one's interpretation of "Each man had his work and each woman her home," who would not long for such a world?

To modern man, as to Sun Yat-sen, the *Ta t'ung* may well sound "socialist," but that is historical accident. In reality, it was, as K'ang Yu-wei protested all along, Confucian. Great Unity, One World, governors "chosen from the able and the good"—these were ancient Chinese dreams, aspired to by Confucius and approximated in his name, if not in his Way, in the unity of the Han, and pursued forever after. The Middle Kingdom, after all, was supposed to mean the middle of One World, the hub of the universe, the hub of Great Unity. And the *Ta t'ung* vision of a society based on an extension of Confucian family relationships—that too was truly "Confucian" for, although Confucius may indeed have been a man who believed in putting first things first, he never meant to stop there. Just as self-cultivation was simply the first step towards peace on earth, so filial piety and parental love were only the natural beginnings of universal brotherhood. They were to be exalted and cultivated precisely in order to be extended. They were the hope of the world, not the end of it.

But what of the fact that the *Ta t'ung* lay in the past? It was in the past, yes, but it was not forever lost there. It was easy to see in the *Ta t'ung* a once and future kingdom, a once realized ideal ever waiting to be realized yet again. One of K'ang Yu-wei's disciples, in what must have been one of the first Columbia University PhD theses written by a Chinese, maintained that his master's doctrine of the Three Ages and the *Ta t'ung* "is the principle of progress; [it teaches us that] we must look for the Golden Age in the future."[69] But in terms of the present, the Golden Age's existence in the past or in the future is really somewhat of an academic question, as long as it is attainable, or reattainable, as Confucius, despite his sighs, persistently maintained it was. As a spur to action, a belief that one can progress back to a goal once reached is as good as a belief that one can progress to a goal never yet attained. Time being what it is, men have to go forward even to get back to where they were.

But there is a psychological difference. Might not Confucius's sense of progress back to the glories of the past lead one to progress only with a lessening sense of inferiority towards the ancients rather than an ever-growing sense of superiority over them? Could climbing back to the peaks from which we fell ever be as exciting as scaling new heights? It might be easier to become defeatist and quit.

In one sense, the Golden Age of the past might seem to be a closer goal than that of the future. If one believed in it, one at least had historical "proof" that it was possible. Moreover, as its attainment seemed to depend on an ever possible moral change of heart rather than any long-term development of history, it should have been available for any generation willing to turn over a new leaf, and not reserved only for the present generation's descendants' descendants. But it was not that simple, of course, for Confucius. The possibility of attaining the goal did vary according to the times. Moreover progress, and even attainment itself, were only possible for a time, for the times were cyclical. Confucius promised that men could again ascend the heights, but he did not promise they could stay there. Thus, later, the orthodox Confucian faith held to the doctrine of *i chih i luan* (a period of order, then one of chaos).[70]

If one believes in *i chih i luan,* however, can one ever have as high hopes as those who believe in linear progress? Perhaps not. But, in practice, do those without such high hopes struggle less for order if they are in chaos just because they know that order will not last? Perhaps they struggle harder than those who know that the fruits of their labor can only be tasted by others. Still, if the cycles are long enough, that latter fact is as true for the cyclist as for the progressionist, just as, on the other hand, is the ego-lifting sense that one is doing better than at least some of one's forefathers.

Historians of China, both Chinese and foreign, have often pointed to the negative aspects of cyclical belief, to the periods of snowballing pessimism brought on by a general conviction that the cycle of society is in decline. But surely progress worshippers should be just as susceptible to depression. For almost all who have professed to believe in linear progress have admitted that the line is jagged at best and capable of backward loops as discouraging as any Chinese cycle. Both beliefs can place men "with" or "against" their times. And yet, historically, neither belief has proven enough in itself to force all men to give up in the face of bad times. Hope of success is obviously not the only source of human action. There have always been men like Confucius of the *Analects,* "the one who knows he cannot make it and yet goes on."[71]

For Confucius, it was enough to know that the *Ta t'ung* should be; for K'ang Yu-wei, what mattered was the reassurance that it would be. K'ang Yu-wei wanted that reassurance so badly that he extracted from Confucius a promise he never gave. Confucius had pointed to the *Ta t'ung* as a possibility, not a certainty. But K'ang Yu-wei insisted that Confucius had written the "Li yun" as good news, as glad tidings of what would surely come to pass. Confucius, great and stern moralist that he was, had actually never cried anything but, "Be good and all will be well," but K'ang Yu-wei claimed that he had said, "Take heart. All *will* be well. Everything is bound to come out right. 'Things' are moving towards the *Ta t'ung.*"

Had K'ang Yu-wei been Jewish or Christian, he would have sided with those who insist that God's will *will* be done on earth as it is in Heaven, albeit in His own time. As it was, he insisted that the *Tao would* prevail. He thus came, by his subtle shifting of "the Great Way of Confucius" from what should be and could be to what would be, to be one of the first to view the *Tao* as something very close to Hegel's "Spirit," although, of course, he had never heard of Hegel. The *Tao,* for K'ang Yu-wei, became no longer just the heavenly Way of virtue, that could or could not be realized depending on who ruled and how, nor the changeless Way of the universe, that supported all things and would support all men, if they would only "get with It." It became a *Tao* with a purpose, a *Tao* that would have its way. The *Tao* still *was* the Way, but it was the Way in which the will of Heaven worked *its* way toward its supposed goal of a perfectly harmonious human society. The Way was the predestined course of human history.

With Hegel K'ang Yu-wei believed that the real meaning of history lay in its end, and that there was something at work in history bigger than the individual wills or even the combined wills of human beings pushing men towards that end. But K'ang Yu-wei's something was not quite Hegel's. Hegel had made his God or Spirit into the ultimate artist, seeking the ultimate in self-expression and self-fulfillment. "The Spirit," he proclaimed, ". . . has world history as its stage, its property, and its field of actualization.[72] . . . The infinite mass of wills, interests, and activities are the tools and means of the world Spirit to accomplish its end . . . to come to itself and to contemplate itself as actual reality."[73] Hegel made all creation and all human history not just a mirror of God but a mirror for God, in which, after eons of polishing, God could see his own true image reflected ultimately in the eyes of men. The end of history may have been human knowledge of God, but "the Spirit of man knowing God," said Hegel, "is only God's own Spirit."[74]

K'ang Yu-wei's Heaven was not so narcissistic, and therefore K'ang Yu-wei's teleology was much more down to earth. There was a heavenly plan at work in human history, but it was working for man. The *Tao* was trying to prevail for man's sake not its own. Heaven was "providence." Only nominally non-anthropomorphic, It would provide, only It was bent on providing for the *future* of all mankind, not for the present of believers.

Hegel, for his part, would have agreed with K'ang Yu-wei that the windings of the *Tao* were observable in past history, but he was not as confident that, in tracing the progress of the *Tao* to date, one could predict where it was going. K'ang Yu-wei *knew* where it was going, thanks both to history and to his new doctrine of Confucian revelation: "As Yang Tzu said, 'The Sage is Heaven's voice.' Heaven ordained that Confucius create his system and establish his morality. For Heaven cannot speak, and so caused Confucius to speak for It. The words of Confucius, therefore, are not the words of Confucius, but the words of Heaven."[75]

K'ang Yu-wei thus transformed Confucius into a prophet as Heaven-sent as any in the West's Old Testament. What Heaven was K'ang Yu-wei could never really say. It was "the great father of all things,"[76] and it was natural law, the subject of active verbs, and the way of the physical universe. But in his very confusion over the term, K'ang Yu-wei showed how really close he was to the many European prophets of progress. For they too, through their wildly indiscriminate use of the words *God, Providence,* and *Nature* led many to the logically baffling belief, nominally quite atheistic, that "nature" had a purpose.

K'ang Yu-wei claimed to know that purpose. It was the *Ta t'ung,* the Way of Heaven revealed through Confucius, that would "arrive with the necessity of natural law."[77] He thus became modern China's first determinist, a Confucian determinist oddly reminiscent of Christian determinists, who was to help others become all sorts of Darwinian and later Marxist determinists. In so doing, he sought, in what seemed to many China's darkest hour, to reassure himself and his countrymen that China would survive. He did not, however, mean for them to sit back in confidence. He wanted to stir them into action and self-sacrifice by giving them confidence in ultimate victory. Thus, he was perhaps also modern China's first determinist-activist.

Actually, like all determinists, he was forced in the end to find the forces of history somehow working through human action and, not surprisingly, he, "the Sage Yu-wei," chose as the executors of the *Tao* the Sages. He came, indeed, to a pre-Darwinian doctrine of evolution by Sages, or evolution by Sage-power.

This theory was quite in line with his whole interpretation of the "Li Yun," which easily lent itself to such interpretation. It was, after all, a story of evolution, the evolution of the *li,* those rules of propriety that governed everything from table manners to the highest rituals of state, and distinguished civilized men from savages. The *li* were rules of social conduct, "polite laws" of self-restraint that, if fully followed, would render the harsh, unpleasant restraints of the state unnecessary. Confucius, in the second important passage of the "Li yun," had related that it was through fostering of the *li* that an age of *hsiao k'ang* (lesser tranquility or relative peace) had been maintained by the great kings Yü, T'ang, Wen, Wu, Ch'eng, and the Duke of Chou, even after the Great Way had been lost. True, a world ordered by *li* was not equal to the *Ta t'ung,* in which eventually goodwill and unselfishness would be so prevalent that all rules would be superfluous, and all men would be able to live their lives as Confucius could at seventy, able "to follow my every desire without transgression."[78] Still, a world ordered by *li* was far better than the world as it had been without *li,* as described in the third important passage:

> Long, long ago, before the first kings had their palaces, men lived in the winter in earthen lairs, and in the summer in huts of branches. They had no fire for cooking, but lived on fruits and berries and the raw flesh of birds and beasts—blood, fur, and all. They had no hempen cloth or silk, but clothed themselves in skins and feathers. It was only afterwards, after later Sages emerged and showed men how, that men

> learned the use of fire, how to work metals, and fire clay, how to build towers and pavilions and palaces with doors and windows, how to bake and roast and boil and broil, how to ferment wines, and weave hemp and silk for cloth and satins. The ways of nurturing the living and giving funerals to the dead, of serving the gods and spirits and the Lord on High, all of them began thus.[79]

Here is still more proof that the "Li yun" is a most remarkable text for here is proof that the ancient Chinese, already "ancestor worshipers," were yet unashamed to admit that their ancestors were savages. They actually knew that their ancestors were cave-men, not from scientifically studied archeological evidence, but from a staggeringly distant line of human memory, somehow handed down quite literally from beyond the pale of civilization. Here was no Garden of Eden, no panoply of gods, no Prometheus–only wild men and later wise men, an ancient un-mythologized account of human progress.

The "Li yun" was actually so interpreted by one of the earliest thinkers in whose writings its influence can be found. Wang Ch'ung (A.D. 27–100?) of the Later Han was obviously thinking of this text when he ridiculed certain of his "Rousseauian" compatriots for glorifying the noble savage. Taking virtually the very position that Yen Fu would so heatedly take almost two millennia later, he deplored the fact that so many of his contemporaries "revere the past and belittle the present."[80] There are those, he wrote, who protest that "the men of far antiquity were simple and easily swayed by virtue, while those of later ages were extravagantly civilized and hard to rule. . . .[81] They see that in ancient times people ate fur and drank blood, and had not the five grains to eat, while in later ages they dug wells, plowed fields, and planted grain. . . . They see, too, that in antiquity men lived in caves and clothed themselves in the skins of birds and beasts, while in later ages they changed to houses and adorned themselves in cloth and silk. From this they conclude that the ancients were simple and pure and later men fancy and superficial. Actually, implements and occupations change, but man's nature and behavior are no different."[82] He meant that material progress did not necessarily mean moral decline. Material progress was a good thing. Like Ho Hsiu, Wang Ch'ung was filled with pride for the advances of his dynasty. He was writing, indeed, to assert Han progress: "The barbarian Jung and Ti are the Chinese of today. The naked men of old now wear official robes. Truly, if you compare the prevalence of virtue, then the Chou cannot surpass the Han. Militarily the Han is stronger than the Chou. In territory the Chou was smaller than the Han. Who says then that the Han is not as good as the Chou?"[83]

Once again we find that K'ang Yu-wei was impressed by a text that had also helped Han thinkers believe in progress; but, where they looked back in satisfaction at how far they had come, K'ang Yu-wei looked back only for reassurance that the Way led forward. When he straightened out the three descriptions of the "Li yun," that just above, that of the *hsiao k'ang,* and that of the *Ta t'ung,* he had a chart of human history and human destiny, a picture of man's emergence, if not quite out of the trees, at least out of the caves and then on up along the slow road to civilization, moral order, and eventually perfect happiness.

Granted, the steady progression from cave to *hsiao k'ang* to *Ta t'ung* was a distortion of the "Li yun." For Confucius, the *hsiao k'ang* already signified a falling off. It was an attempt to put things back together again, to tie a selfish people back together by the *li.* But to draw the principle if not the promise of progress from the "Li yun" was not a distortion, for by his description of primitive man even Confucius had admitted that man had once progressed from *li*-less savage to *Ta t'ung* citizen. Actually, K'ang Yu-wei's progression was just as logical as and much more satisfyingly simple than that of Confucius. For K'ang Yu-wei, man's history and destiny were simply one great course of learning. First taught basic technological skills, then rules of conduct, and finally the truth of universal brotherhood, mankind, raised by the Sages, would grow up to the *Ta t'ung.* Mankind would come of age.

GETTING THINGS GOING. Such was K'ang Yu-wei's first idea of progress, before he knew of Darwin, or even knew of a term for progress. In an un-Darwinian sense his concept was evolutionary; it was also providential. But it was a doctrine that as yet had no essential mechanism of struggle, no absolutely necessary dialectic. For K'an Yu-wei, evolution was ideally a "polite" evolution, like the evolution of the *li* themselves. Like that of the Chamberses, his evolution was enlightenment, and its proper mechanism was learning.

But this did not mean that K'ang Yu-wei was a pacifist, or that struggle had no place in his system. Struggle was not a good in itself. It was not the Way. It was not even a strictly necessary evil, for progress was supposed to flow without it or rather at its expense. Progress was supposed to mean a steady reduction of strife and struggle of all sorts. But struggle became necessary when this whole process was thwarted, when something or someone stood in the way of the Way, when obstacles blocked the *Tao.*

K'ang Yu-wei, one of the first in modern China to preach progress, became also one of the first to address the problem of China's "arrested progress," and he explained it in the seemingly simplest possible terms—as the result of villainy. Convinced that Confucius had been on the right track, he understandably concluded that progress ceased when that track was lost, when Confucius's Way was misconstrued. The villain was one Liu Hsin, who at the end of the Former Han, with the backing of the usurper Wang Mang, first established as the official versions of the Six Classics the newly discovered "ancient texts." These texts, said K'ang Yu-wei, were forgeries, and the harm that ensued from their acceptance stunted China's growth from that day to this. Liu Hsin and his ilk "seized Confucius's Classics and attributed them to the Duke of Chou, thus turning Confucius into a simple commentator. They swept away Confucius's holy plan of reform [the *Ch'un ch'iu*] and viewed it as 'a senseless chronicle of state.' The Six Classics were mixed up and hopelessly bastardized. The Sagely Way was buried and lost in a fog. Heaven and Earth went awry. Sun and Moon changed countenance."[84] The whole subsequent development of Chinese history was thus a mistake. "Confucian China" was not Confucian. Confucius, cried K'ang, had tried to change China and set her on a steady course of further change, ever changing towards the *Ta t'ung.* The rules of Confucius, set for his day, were only for his day, and to adhere to them slavishly in this was to miss his point. He had tried to show the Way, and still did. There was now only one way out: clear the Way, change the institutions, tear down the obstacles, *pien fa.*

Others had claimed Confucius as a reformer. Indeed reformers throughout the ages had insisted as Wang T'ao had in the 1880s that, "if Confucius were born today we may be certain that he would not stubbornly believe in antiquity and oppose making changes . . ."[85] But no one had as yet tied Confucius to a belief in a great impatient, pent-up *Tao,* stymied by human folly, but pressuring to get on. K'ang Yu-wei's passion for reform thus came not simply from a desire to bring Chinese institutions back into harmony with the times. He did not just seek peace and order. He wanted to open the door for the *Tao,* for history, so that it could get going again.

Thus did a philosophical belief in progress deepen K'ang Yu-wei's rationale for reform. His rationale in its peculiar entirety was quite his own. It was Chinese not Western, of classical inspiration, as yet un-Darwinian. It helped prepare the way for Darwin. But K'ang Yu-wei had very little time to spread his ideas before those of Darwin were introduced by Yen Fu. And so, it turned out, the ideas of Darwin and the *Ta t'ung* spread together–and were naturally confused.

In 1890, K'ang Yu-wei took disciples, most significantly Liang Ch'i-ch'ao. In 1891, he started a small academy at Ch'ang Hsing Li in Kwangtung and published the first step in the "proof" of his doctrine, *Hsin hsueh wei ching k'ao* (A study of the forged Classics of the Hsin period). To a selected few of his disciples he expounded on the *Ta t'ung.* More generally he preached his doctrine of Confucius as a reformer, and in 1892 he enlisted several students to help compile the book of that name (*K'ung-tzu kai chih k'ao*), which he had begun in 1886 but did not publish until 1897. In 1894, he started writing his *Ch'un ch'iu Tung shih hsueh* (Tung [Chung-shu]'s study of the *Ch'un ch'iu*), which gave the best account to date of his theory of progress. But it was not published until three years later.

Thus, in 1895, the year Yen Fu first wrote of Darwin, K'ang Yu-wei's Chinese proof of progress was not well known. Indeed, if there was any stated proof of progress that was well known, it was a Western one, expressed in Timothy Richard's translation of Robert Mackenzie's *The 19th Century* (translated as *T'ai hsi hsin shih lan yao* [The essentials of recent Western history]), widely read in China in 1894.

Coming with perfect timing, in the middle of the Sino-Japanese War, *The 19th Century* was the perfect work to whet the interest of men who in the very next year would be impressed by both K'ang Yu-wei and Darwin. It was perfect background reading for the intellectual endeavors of the next twenty years. Written with a goodly measure of British self-satisfaction, tempered though it was with British good humor, the bulk of the book was a most colorful description of the fantastic forward strides England (and even other places) had made in the last century. Ebulliently confident that Britain now stood first on the threshold of an undreamt-of new level of civilization, Mr. Mackenzie reveled in depicting the darkness out of which she had so recently come, for the darker that darkness the more brilliant shone Britain's progress—and, by extension, the progress of mankind.

It was, indeed, progress that *The 19th Century* was meant to prove: human history is a record of progress—a record of accumulating knowledge and increasing wisdom, of continued advancement from a lower to a higher platform of intelligence and well-being. But as Mr. Mackenzie finished his book with that happy conclusion, he introduced, almost as an afterthought, one elaboration that was not so happy:

> The rate of progress, as the eye of man deciphers it, is irregular and even fitful. Now it seems to pause, and the years seem to repeat themselves unalterably. Now it bursts forth in sudden ameliorations, in the violent overthrow of evils, which had been quietly endured for generations.

No one in China had yet said that progress required violence. And no one had yet said that the enemy of progress was despotism. But Robert Mackenzie said so very clearly:

> The nineteenth century has witnessed progress rapid beyond all precedent, for it has witnessed the overthrow of the barriers which prevented progress. . . . Despotism thwarts and frustrates the forces by which providence has provided for man; liberty secures for these forces their natural scope and experience. The nineteenth century . . . has seen all that the most ardent reformer can desire. . . the removal of artificial obstacles placed in the path of human progress by the selfishness and ignorance of the strong. The growth of man's well-being, rescued from the mischievous tampering of self-willed princes, is left now to the beneficent regulation of great providential laws.[86]

Gentle as Mr. Mackenzie might seem, for China in 1894 that was revolutionary rhetoric. But the Chinese never heard it. Timothy Richard's translation had no words for *progress* or *violent overthrow,* or *despotism,* or *liberty.* In Chinese, Robert Mackenzie simply said, "The life of man begins in infant weakness, grows to maturity, and then declines. And so it is for countries; as they flourish so must they decline. Only the earth [*ti-ch'iu*] grows without getting old and flourishes without declining. . . . [For] the ways [*fa*] of former times have been passed on to later people who have passed whatever new ways they could add to them on to people later still." Human institutions might "sometimes appear stuck." They might sometimes be "like a chronic illness of many years that cannot be healed." But Mr. Mackenzie gave no Chinese prescription for cathartic violence. Instead he simply said that "sometimes with a sudden change all is clear and a new world forms." For "there is a great Way encompassing all nations," and "it is the will of Heaven that men should become ever better than before, until they gradually approach the perfect virtue of Heaven above."

The Chinese text calmed Mackenzie's progress. Yet the Chinese text still had a force of its own. Certain of its phrases could provide ample food for thought for the despot, for "the most ardent reformer" (although that term too was missing)—or for the revolutionary:

> Of old, those who wielded power and influence long obstructed the people and did not let them put forth any new ways. . . . In this last one hundred years, the great nations of this earth have already gotten rid of the great evil of power and privilege, and their people enjoy great peace and prosperity. . . . Therefore the greatest evil in the world lies in men without vision, who know only that they themselves exist, and who can thus use their power and privilege, with perfect equanimity, to oppress others, and keep the common people from maturing and advancing. But the people of the future, once they have escaped the bitter bonds of imperial and royal authority, will move in accord with Heaven, and all enjoy boundless bounty."[87]

Even in translation, *The 19th Century* gave powerful support for K'ang Yu-wei's contention that, one way or another, men must clear the way for the *Tao.*

On the eve of Darwin's day in China, there was explosive political power in that contention, far more, indeed, than K'ang Yu-wei imagined or desired. But there was also powerful philosophical confusion, the confusion ironically caught perfectly in the last two words of *The 19th Century,* which Richard had failed to translate, most probably, and understandably, because neither he nor his assistants knew how. There was a conundrum in those words, one that lay, in Christian terms, at the heart of Mackenzie's thinking, and in Confucian, Taoist, and Buddhist terms at the heart of K'ang Yu-wei's. And it would lie at the heart of the thought of Yen Fu and of almost all of China's Darwinists and Marxists yet to come. It was the conundrum inherent, but unrecognized, in that most confused and confusing notion, "providential laws."

TWO

Enter Darwin

Defeat in the Sino-Japanese War left China afraid not of Japan but of the West. The back-stabbing gall of the Japanese, in their first full-fledged attempt to play at Western imperialism, was enough to make Chinese patriots weep with rage and frustration, but it was not enough to make them fear that Japan might conquer China. No one could yet believe that China could not have defeated Japan in an all-out war if she had wanted to, and many were therefore furious that the court had chosen to buy such a despicable peace.

The court, however, may not have been as stupid and self-seeking as K'ang Yu-wei and the 1,300 odd signers of his "ten-thousand-word memorial" thought, for a protracted war with Japan might indeed have invited the very foreign aggression K'ang Yu-wei believed would ensue from capitulation. There was no good way out through either peace or war, and the court, although choosing peace, must have been agonizingly aware of the truth of K'ang Yu-wei's contention that, "if Japan can so terrify us with big talk that she can grab the whole island of Taiwan without loosing a single arrow, then all the barbarians will think China is easy to bully. The French will ask for Yunnan and Kwangsi, the English will ask for Tibet and Kwangtung, the Russians will ask for Sinkiang, and the Germans, the Austrians, the Italians . . . the Portuguese and the Dutch will all connive to join in."[1]

This was no exaggerated fear. For at least another decade, the greatest threat to China was still the West, not Japan. There was every reason in the world for the immediate response to the Sino-Japanese War to be the frantic cry, "If we cannot even crush the puny Japanese, how can we repel the wretched Westerners?"

WHENCE STRENGTH?

The basic problem in 1895 was the same problem that had plagued China since the Westerners first arrived. Before that time, wrote Yen Fu, "We thought that of all in the human race none were nobler than we. And then one day from tens of thousands of miles away came island barbarians from beyond the pale, with bird-like language and beastly features, who floated in and pounded on our gates requesting entrance and, when they did not get what they asked for, they attacked our coasts and took captive our officials and even burned our palaces and alarmed our Emperor. When this happened, the only reason we did not devour their flesh and sleep on their hides was that we had not the power."[2]

Power was the problem—by far the more immediate half of the goal of "wealth and power" already voiced in a thousand essays. And the corollary question, now asked with more urgency than ever before, was quite naturally, where did the Westerners get their power? What was their accursed secret?

The Chinese, of course, had long since admitted that something would have to be changed if China was to handle the Westerners. They had admitted that much, although reluctantly, after they had failed to change the Westerners themselves and make them act as respectful vassals according to the protocol of the Middle Kingdom. So they tried the traditional method of changing their own men, and in so doing found in time a team of formidable men indeed. But they also found that new men were not enough, and the new men themselves demanded further change, a change of weapons and technology. It was when that too proved not enough that the Chinese turned in desperation to institutions.

That is where they were in 1895, when K'ang Yu-wei and hundreds of others took up the cry of *pien fa.* And it was to support that cry that Yen Fu first dared step into print. When he did, however, his very first steps placed him in a position which, although not immediately recognized as such, was infinitely more radical than that held by any of the ordinary advocates of institutional change. For, although he too addressed himself to the secret of Western power, he found that secret ultimately no more in institutions than others had in weapons. For Yen Fu, it lay beyond institutions, far deeper, in men's minds. The Westerners' secret was in their attitude, their philosophy.

Without vaunting his expertise, and yet speaking unavoidably with the rare authority of one who had sought for the Westerners' secret in their own land and from their own books, Yen Fu pronounced that "the greatest and most irreconcilable difference between Chinese and Western thinking is that the Chinese love the past and neglect the present, while the Westerners strive in the present to surpass the past. The Chinese believe that to resolve from order to disorder, from ascension to decline, is the natural way of heaven and of human affairs. The Westerners believe, as the ultimate principle of all learning and government, in infinite, daily progress, in advance that will not sink into decline, in order that will not revert to disorder."[3]

So the Westerners' secret was belief in progress. Chinese believed in cycles and got nowhere. Westerners believed in progress and progressed. But how could this be? Was this the power of positive thinking? Did the West get ahead just by saying, "I think I can, I think I can"? Were they number one because they tried harder? Many were to come to think so, and Yen Fu seemed to. The West was stronger, so he said, because "China trusts to fate; the Westerners rely on human strength."[4] But it was fate that ruled the world. Yen Fu had just said so in his opening sentences, his very first words to China: "Change in the world—none knows whence it comes. For want of any better name, we call it destiny. But, once destiny is set, not even a Sage has any power over it. For even a Sage is an entity within destiny and, being so, can obviously not shape its course."[5] If not a Sage, how much less so can the rest of us, and yet, if we cannot, how can it be our beliefs and actions that make us strong or weak?

Yen Fu's very first step thus took him in over his head. We should not, of course, fault him for this, for being in over our heads is perhaps our natural condition, but we must note that Yen Fu's particular contradiction, here at the very beginning of his work, is one of the most persistent confusions in all modern Chinese thought, one that runs from K'ang Yu-wei to Mao Tse-tung. It is a confusion "fated" to all would-be determinists who want also to believe that human determination is a determining factor in human history.

Yen Fu, perhaps like most serious thinkers, was tormented by the desire both to have his cake and eat it too or, in the Chinese variant, to have both fish and bear's paws.[6] He did believe that the Westerners' attitude made them strong, and he believed it not in the relativistic way in which K'ang Yu-wei for a time believed that the Westerners' strength lay in their faith in Christianity, and that Chinese could be as strong if they would only believe in Confucianism with similar fervor. K'ang Yu-wei, in his own confusion, and quite to the detriment of his contention that Confucius's Way was true, fell into the trap of lauding the power of religion not as truth but as a force of social unification, which in the end meant that it was not a particular belief but simple solidarity that was strength.[7] Certainly the possible psychological power of any faith is easily discernible (if unexplainable) in history, however many have, like the Boxers, learned that one cannot survive on faith alone. But the pyschological power of faith was not Yen Fu's point. The Westerners' belief, not in Christianity but in progress, gave them power because it was based in truth. There was progress.

But here he got into trouble. Yen Fu wanted to laud the Westerners for holding the correct world view, and for acting upon it. He wanted to commend their endeavor to act realistically, but he also wanted to believe that it was actually the universe that ran everything, so that those who endeavored to act unrealistically were doomed to despair. The real power of human action, which the Westerners supposedly relied on so successfully, was thus, from the beginning, thrown into most serious question—and so was the force of the universe.

Conflicting pulls of determinism and "determinationism," strained Yen Fu's emotions and his logic, nowhere more clearly than in his heated denunciation of "the Sages." He began his essay "On the Race of World Change," after all, by claiming that the Sages had no power over destiny whatsoever. The Sages only seem to effect change, he began: "They simply know the course of destiny, and can predict its destination . . . hence they can wait for Heaven and grasp Heaven's time—and they can also act before Heaven, and Heaven will not go against them. It is thus that they seem to bring the world to peace. Later men look at their successes and think that Sages really can change the course of destiny. They do not realize that the Sages actually have nothing to do with it."[8]

Sage power was *not* the motive force of human progress. And, in a second essay, "P'i han" (In righteous refutation of Han Yü), Yen Fu tried to make that even clearer. He pounced on the famed T'ang literatus Han Yü for taking a position that was almost precisely that of the much earlier "Li yun" chapter which so inspired K'ang Yu-wei. "In Han Yü's essay, 'Yuan *Tao*' (On the source of the *Tao*)," he wrote, "it says that, 'in ancient times, the dangers that beset mankind were many. But Sages arose and as men's lords and teachers taught them the ways of living together and supporting each other. . . . If of old there had been no Sages, then mankind would long since have perished. Why? Because men had no fur or feathers, no scales or shells with which to withstand heat and cold, no claws or fangs with which to fight for food.' But, if Master Han's words are true, then the Sages and their fathers and their grandfathers could none of them have been people. Indeed they must have all had feathers and fur and scales and shells and claws and fangs . . . else they themselves would have succumbed in their infancy to the ravages of the elements, to cold and hunger, and to the attacks of insects, serpents, birds, and beasts."[9]

We may well wonder how Yen Fu, already excited by Darwin,

could possibly have passed up such a chance to shout out that the Sages' ancestors, and even our own, probably did have fur, if not feathers, and only began to lose it after they stopped needing it. The fact that he did not is our first proof that, from the beginning, his excitement over Darwin was political and sociological, not anthropological or biological. Later, others who were more curious about man's flesh-and-blood beginnings would look again at Han Yü's ill-fit and ill-feathered friends, but, in this essay, Yen Fu had not the time, for his purpose was quite political. He wanted to discredit the Sages because of the harm he felt belief in them had inflicted and was inflicting on China. He wanted to destroy forever the idea that progress had been spoon-fed to mankind by near supermen, for he wanted to prove that progress comes out of the people.

But here, arguing back and forth, he outran his logic in two directions. First, when he turned to destiny to crush the myth of Sage-power, he grasped a force that was logically just as devastating to people-power. But then, scarcely three paragraphs further on, he suddenly devastated destiny, by giving back to the Sages the awesome negative power of being able to hold back destiny for two thousand years. Not content to take away the Sages' credit for China's progress, he insisted on heaping blame upon them for China's lack of progress. But, in so doing, he empowered them anew.

Who were these Sages anyway? In context, they suddenly seemed to be not only the philosopher-kings and the philosophers but all China's rulers, so that Yen Fu's outburst was not only Confucian blasphemy, but something perilously close to treason. Chou unity, the Ch'in's melting down of weapons and burning of books, and in the same breath the great examination system, all were "sagely" ways of "preventing conflict," and thus progress. "Alas!" cried Yen Fu, "here indeed we see the artistry with which the Sages bridled our empire, put down contention and quelled disorder. And it is because of such artistry that our people's knowledge has steadily decreased and their power day by day declined.

The Sages never realized that in the end they might render us incapable of contending with foreign nations for another day's existence."[10]

The Sages did us in. They told us not to struggle, said Yen Fu in his first protestation that struggle was necessary, natural, even good for one. And the reasons he gave were Malthusian, although he failed to mention Malthus. "The produce of Heaven and earth is limited," he announced, "but human desire knows no bounds. Men keep multiplying, cultivation is ever extended, and in the end there is not enough. And, when there is not enough, men contend. But contention was a great 'crime against humanity.' So our Sages chose instead to teach contentment with bare necessities, so that everyone would be satisfied with low and ignorant simplicity, and till the soil to serve his betters."[11]

Small wonder that Yen Fu looked so coolly on K'ang Yu-wei's later shoutings to "preserve Confucianism" (*pao chiao*). For here he made an almost Nietzschean attack on Confucian morality, or a part of it, as the cause of China's backwardness. He dared denounce the ancient virtue of *shou pen fen* ("Accept your lot," or "Keep in your proper place") for having kept China too well in one place. Before he ever preached of Darwin, China's first Darwinian had already had doubts about Confucian goodness that were strangely close to those in Nietzsche's stabbing question, "What if the 'good' man represents not merely a retrogression but even a danger, a temptation, a narcotic drug enabling the present to live at the expense of the future? More comfortable, less hazardous, perhaps, but also baser, more petty—so that morality itself would be responsible for man, as a species, failing to reach the peak of magnificence of which he is capable?"[12]—or for China failing to retain her rank as the Middle Kingdom.

Dark doubts indeed, held a quarter of a century before it became "the thing" to scream, "Down with Confucius and Sons!"[13] But for now, let us simply note in what an awkward position Yen Fu finally left the Sages and, by extension, all mankind. On the one hand, he quoted Kuo Sung-t'ao, with whom he had first pondered the Westerners' secret in England: "Once the workings of Heaven and earth have started, no one can stop them. The powerful in their selfishness may try . . . but no one has ever won."[14] This was perhaps the first modern Chinese cry of that hackneyed Western phrase, "You can't stop progress." But, on the other hand, Yen Fu claimed that the Sages had stopped progress, so much so that China's very survival was at stake. As so often has been the case, "You can't stop progress" was angrily shouted at those who seemed to have done so, at those whose maddening presence proved that human beings, supposedly "but entities within destiny," had yet the power to thwart it, if not forever, at least for far too long. Thus Yen Fu, about to help send off his countrymen in glad pursuit of a new, "scientifically" naturalistic philosophy, still slipped man into a semi-supernatural position—at least far enough above the works to throw in wrenches.

So Yen Fu was bafflingly inconsistent. The Sages had first no power and then negative power but not positive. They only seemed to order the world, because through foreknowledge they could keep in step with it, and yet their seeming power was the power blamed for throwing China out of step. The Westerners, on the other hand, were deemed positively powerful for doing the same thing the Sages seemed to do. They had the power to *make* progress because they *knew* the world was progressing. And yet, to add one more confusing element, their knowledge-power worked only because they were "free": "You can't stop progress," but "what makes the difference is freedom."[15]

There is no logical way out of these contradictions in Yen Fu's early essays. But never mind. Confusion was a deeply important part of Yen Fu's influence. And it was with just the above flourish of confusion that he set the stage for Darwin. It was with infectious excitement, not logic, that he, as had K'ang Yu-wei and the translated Mackenzie, painted progress as the promise of mankind. But progress, of course, set only half the stage. The other half, which seemed to loom larger, was painted with the dark colors of potential doom, all too obvious to all—the picture of whole countries and peoples "going under."

Darwin, for many, would pull these two pictures together. In China as in the West, both would be part of the Social-Darwinian vision. But consider how easy it was in Chinese terms to paint the second Darwinian picture without mentioning Darwin, indeed how easy it would have been for Chinese to have invented Social-Darwinism without Darwinism proper.

In the disheartening winter of 1895, K'ang Yu-wei, losing sight momentarily of the light of the *Ta t'ung,* wrote an impassioned introductory editorial for the Society for the Study of Strength's new journal, in which he drew the following warning from the world of nature:

> The largest animals are the camel, the elephant, the mule, the horse, and the ox. All are of tremendous size, all several times as big as man. And yet men corral them, bridle them, break them, and ride them. They even slaughter and roast them. Camels, oxen, and horses try to voice their suffering . . . but, despite their cries, no one saves them, no one redresses their wrongs. Why? Because they are weak. Oxen and horses are guiltless, they are innocent, yet, however hard they toil and slave, they cannot escape the butcher. Why? Because they are stupid. The *Book of History* says, "They attack the stupid and annex the weak." If one is stupid and weak and so invites annexation and attack, then what can one expect?[16]

K'ang Yu-wei then delved into Chinese history with its wealth of stories of conqueror and conquered. What difference between the conquered and the camel? Just look at the West, from Alexander's far campaigns to the recent, brutal cutting up of Africa: "Different tribes and different races have all been gobbled up through their own weakness and stupidity."[17]

The moral was that "Heaven is unknowing. It blesses only the strong." The consequences were obvious: "Does China want to fall before a foreign race?" If not, if she wants to preserve her race and her religion, and avoid the fate of elephant and ox, then there is only one way out: The *Book of Changes* tells us to "strengthen yourselves without ceasing." The *Great Example* says that, of the six extremes, weakness is the worst: "But strength and weakness are matters of fact. Is it not true that even Sages must follow

Heaven? Then there is only one thing to do, strengthen ourselves."[18]

There were two kinds of strength, mental and physical, and brain could beat brawn (that was why men could conquer tigers), so China must study. There was one other secret, however, revealed by Hsun Tzu: "Animals cannot organize as groups (*ch'ün*); only men can organize as groups." (Otherwise, said K'ang Yu-wei, even insects would be terrifying, not to mention elephants.) So China must realize that "all creatures alone are weak, united strong." China is huge but her people "hold aloof and will not group; they are stupid and will not study." So "in study there is strength; in grouping there is strength."[19]

Here was a Social-Darwinesque slogan in pure Chinese, "Heaven blesses only the strong," immediately coupled with an ancient Chinese version of "solidarity is strength," two powerful ideas for any drive for social reorganization. But K'ang Yu-wei's purely Chinese version did not prove as powerful as Yen Fu's Darwinian one. K'ang Yu-wei took his examples from nature and human history—indeed he virtually claimed that his slogans represented natural law—but his Chinese natural law had not the authority of Western natural law. Already, thanks to the show of Western technology and to the efforts of the treaty ports' early magazines and papers, Chinese were beginning to believe, however reluctantly, that Westerners really knew the laws of nature. They were beginning to believe in "science," although they still had only a borrowed Neo-Confucian term for it.[20] That is why Darwin proved so powerful. Darwin was a "scientist," and what a scientist said was probably true.

Nonetheless, the similarities between K'ang Yu-wei's early approach and Yen Fu's were striking, and, after all, in 1895 their cause was still the same. K'ang Yu-wei brought forth "the largest animals" for the sake of the Society for the Study of Strength. Yen Fu brought forth Darwin to answer the title of an essay, "Yuan ch'iang" (Whence strength?). Yen Fu claimed that Darwin knew.

ASK DARWIN

"Do those now madly talking about Western learning and foreign affairs have any idea what it is Westerners have been ceaselessly pursuing for the last fifty years, that in accord with which they are able not only to order and protect their own lives, but also to govern their countries and enrich their people?"[21] Yen Fu challenged. What Westerners had been pursuing were ways to fit the truth of Charles Darwin:

> Darwin was an English scholar of animals and plants, who as a young man, carrying on the scholarship of his family, sailed round the globe, collecting in great profusion the most rare and exotic of flora and fauna. Then, after infinitely careful research, extending over many decades, he wrote a book entitled *In Search of the Origin of Species.* After it came out, it was soon to be found in almost every home in Europe and America, and Western scholarship, government, and philosophy all drastically changed. Someone has said, and it is no idle boast, that Darwin's theories have given men new eyes and ears and have changed their thinking more even than did the science and astronomy of Newton.
>
> Darwin's book says that species in their multiplicity originally were one. The fact that they have come down in such infinite differentiation is largely due to the varying influences of earth and sky. Nature tending towards slight variations, species have so spread from their source that they are now separated by great gulfs and can no longer return to one. But this is all the result of development. Things have come out as they are gradually and in accord with nature. Life was not set this way in the beginning.[22]

So was Darwin introduced to China. And how Chinese he seemed! Continuing the work of his fathers (at least of his grandfather), he had the patience of a Confucian compiler. He spoke only after long and painstaking study. And how Chinese was his first conclusion, that the many are descended from one. Lao Tzu had said, "The *Tao* gives birth to one, and one to two, and two to three, and three to the ten thousand things," which meant, explained the third-century thinker Wang Pi, that "the ten thousand things, the ten thousand forms, all go back to one."[23] This was an idea almost universally accepted in Chinese thought, certainly at least by the Neo-Confucianism that was still the official doctrine of the day. Of course, Chinese philosophers had not meant by Wang Pi's statement, or the Neo-Confucian dictum that "from one principle come all forms,"[24] what Darwin meant when he said that "probably all the organic beings which have ever lived on this earth have descended from some one primordial form, into which life was first breathed."[25] Nevertheless, as Yen Fu stated it, Darwin's belief struck a resonance in Chinese terms that was a far cry from the thundering dissonance his original sentence, despite its cryptic final phrase, produced in Christendom, where the faithful had thitherto held as an inviolable pillar of their faith the gospel that "each species has been independently created."[26] When Darwin attacked that doctrine, he expected in so doing to "get more kicks than half-pennies."[27] But Yen Fu, on this score at least, could report Darwin's theory without any such fear. For although there were, to be sure, passages in the Chinese Classics that upheld the idea of the immutability of species,[28] there was no great faith, and certainly no church, that saw itself standing on such a foundation.

For the same reason, the idea that man had animal ancestors produced no immediate shock in China, despite the prevalence of "ancestor worship." It is true that Yen Fu did not stress man's family tree in his first essay on Darwin—he did not cite Darwin's portrait of man's ancestor as "a hairy quadruped, furnished with a tail and pointed ears, probably arboreal in habits"[29]—but he did say that "men were of a class with animals."[30] So stated, however, that idea offended few. For, in most Taoist and Neo-Confucian thinking, man had always been called a "thing" (*wu*), one of the "ten thousand things" (*wan wu*), and related in principle to the lowest of them.[31] Lacking a hypersensitive clergy, the Chinese did not seize upon the lurking philosophical implications of Darwinism until later.

STRUGGLE AND THE CH'ÜN. What was shocking in Yen Fu's presentation was the Darwinian doctrine of the "struggle for existence" and its somber corollary, "natural selection." Yen Fu created two literary Chinese terms to convey these concepts; *wu ching* ([living] things contend) and *t'ien tse* (Heaven [or nature] chooses). "'Things contend,'" he explained, "means that things struggle to preserve themselves. 'Heaven chooses' means that only the fit races are preserved."[32] Putting the two together, *wu ching t'ien tse,* he created a new idiom that in four stark syllables captured two of the most potent of Social Darwinian slogans.

He went on to explain them more fully: "Men and all other living things are born on the earth in great profusion, and together they share the advantages of nature. But they band together, and each people and each species struggles to preserve itself. Thus, in the beginning, species struggles with species, and group struggles with group, and the weak are devoured by the strong, and the stupid enslaved by the wise, so that, in the end, those who survive and have descendants are perforce the hardy and the tough, the outstanding, the able, the skillful, the intelligent—those who at that moment are most fit for their time, their place, and their human situation."[33]

But struggle (*cheng*) was a bad word in Chinese. The Confucian Hsun Tzu had protested that "struggle (or contention) is disaster."[34] The Taoist Lao Tzu preached "the virtue of non-struggle."[35] So, in praising struggle, Darwin did go against the Classics. And Yen Fu realized it, for he himself challenged the Classics on just that issue, anticipating conservative attack. In lambasting the Sages, he had already specifically attacked the words of Hsun Tzu, although he rather cleverly refrained from mentioning Hsun Tzu's name. Now he took up the challenge of the Taoists, attacking the view that the true Way lies in passivity, that as "everyone knows . . . weakness conquers strength."[36]

The theoretical Taoist objections to Darwin's "message" that Yen Fu anticipated were not moral objections. They were tactical, for Taoism too was a "survival" philosophy. Taoism simply took the side of La Fontaine's reed instead of the oak. Philosophically, Lao Tzu was a devotee of *judo.* His way was the *jou tao,* the soft way, but it was still the way to win, the way to live. Therefore the Taoist argument that Yen Fu answered was not that aggressive struggle and strength were bad, but that they were bad for you: "You will object," wrote Yen Fu, "that 'strength leads to death,' but you do not realize that [if it does] . . . there are reasons. . . . If one is good at preserving one's strength, then strength is precisely what insures one's survival. If one is not good at using one's softness, then softness is exactly that which will speed one's death."[37]

"Lao Tzu's talk of using the power of female passivity," he went on, "is a masterful technique for sagely wisdom, but that does not mean that one should 'not apply oneself, but let things come out as they will.' . . . For has Heaven ever lessened the cold for those who will not weave, or lessened the hunger of those who will not plow?"[38] So, with apologies to Lao Tzu, Yen Fu's exhortation, backed up by Darwin, was for China to get in there and fight.

That message, even with its "scientific backing," was shocking, but it was also electrifying—because many Chinese intellectuals wanted to fight. If Darwinism had been introduced to China in an age of peace, in a secure age of inner and outer harmony, there might have been an immediate moral hue and cry somewhat more akin to the instant *odium theologicum* so loudly voiced in the West.[39] But in the charged atmosphere following the Sino-Japanese War, traditional moral aversion to the word *struggle* was momentarily put aside.

Still "Get in there and fight," in Darwinian terms, was no simple message. Even in this first introduction, Darwinism, once let in, proved a Pandora's box of deep and disturbing questions. The amazing thing was that Yen Fu in "Whence Strength?" touched on so many of them, even though he failed to ask them outright. Benjamin Schwartz has well called "Whence Strength?" a prolegomenon to the whole of Yen Fu's thought.[40] It might equally well be called a prolegomenon to Chinese thought in general for the thirty years that followed. For, with remarkable prescience, Yen Fu raised almost all the major themes his contemporaries were to argue about and finally fight about so heatedly.

The first and most important theme was solidarity. "Whence

strength?" asked Yen Fu—"Solidarity is strength." But he expressed this conviction not in the later leftist translation of that Western phrase, *t'uan-chie chiu-shih li-liang,* but in a most subtly twisted paraphrase of a passage from Hsun Tzu, which, as we have seen, K'ang Yu-wei also used, and which hundreds of others were soon to use as well: "Hsun Tzu says," said Yen Fu, "that man is nobler than all the animals because he can *ch'ün,* " because he can group.[41]

Now this was not what Hsun Tzu really said. He said man was the noblest living thing on earth because he had morality (*yi*). He was stronger only because that morality enabled him to unite, to band together in a tight, well-knit group.[42] Hsun Tzu's argument was first and foremost a moral argument. Like all true Confucianists, when he talked of strength he talked of "moral rearmament." Still it is perfectly true that solidarity, although not the source of strength, was yet the manifestation of it. The *ch'ün* was indeed a *sine qua non* of human life,[43] and man's ability to *ch'ün* did set him apart. Apparently Hsun Tzu was unimpressed by the amateur naturalist's observation that "birds of a feather flock together," for he seemed to believe that only man could truly "flock." This was the argument K'ang Yu-wei borrowed for his self-strengthening editorial, and the classical argument Yen Fu chose to bolster Darwinism.

But it was so much more than that. When Yen Fu seized upon Hsun Tzu's magic word, *ch'ün,* he grasped with uncanny appropriateness the perfect watchword for Chinese Social Darwinism as a whole, for its unique and distinctive genius. For if there was anything, and there was not much, that almost every Chinese Social Darwinist shared, it was the predilection for the *ch'ün* as *the* important element in the struggle for existence.

Of course, the *ch'ün* was important elsewhere. Western Darwinian justifications for nationalism, racism, and imperialism are well known. Indeed, they played all too great a role in reinforcing China's primacy of interest in the *ch'ün.* But, however overshadowing, they were still the branches, not the root, of Western Social Darwinism, at least in Britain and America. For there, when people first caught the Darwinian fever, they looked inwards. They looked first at the struggle within species, at the struggle within their own societies. And they came up with another reason to worship "rugged individualism." Herbert Spencer was in America hailed first and foremost as "the shining light of evolution and individualism";[44] it was no coincidence that one of his "two best American friends" was Andrew Carnegie.[45]

How different it was in China. Although the struggle within species was for Darwin the ultimate cause of the transformation of species, Yen Fu did not mention it at all. "In the beginning," wrote Yen Fu, "*ch'ün* struggled with *ch'ün,*" group with group—not individual with individual. As for Herbert Spencer, Yen Fu introduced him to China as being "like Hsun Tzu,"[46] because he felt that man was man because of the *ch'ün.* Consequently Spencer's science became *ch'ün-hsueh,* the "study of the group," a translation of "The Study of Sociology," but which in context really meant "the science of group strength." For *ch'ün-hsueh* was the study of "that by which a race is strong and that by which a group can stand."[47]

To make Spencer into the shining light not of individualism but of "*ch'ün*-ism" was not, of course, an altogether gross misrepresentation. Spencer could "shine" in either direction. To Yen Fu he obviously was most dazzling in his (pseudo-) scientific explanation of just why it was that solidarity was strength, that the strength of the group lay in its ability to group. This blinding light lay in that most devastating metaphor and myth, the "social organism." Spencer, the social scientist, realized, wrote Yen Fu, that "when a *ch'ün,* a society, is formed, it is in body, function, and capability no different from the body of a living thing. . . . If we know what keeps our own bodies alive, we will know what makes a *ch'ün* secure."[48] So too "a country is like a body,"[49] a true "body politic." That is why China would be strong if only it would "organize," if only it would act as a social organism, if only it would pull itself together and act as one body.

The consequences of not acting as one body were deadly serious:

"If a people is stupid and afraid, with each individual out for himself, then its *ch'ün* will fall apart, and, if a *ch'ün* that is falling apart meets a people that is fierce and has much knowledge, that loves its country and protects its race, then at very best it will be enslaved; at worst—it will be exterminated."[50] Thus did Yen Fu, only a few short paragraphs after his introduction of Darwin the naturalist, leap immediately to the realm of Social Darwinism: the world not just of warring states, but of warring organisms, warring social organisms, of which only the fit would survive.

Then how was one's social organism to be made fit, to be made healthy and strong? The only way lay in cultivating the three people's strengths, *min li, min chih* and *min te,* "the people's physical prowess, the people's knowledge and intelligence, and the people's virtue."[51] These three, of course, represented Yen Fu's version of Spencer's three energies, the physical, intellectual, and moral,[52] all of which had to be developed before a people could be secure.

At first glance, it would seem that Spencer's facile inclusion of morality in his scientific analysis of strength would make it a most suitable doctrine for "Confucian" China. Indeed it is odd that the Confucian revivalist K'ang Yu-wei should have neglected to include morality in his own analysis of strength, recognizing only "physical strength and intellectual strength."[53] It is very odd, for his analysis was obviously based on the above-mentioned passage from Hsun Tzu, in which man's ultimate source of strength was said to lie not in his physique or "know-how," but in his morality. That, of course, was *the* Confucian argument, that right was might, that virtue worked, that it was the Way to peace and prosperity. With virtue, Confucius, Mencius, and Hsun Tzu each insisted, the smallest and weakest state could hope not only to survive but to conquer.[54] By virtue alone, it could hold the world in sway.

The reason neither K'ang Yu-wei nor Yen Fu quoted this part of Hsun Tzu's argument was probably because they no longer believed it. They were, in effect, making precisely the opposite argument, that morality in itself was not enough. Physical and mental strength were also essential. It was no time to sit back and rely on the virtuous power of non-action (*wu-wei*). It was even less a time to quote the more pessimistic and slightly contradictory Confucian dictum: "If one cannot have both, then give up life and keep morality."[55] For survival was *the* issue, *China's* survival that is. It was all right, indeed it was highly moral, for the individual to give up his life for China. But for China to give up her life for morality, to hold to righteousness at all costs, to cling to her integrity and let the barbarians do their worst—was unthinkable. At least Yen Fu did not think of it, for he did not conceive of such a dilemma. China's survival had also become the moral issue. It was that which defined, and limited, the very concept of morality.

Yen Fu used the word *virtue* to mean public-spiritedness, that selfless dedication to the group that was essential to its preservation, that which alone allowed physical and mental strength to be effective. What was right, therefore, meant what was right for one's group. And this was only "natural," for, as Yen Fu said, "Western theorists of political philosophy hold that the first law for living organisms is for each to protect its own life. The second is to protect its species. But, when it is a question of one's own life or the life of the species, then one should sacrifice oneself to preserve the species."[56] Just why this should be true, if self-preservation was the first law, he did not say, but such self-sacrifice was a "natural law," and it was also meet and right and one's bounden duty.

So convinced was Yen Fu that China's survival was both a natural and a holy cause that he could not conceive of the possibility that in following that cause the group might go wrong, and force the moral individual to stand against it. That is why Yen Fu could so easily say that China need only "take the people's might, knowledge, and virtue as our standard. Whatever will advance these three we must carry out with all our strength. Whatever is to their detriment we must abolish."[57] He mentioned no possible conflicts among the three. He did not even mention the all-too-common rub between might and virtue. At least he did not admit that there might be policies that could help the one and hurt the other. And yet he was not shouting "My country right or wrong." He did not conceive of his country's being wrong.

Yen Fu probably would have been shocked if someone had accused him of claiming that the end justifies the means, that the need for survival could justify anything. By temperament Yen Fu was no more a ruthless and callous Social Darwinist than Darwin. He would have shared the sense of shock, if not the amusement, that Darwin revealed in a letter to his friend and mentor, Sir Charles Lyell: "I have received in a Manchester newspaper rather a good squib, showing that I have proved 'might is right,' and therefore that Napoleon is right, and every cheating tradesman is also right."[58] Yen Fu would not have believed such a "squib" any more than Darwin did, but, like Darwin, he would have been hard-pressed to prove Darwin had not proved what the squib said he had.

Actually, Yen Fu had simply not yet seen or addressed himself to the philosophical problem inherent in the unpleasant possibility that, although right might be might, and might might be right, right also might not be might, and might might well be wrong. Yen Fu's enthusiastic acceptance of Spencer's inclusion of morality in the make-up of group strength revealed no appreciation of the perplexing problems that morality raised for Social Darwinism. Morality as the great stumbling block of Social Darwinian thinking as yet lay unforeseen.

THE RACE AND DEMOCRACY. Yen Fu's passion for the *ch'ün,* and his excitement with the concept of the social organism, did, however, lead him on to two of the other major themes that were to engross Chinese thinkers in the early twentieth century, the themes of race and democracy.

The problem of race he only touched upon, but the ways in which he did so helped open the door for a generation of unpleasant racial thinking. Originally, the problem came up in answer to an obvious question: If "the group" was so all important, which group was the group—the immediate answer, of course, being China—but China as country, people, or race? The answer was never made perfectly clear, but Yen Fu laid great stress on the word *race,* in order to refute an imagined critic who protested that China was not in danger of extinction. She was not in danger, because no conqueror could kill her. She had been conquered by alien races before and yet still survived, for inevitably her culture conquered her conquerors. It swallowed them up while they were still in power. To this old chestnut, Yen Fu replied, "Yes, yes . . . [but] you are, as they say, 'clear about the past and muddled about the present'." The question is what is an alien race? "Now on earth there are only four great races: the yellow, the white, the brown, and the black. . . . The Manchus, Mongolians, and Han Chinese of today are all of the yellow race. . . . Therefore China from of old has been ruled by one race only. It has never actually fallen to an alien kind."[59]

This statement had two implications. The first was that interracial conflict was much worse than intra-racial conflict. For China, as a segment of the yellow race, to fall to a people of the white race would mean either death or a slavery far worse than death, a slavery that indeed would mean only a slower death, that would keep the yellow race, like the black and the brown, hovering always on the verge of extinction.[60] In warning his countrymen of the Western threat, Yen Fu thus raised the specter not only of international warfare but of racial warfare, which, in the Darwinian context in which he placed it, was an ominous specter indeed. He sounded a warning of the "white peril."

The second implication, in perfect keeping with Yen Fu's conviction that solidarity was strength, and that a social organism to be healthy must act as one body, was that the Manchus and the Chinese were of one race and hence should fight as one against the whites, not against each other. Others, of course, would soon protest that the Manchus were a different race, and that the "Chinese race" could not resist the whites until it had freed itself from the Manchus. But Yen Fu refused to say so, despite the fact that, by extension, he did indeed speak in dangerously disparaging terms about the Manchus. The Chinese, he said, were an "advanced people."[61] Those who conquered them were wild, unadvanced nomads, who conquered, as the poet Su Tung-p'o had long since said the Hsiung-nu conquered, thanks not to their civilization but to their lack of it.[62] But, in the end, said Yen Fu, now adopting his fictitious critics' argument, all such peoples succumbed to Chinese culture, and were, in fact, ruled by it. Thus, the Manchus, China's rulers, were in all that counted Chinese. They were part of the *ch'ün*—which white rulers could never be.

Yen Fu again looked outside China for the true field of Darwinian struggle. Even though it had been China's defeat at the hands of another member of the yellow race, Japan, that had prompted him to write in the first place, Yen Fu looked beyond struggles within his race to a struggle that for him was far more frightening, the "ultimate" struggle between races of different colors.

But why in foretelling such a struggle did he mention only the yellow and white races? He did not say much about the others except to suggest that they were virtually already out of the struggle. But in his very lack of concern in what he did say, there lay an ominous hint of contempt, that in the writings of other patriots would soon develop into a full-fledged intellectual prejudice—against any race darker than the Chinese. Yen Fu, in his list of races, listed the brown race and the yellow and the white, and then said, "lowest is the black race . . . the so-called black slaves."[63] *Lowest,* for Yen Fu, meant "furthest south." For others, however, it was soon to mean "lowest on the scale of human evolution."

But Yen Fu was not yet interested in any scale of human evolution. He was interested in China's immediate struggle for existence, and he had not yet entertained the chilling thought, which would soon strike others, that the outcome of that struggle might already be determined, that races might have evolved so unequally that eventual white supremacy was inevitable. For Yen Fu the fight was still free. All China really needed was the *will* to struggle. Without asking how will could possibly fit into the Darwinian scheme, Yen Fu seemed perfectly convinced of the faith later so strongly held by the would-be materialist Mao Tse-tung, that "if there's a will there's a way." It was this belief, together with his conviction that the real struggle was not between governments but between social organisms, that made Yen Fu look so excitedly towards democracy. Democracy would unify the social organism; it would instill in the whole people the will to survive and the will to fight.

For national strength, after all, was still the issue. Democracy was the way, because it was the way to strength. "Westerners who talk of government," explained Yen Fu, "say that a country is the public property of the people. Kings, dukes, ministers, and generals are all the country's servants. Chinese loyal to their king say that to the Son of Heaven belongs all within the four seas, and all the millions are his servants. . . . If, then, there be war, the people of the West will be fighting for their own public property, in their own interest, for themselves; the Chinese will only fight as slaves—for their master. But when slaves are driven to fight nobles have they ever not been defeated?"[64]

Thus there was only one thing to do: "If we want to advance our people's virtue so that they will consolidate their strength, unify their will, and with one spirit resist our foreign enemies, we must make each Chinese take China as his own." How could one do this? "Establish a parliament at the capital, and let each province and country elect its own officials."[65]

This was democracy. But note its motivation. The people would rule not so that their will would be done, but so that the People would be strong. Democracy was of the people, but *for* the *People,* the hope being that, by letting the people have the country, they would be willing to give their all for it. In the harsh Darwinian world that China "lived" in, it was the People's life not people's lives that counted most.

Thus Yen Fu, in this early Chinese cry for democracy, was actually already crying for a "People's Democracy" in a sense that makes that later tautology, if not altogether unamusing, at least somewhat understandable. The people were subordinate to the People. The People came first. The People was the race, the race was the *ch'ün,* the *ch'ün* was the social organism, the social organism was the country. The individual was almost absent from the argument.

Even individual freedom seemed primarily intended for the good of the group. "Freedom" (*tzu-yu*), and it was never stated what this was really to mean for the individual, was necessary to liberate the people's energies, so that they could have strong bodies, strong minds, and strong wills, all to defend the race.[66] Freedom would allow a pooling of talents that was not yet taking place. Freedom would work for China as it worked for the Westerners: "They have no forbidden topics; they rid themselves of petty, restricting formalities; they break through constricting molds of thought. Everyone can have his own mind. Everyone can voice his view. There is no great gulf between men of high and low position. The ruler is not too much esteemed, the people not demeaned. All join as one body."[67]

There was, of course, a considerable leap of faith in the assumption that everyone, "given his own mind," would willingly "join as one body," but Yen Fu had that faith. He did not expound on the utilitarian doctrine of "the natural identity of interests,"[68] whereby each man "out for himself," as British and American Social Darwinists were wont to think he should be, would almost, despite himself, help bring about the common good. Yen Fu's belief, at least as expressed in "Whence Strength?" was much simpler: If only each individual could be led to believe the group was indeed his own, then he would "naturally" identify his own interest with that of the group. He would see no conflict—no more so than did Yen Fu.

The reason Yen Fu saw no conflict between the general interest and the individual's was that he was writing in a time of national crisis. His philosophy was a crisis philosophy. It seemed to him obvious and unquestionable that the individual's interest now lay with the group's, for the group's life was in danger and, if the group perished, what hope had the individual?—What hope, that is, if the individual was indeed only a cell in the social organism? It was all perfectly simple—and simplistic.

Nevertheless, freedom and democracy, however defined, were extremely revolutionary concepts, and Yen Fu, in his outspoken espousal of them, came the closest he ever came to being a revolutionary. Indeed his arguments, especially his virulent attack on traditional Chinese rulers and on monarchy itself, undoubtedly did much to inspire the real Revolutionaries, whom he later so steadfastly opposed. Yen Fu brazenly quoted Mencius's famous statement: "The people come first, then comes the country, and the ruler comes last," and he happily quoted Chuang Tzu: "He who steals a belt buckle is beheaded; he who steals a country is enthroned."[69]

He lashed out at all China's emperors, Chinese or barbarian: "Ever since the Ch'in dynasty [the first empire], China's rulers have always been the most strong, the most harsh, those most able to conquer and oppress."[70] Here was Darwin struggling with Napoleon again. Did not Chinese rulers rule thanks to the principle of "the survival of the fittest"? If Yen Fu thought so, he did not seem to think the fittest were the best. He seemed momentarily to have forgotten his Darwinism. At least he would have confused many Social Darwinists with his angry rhetorical questions, "Was this the will of Heaven? Was this the true Way?"[71]

Yen Fu explained the existence of rulers in social-contract terms. Rulers were originally "suffered" for the people's protection, but only as a last resort. They were necessary because society "was not yet far enough advanced, and people were not yet perfect. Therefore rulers only exist alongside evil in the world; they are not related to the world's good."[72]

Thus Yen Fu came perilously close to blasphemy and treason. But, when he got to the final question, he made one all-important qualification: "Should we, then, now throw away all loyalty to our ruler? We most certainly should not! Why not? Because the time has not arrived. Our ways are not yet suitable. Our people are not yet ready to rule themselves."[73]

Here Yen Fu's notion of Darwinism came back both to give Yen Fu himself a political position and to give China the basic argument for what would become a virtual party of constitutional monarchists. Darwin and Spencer, said Yen Fu, show us that evolution moves in stages. How dear to the heart of K'ang Yu-wei this must have sounded. But if evolution, and progress, move in stages, then the proper means to a proper end must fit the stage. "It is a recognized principle," said Yen Fu, "that if a man is idle he will be weak, and if he labors he will be strong. But if you force an invalid to seek strength through daily devotion to violent exercise [lit. "leaping about"], you will only hasten his death. Now, is not China today like such an invalid?"[74]

So, China, beware of great leaps forward. Progress takes time, said Yen Fu. And reform must fit that time. It cannot succeed by decree alone. He quoted the Sung poet Su Tung-p'o's tenth-century version of "the mass line": "There is no greater disaster than for the leaders to move and the followers not to follow." The people must be ready. The success or failure of reforms "depends simply on the people's feelings and ways of life at the time, and on whether or not the reforms are fitting." That reforms must fit the times was an ancient Chinese idea, but now it was Darwin who said why: "For Darwin says, 'All things struggle for existence, and the most fit survive.'" It was Spencer, however, who told Yen Fu what to do in such a world: "Wealth and Power can not be forced . . . [one must] nurture their roots."[75] Why nature needed nurturing Spencer did not say, but neither did Yen Fu ask. For him all was clear: China's hope for progress, and indeed survival, depended on nurturing the three powers of the people—physical, mental, and moral. The people must be "raised." China must "grow" strong.

For those dismayed at the proposal of such seemingly slow progress, Yen Fu offered encouragement. He became one of the first to venture a prediction of how long it would take China to catch up to the West. Using an argument destined to become a favorite of Sun Yat-sen's, Yen Fu pointed out that the superior power of the West was actually all the result of, at most, the last two hundred years, and that the most startling advances had taken place only in the last fifty.[76] The most important fact, however, was that, since the West had shown the way, "what for them was difficult, will for us be easy." Therefore China could hope to vie with the West in only sixty years, and would already show tremendous progress in thirty.[77]

But many were dismayed at being even thirty years behind. Many despaired of China's being able to live that long. To these, Yen Fu offered two related words of encouragement, words that were perceptive, but also most psychologically revealing. Despite all he had said in optimistic praise of progress, he now subtly intimated that progress, at least technological progress, might be a mixed blessing. Steamships and railroads, for example, "although they are of benefit to communications and the people's livelihood, are also of great advantage to the monopolizing efforts of robber barons (*chien-hsiung*)."[78] Although this argument left at least an aftertaste of sour grapes, it hinted at a most serious philosophical problem, one that would soon deeply trouble the eccentric revolutionary Chang Ping-lin, and lead him to become one of the first of Chinese intellectuals to question Darwinesque notions of progress and evolution.[79]

Yen Fu's mixed-blessing argument, however, merely led into a second argument that would become a favorite of Sun Yat-sen's. Strength, said Yen Fu, was relative. Although the West might look formidable compared to China, it was far from perfect. The Westerners had not reached their ideals of freedom and equality. On the contrary, their technological advances had simply increased the distance between rich and poor. This fact would give rise to "cries of inequality and determination to struggle . . . the causes of great rebellions." Already, said Yen Fu, "parties seeking to level rich and poor are rising up."[80] In short, as Mao Tse-tung would later put it, the West was infected with "internal contradictions" that would sap its strength from within. Indeed Yen Fu actually used Mao Tse-tung's "paper-tiger" argument, although Yen Fu's metaphor, which he attributed to certain "very knowledgeable scholars," was that of the "hothouse plant" that might bloom early, and look magnificent, but would not stand up against the elements.[81]

Now, both hothouse plants and paper tigers are Social Darwinian metaphors: they both display a desperate concern with the struggle for existence. For if the political world really was not only similar to but actually part of "nature red in tooth and claw," then paper tigers (like dinosaurs) were the underdog's only hope. If China accepted Darwin, she *had* to believe in paper tigers for, if the world was only a great arena, and the West was *not* a paper tiger, then China's life was over.

Yen Fu did not think China's life was over. He did not, of course, suggest that China could afford to sit back and wait for the West to wither; he only used his hothouse plant to gain time, to suggest that the West was just enough of a paper tiger so that China, if she would only pull herself together and struggle, had a chance of catching up. But he did thus offer China hope: the world was a savage place, but the struggle for existence was a long one, and those who seemed the fittest now were not necessarily as fit as they might seem.

BEWARE, BUT TAKE HEART

Thus Yen Fu called on Darwin both to frighten China into reform and to encourage her to believe that through reform there lay a way to wealth and power. Charles Darwin's first words to China were extremely mixed. He seemed to be saying at one and the same time, "Beware, doom looms,"[82] and "Take heart, for all shall be well."

Darwin's warning should have filled the Chinese with terror, if not despair. After all, it would have been bad enough just to be told that the world was a world of warring states, a fact unfortunately all too easy to believe. But this scientist, Darwin, insisted on saying that the world was *naturally* a world of warring states. The Chinese had thought they had already had their Warring-States period—from 403 to 221 B.C. The resulting Chinese Empire had, with only occasional lapses, supposedly left all that behind. Thus the idea that China must now contend for its life in a new warring-states period on a global scale was already a horrifying prospect. But that this should be "only natural" made it far worse. Confucians had always said that the Warring-States period was the result of a moral breakdown, and that unity and peace could hence be rewon through moral regeneration. But Darwin now said, or seemed to say, that the "evils" that beset China were not evils, but simply a part of the natural order. Indeed disorder *was* the natural order.

Now it should have been every bit as cataclysmic for a Chinese scholar like Yen Fu to be told that the order was disorder and that struggle was the Way, as it was for Bishop Wilberforce to be told, or to think he had been told, that man owed his existence to an ape, or that the world and all its wonders had come into being by chance, not design. For nature was order in Chinese, and disorder was "unnatural," perverse, and evil. And yet Yen Fu was not unduly shocked by Darwinian disorder, first because the obviousness of world disorder in the nineteenth century, and the intense desire among most Chinese intellectuals for China to struggle in that disorder, made a cosmic theory that justified such action attractive, and second because, from the very beginning, Yen Fu had fused and confused in his mind the concepts of evolution and progress. Just as Darwin himself had done in *The Origin of Species*,[83] and as so many other Westerners had done after him, with equal lack of scientific evidence, Yen Fu assumed that evolution was for the good, that things evolved upwards. And therefore he could believe with Andrew Carnegie that "all is well since all grows better."[84] He could believe that, for all its seeming disorder, evolution still was natural order, that it was a great and good process, that it was indeed the Way.

Yen Fu, therefore, almost did just what the first Christian converts to evolution did: he endowed evolution with design, although in age-old Chinese fashion he managed to conceive of design without a designer. Just as Asa Gray, believing that "a fortuitous cosmos is simply inconceivable," could "insist that, for all that yet appears, the argument for design, as presented by the natural theologians, is just as good now, if we accept Darwin's theory, as it was before that theory was promulgated,"[85] so Yen Fu could insist that the Tao in all things still prevailed. With almost religious excitement Yen Fu seized upon Spencer's "the unknowable," that mystery which Spencer allowed into his system and then ignored, which Yen Fu translated with an ancient Buddhist term, *pu-k'o-ssu-i* (the inconceivable), as the mystery behind evolution; but he saw that mystery, which was the *Tao,* working *towards* something, and something in earthly terms. This was admittedly a new vision of the *Tao.* Evolution could change men's concepts of the *Tao* as it changed men's concepts of God, but it denied neither. That is why Yen Fu could later say with full sincerity, "[In all of early Chinese thought], only the views of Lao Tzu are compatible with the views of Darwin"[86]

"Darwinian Taoism," however, was for Yen Fu, as we have seen, decidedly not an invitation to non-action. Yen Fu would have accepted the Taoist idea of non-action (*wu-wei*) only in its proper sense of no action against the *Tao,* which now meant no action that would hinder evolution. For seeing Heaven as nature, Yen Fu could readily accept a Taoist and a "Darwinian" interpretation of Mencius's famous statement, "He who goes with Heaven shall survive; he goes against it shall perish."[87] But Yen Fu was more than a negative activist. He favored much more action than Lao Tzu or Chuang Tzu would have prescribed, for he believed that human action, indeed human struggle, was (somehow) a necessary part of the process. He was, like K'ang Yu-wei, a determinist-activist, a determinist and a determinationist all at once. To be sure, he was a gradualist—he believed progress moved slowly and through necessary steps—but he was convinced that each step required action. He would thus have agreed with Mencius that one should not try to help plants grow by pulling them up,[88] but he would have agreed just as strongly, to change stories, that neither should one leave one's garden up to God.

It was this activism that let Yen Fu hear Darwin's warning as fair warning. Darwin, the Western wise man, told China she was in danger, indeed from the West, but he also told China, so Yen Fu thought, how to get out of it. At least Spencer, Darwin's great "interpreter," did. Yen Fu, making the same subtle revision that Spencer at least came closer to making, heard in the sentence "The fit survive" the dictum "They who *make themselves fit* survive." Drawing from within him on one of China's most hoary and noble prejudices, Yen Fu added to evolution his faith in "self-cultivation" (*tzu hsiu*). China could make herself fit by cultivating her people's strength, knowledge, and virtue.

By extension, working *back* into biological evolution, Yen Fu seemed to be giving the leopard credit for his spots. Why, we may wonder, did no one ask him just how strong, knowledgeable, and virtuous the arctic hare was? No one did ask him, and he did not ask himself, for he was not interested. Spencer saw survival resting on strength, intelligence, and virtue, and he ought to know. As Yen Fu read Spencer, with knowledge of the way the world worked and the will to work in the world's way, one could work out one's own salvation.

It is remarkable how differently people could interpret Spencer. Richard Hofstadter quotes Edward Livingston Youmans, one of Spencer's most ecstatic American disciples, as replying to Henry George, when asked what he intended to do about the social evils of New York, "Nothing! You and I can do nothing at all. It's all a matter of evolution. We can only wait for evolution. Perhaps in four or five thousand years evolution may have carried men beyond this state of things."[89]

Not even in his most conservative old age would Yen Fu have accepted gradualism, or evolution, at that price. Yen Fu was for reform. He wrote of Darwin and Spencer in the interests of reform. It was therefore a good thing he read Spencer and not Youmans. For, as Benjamin Schwartz has so well proven, Yen Fu had just as logical a right as Youmans to call himself Spencerian.[90]

This is the way Yen Fu introduced Darwinism to China. The full theory of evolution, even as it existed at that time, was only barely touched upon. Much was misstated, more was unstated. But there were enough ideas expressed or latent in Yen Fu's writing to inaugurate a new age of philosophical Chinese thought.

We must pursue Yen Fu's specific influence much further. But two things should already be clear. In so emphasizing progress, the people, struggle, the "hothouse" strength of the West, and the irresistible power of knowledge, will, and virtue, when backed by the forces of the universe, Yen Fu was, if not sowing, at least cultivating, many of the dragon's teeth that would bloom into the thought of Mao Tse-tung. Indeed his very emphasis on the power of ideas would help make it easier for Chinese one day to believe—for a while—that "the Thought of Mao Tse-tung, when fully grasped by our great masses, will become an infinite power . . . an incomparably powerful spiritual atomic bomb."[91]

And yet, with double irony, Yen Fu, famed introducer of Western ideas, would make that belief easier by making certain Western ideas in his well-meaning introductions very difficult to understand. He would make it very difficult, for example, for many to understand what the Western liberal tradition meant by "individualism," because his own understanding was so clouded by his "*ch'ün*-ism," his militant belief in the primacy of the group. It was not that Spencer had single-handedly led him to believe that the individual was just a cell in a social organism. Yen Fu had seized upon Spencer's notion of the social organism "naturally," because something deep in the Chinese tradition had predisposed him to, something deep enough, probably, already to have so predisposed Hsun Tzu, some two millennia before.

So it was not strange that Yen Fu in introducing Darwin should overlook one fact of life that the discredited proto-Darwinian and "natural theologian" William Paley did get right: "Although we speak of communities as of sentient beings, although we ascribe to them happiness and misery, desires, interests and passions, nothing really exists or feels but individuals."[92]

Part Two

Darwin for Reform

THREE

Darwin and The Chinese Progress

The first movement in the Middle Kingdom that Charles Darwin took part in was the Reform Movement of 1895–1898, a movement ill-fated but of great import. Darwin's part in it was also of great import, of unique import, because this first movement in China in which Darwin took part was the last in which he took part on one side only. His two-decade "day" was not really to begin until the Reform Movement had failed, and the great argument between the Constitutional Monarchists and the Revolutionaries had begun, but by that time his banner was waved by both sides. In the Reform Movement of 1895–1898, only the Reformers called on Darwin for support. And so China first saw Darwin standing unequivocally for reform.

Yen Fu's introduction of Darwin in "Whence Strength?" had not made Darwin's name a household word. It was not to become one until after the first widespread publication, in 1898, of Yen Fu's most famous work, which would also make his own name a household word (in intellectual households), *T'ien-yen lun* (The theory of evolution), his brilliant paraphrastic translation, with commentary, of T. H. Huxley's *Evolution and Ethics.* Before that time, Darwin's name spread only slowly, largely thanks to the efforts of another man, destined to become a veritable "Saint Paul"[1] of Chinese Darwinism, one of the chosen few who read *T'ien-yen lun* before it was published, K'ang Yu-wei's wavering disciple, Liang Ch'i-ch'ao.

LIANG CH'I-CH'AO, JOURNALS, AND EVOLUTION

Liang Ch'i-ch'ao read *T'ien-yen lun* precisely at the time when, still unbeknownst to himself, he was finding his true vocation. In spring 1896, after several euphoric months of practice, writing and editing the Society for the Study of Strength's first ill-fated efforts at journalism, followed by several months of increasing frustration and discouragement, as justly fearful newsboys refused to deliver those efforts, angry officials impeached the Society, and imperial edicts forbade it, Liang Ch'i-ch'ao, encouraged anew by the support of the influential scholar-official-diplomat-poet, Huang Tsun-hsien, went to Shanghai to create his first great journal, the *Shih-wu pao* (Current affairs). Strictly speaking, the journal was not his own. Various officials gave most of the money, and Wang K'ang-nien, manager, contributor, and later editor, really set it up. But Liang Ch'i-ch'ao was the first Editor-in-Chief, and chief contributor, and very soon, in his readers' eyes, any paper he wrote for was his.

THE CALL TO WRITE. By vocation and avocation, Liang Ch'i-ch'ao was becoming a journalist, the most influential China has yet known. He was a unique kind of journalist. Indeed, that modern Western word, when applied to a man like Liang Ch'i-ch'ao, is almost an insult. For what Western journalist has ever gained by his writing the ear of an Emperor, a price on his head, exile abroad, party leadership at home, and, finally, cabinet positions, semi-official diplomatic missions, and three professorships? And what Western journalist has for at very least a decade been the most influential intellectual in his land? Liang Ch'i-ch'ao wrote so well that the "Liang style" of writing became the style cultivated by almost all would-be polemicists until the *pai-hua* vernacular finally triumphed in the 1920s (at which time Liang

Ch'i-ch'ao switched to his own very powerful *pai-hua*). His *Shih-wu pao* and its successors became models for a new kind of journal that spread in countless imitations to every province.

Of course, the situation did as much to shape Chinese journalism as any individual, a situation so serious that Chinese journalism was on the whole more serious than Western journalism has probably ever been. In the beginning, there was very little commercial interest in providing entertainment, or in satisfying comfortable subscribers' curiosity. Editors were interested not in "all the news that's fit to print" but only in that news pertinent to their main concern, China's welfare. This was especially true of journals like *Shih-wu pao,* published once, twice, or three times monthly, yet much more influential than the dailies. Such journals did carry "pertinent" news, but their strength and fame lay in polemical essays and educational articles, educational, that is, in the cause of China's safety. In none of this was there much humor.

Their influence was phenomenal, and it was something new in Chinese history. Even Chang Chih-tung, last of the great loyalist Reformers, who in the beginning was the *Shih-wu pao's* greatest patron (if also a would-be censor), recognized and welcomed the phenomenon, even though he was often infuriated by the journals' radical ideas and would have been broken-hearted if he could have predicted the eventual outcome of their influence. In 1898, despite his opposition to the radicalism of K'ang Yu-wei and Liang Ch'i-ch'ao, he still generously made this admission:

> After 1895, literary men of patriotic spirit began to publish journals. They translated widely from foreign papers and added sweeping discussions of their own. Their efforts began in Shanghai and spread to every province. Internal politics, foreign affairs, academic knowledge—all were within their scope. Although their arguments greatly varied, they shared a common aim: to spread information, arouse unselfish spirit, wash clean the poisonous apathy of those who seek only their own peace and contentment, and break down the blind arguments of those isolationists who would still lock out the world. As a result, gentry from the most obscure pockets of the realm, and isolated peasants, learned for the first time that there was a China. Ignorant, petty offi-

> cials and the whole multitude of would-be scholars learned for the first time that there were "current affairs." One cannot say that this was not an aid in the education of men who wanted to help their country.[2]

Chang Chih-tung clearly rejoiced in the fact that such journals were spreading his own great patriotic spirit to the people, although he rightly feared that they were not spreading his loyalty to the Chinese tradition. They did not, of course, inspire an immediate mass movement. The vast majority of Chinese peasants, that is, China's people, even those plowing just outside the Emperor's own walls, could not read. But men who *could* read read Liang Ch'i-ch'ao's writings, and talked about them throughout the realm, and ideas spread, and nationalism grew. That is why, in retrospect, the new journals were the most important part of the Reform Movement of 1895–1898, much more important than the final One Hundred Days.

China's earlier attempts at reform, in the Tung-chih Restoration and the Self-Strengthening Movement, had all been official attempts, urged by loyally patriotic officials like Tseng Kuo-fan, Li Hung-chang, and Chang Chih-tung. With the new journals, the initiative for reform switched to men outside of government.

Chang Chih-tung was surely right when he said that the new journals were "an aid to education." For many, they provided the only modern education available. For at least two decades, they were miles ahead of the schools. And almost all China's most famous intellectuals helped write them. Indeed, China's intellectuals became famous as authors not of books but of articles, for, aside from translations, influential new books were relatively few, perhaps because there was no time to write books. Every scrap of new knowledge seemed *too* pertinent, and had to be shared as soon as it was found. This was why the new journals were taken so seriously by both their writers and their readers.

It was in these journals that Darwinism flourished, especially after those "who wanted to help their country" began their violent arguments as to how to help it. For more than twenty years, almost everyone who had anything to say mentioned Darwin, or used his slogans. But no one did so in as many ways as Liang Ch'i-ch'ao.

In 1896, however, when Liang Ch'i-ch'ao first set out on his journalistic career, which was to be so colored with Darwinian metaphor, he did so with grave doubts about the whole business. He was seriously afraid that he had no right to lecture his countrymen, that journalism was presumptuous without more knowledge.[3] Such inner fears, of course, or qualms of conscience, have not kept journalists down for long in any land, but Liang Ch'i-ch'ao's initial doubts were unusually sincere, partly because of the failure of his efforts in 1895, but partly also because of his faith in the visions of his master, K'ang Yu-wei.

This combination of faith and frustration led him to suggest in a letter to K'ang Yu-wei that, as "we take our religion to be central, and as the country's survival has nothing to do with our religion, perhaps we should forget about all else and concentrate on study and teaching, and wait until we are all really well prepared, before we go out to propagate our doctrine."[4] The doctrine, his "religion," was K'ang Yu-wei's progressive, utopian, *Ta t'ung* Confucianism, which did, supposedly, look towards the salvation of the world, not China. Still it seems incredible that, so soon after his first great efforts to arouse nationalistic spirit for self-strengthening, Liang Ch'i-ch'ao could still protest, "I told them that those of us with true motivation should all retreat to the mountains for several years before going out into the world, but Hsu Ch'in and the others laughed at me. . . . They do not understand that our goal is to be missionaries, not politicians, to save the world and all creatures of all worlds, not one country. What has one country's survival to do with us?"[5]

One country's survival had so much to do with Liang Ch'i-ch'ao that he was to spend the rest of his life writing about it, which makes the above sentence somewhat puzzling at best. But it was written in a mood of grave despondency, the very despondency that had made K'ang Yu-wei's vision so attractive in the first place. If China, seemingly bent on self-destruction, insisted on silencing her would-be saviors, and if the country seemed doomed in the international arena, then how comforting to believe that countries did not matter, that the whole abhorent concept of nation-states was backward, passé, itself doomed. K'ang Yu-wei's universalism could lift China out of the whole wretched international picture, which China, so rudely shocked out of her old One-World security, had been so loath to accept in the first place. By redefining the boundaries of *t'ien-hsia* (all under Heaven), international strife could be outlawed. And, even if China then lost itself in a new One World, it would still have been a Chinese, K'ang Yu-wei, who saved that world, with the long-lost vision of a Chinese Sage, Confucius.

Despite such brave talk, of course, Liang Ch'i-ch'ao could not really forget China. And yet, when he said, "I do not understand by what means one can hope to save China, if one's learning is incomplete,"[6] he probably meant it, for throughout his long career he kept insisting that China had to make itself ready before attempting any dramatic step forward. He kept stressing the need for education first and action second. And yet in his own action he obviously compromised, for he was bothered by the argument of the friends who had laughed at him: "Our land is about to perish, and you're ready to go off into the mountains. Who can wait for you?"[7] Thus he went ahead with preparations for his *Shih-wu pao,* although apologizing to K'ang Yu-wei: "I feel my knowledge is not enough, and I have greatly desired to retreat into the mountains for several years, but I cannot yet abandon all that has already begun."[8]

He was still apologizing in his first known letter to Yen Fu, who had also apparently urged him on to further study and more cautious writing, lest he later regret what he now published, indeed lest "a miniscule error, loosed in the mind of the masses, become a thousand-mile misconception."[9] But by this time, after several months of publishing the *Shih-wu pao,* Liang Ch'i-ch'ao had rationalized his defense. He had no illusions that his paper could solve everything, but it was a beginning, however crude, on which greater men could later build.[10]

ENTER T'IEN-YEN LUN. It was in this same letter to Yen Fu, probably written in spring 1897, that Liang Ch'i-ch'ao first revealed that he had read *T'ien-yen lun,* which Yen Fu had finished translating the previous fall. Yen Fu had apparently sent him a draft copy, which he in turn showed to his teacher, K'ang Yu-wei, and to his best friend and fellow disciple, Hsia Tseng-yu. All three, said Liang Ch'i-ch'ao, were infinitely impressed. Hsia Tseng-yu expressed "inexpressable" respect for Yen Fu, and K'ang Yu-wei "said that he had not seen a man of such calibre," although there were, he added, a few of Yen Fu's points with which he could not totally agree.[11]

Actually, Liang Ch'i-ch'ao claimed that he and his fellow students had heard of some of the ideas in *T'ien-yen lun* from K'ang Yu-wei himself, but had never before had the complete story. This again raises the question of whether or not K'ang Yu-wei knew of Darwin before he knew of Yen Fu; again, one can only say that, if he did, he did not know very much. Several years later, in his own description of what he had taught his disciples, he wrote, "I taught them that man gave birth to the horse, and the horse gave birth to man, and that man was a transformation from the ape, and they believed and could prove it."[12] K'ang Yu-wei had not come upon an illustrated Tartar history of Ixion and his brood of centaurs, or any other exotic manual of Western horsemanship. His horse story came from a slightly misquoted line from *Chuang Tzu,* which in the original, only barely less bewildering, stated that an insect (some say a leopard!), not man, gave birth to the horse.[13] But how Chuang-tzian horse and Darwinesque ape fitted into the same story K'ang Yu-wei did not say. Nor, alas, did he reveal the pedagogical secret by which he convinced his students to believe him.

The true depth of his Western learning, and scientific spirit, is best seen in his account of a moment of personal, evolutionary revelation that came to him as he pored over accounts of ancient rituals in an attempt to find suitable rites for Confucian worship in his academy:

> As I was worrying over the fact that the *kuei-ch'ih,* one of Hsun Hsu's twelve flutes, was so long that one's fingers could not stretch far enough to play all the notes, it suddenly came to me that the ancients

> must have had extremely long bodies. That is why they had long flutes. This was because they were not yet far removed in time from the period of the giant beasts, and the heat of the earth was still great. Today, after more than two thousand years, the earth has gradually spun further away from the sun, its heat has gradually lessened, and human bodies have gradually shortened. Therefore, we can predict that, in another two thousand years or more, today's flutes will be equally unsuitable for the people of that age. And in ten thousand years, people will be small indeed. [By which time, Mao Tse-tung, had he lived the ten thousand years his people wished him, would have truly been a giant, a veritable "stork among chickens."] [14]

If such a theory proved that K'ang Yu-wei *had* heard of exciting bits and pieces of current scientific theories, however, it also proved that he was in absolutely no danger of becoming a scientist. It is unlikely, therefore, that K'ang Yu-wei, however spellbinding his lectures, taught Liang Ch'i-ch'ao much about evolution. And there is no proof at all that he taught him anything about Darwin. The first of K'ang Yu-wei's works that really set his contemporaries on fire (either for or against him), *K'ung Tzu kai chih k'ao* (Confucius as a reformer), published in winter 1897, but compiled with the help of his students between 1892 and 1896, contained not a single reference to Darwin, nor did it use any of Yen Fu's evolutionary terms.[15] Neither did his other important work published at that time, *Ch'un ch'iu Tung shih hsueh* (Tung Chung-shu's studies of the *Spring and Autumn Annals*). In the latter, K'ang Yu-wei gave the best exposition to date of his Confucian theory of progress, but he shunned the later term for progress, *chin-hua,* which Yen Fu had used.[16]

Actually, the first time K'ang Yu-wei seems to have used *chin-hua* was in the brief autobiography he wrote in 1899, from which the above evolutionary anecdotes were taken. Even then, however, he used the term only once, when he reported that in 1892 he had inspired his daughter to make a study of customs and institutions of China and other countries "to demonstrate the principle of human progress."[17] Throughout his autobiography, K'ang Yu-wei seemed so intent on demonstrating the genius and independence of his own thinking that one can almost suspect him of being jealous of Darwin, whose fame was just then becoming widespread, for creating a "progressive" theory which took glory away from his own. Certainly, when he did finally take up Yen Fu's vocabularly and mentioned Darwin's name, he did so with far less than one-hundred-percent enthusiasm.[18]

Liang Ch'i-ch'ao's enthusiasm, however, knew no bounds. In thanking Yen Fu for showing him *T'ien-yen lun,* he said that, when he was first told that Yen Fu was working on such a book, his "mouth had watered uncontrollably" in anticipation.[19] Once he had actually read it, he never recovered. Its influence was with him forever.

TO CHANGE OR TO BE CHANGED

As *T'ien-yen lun* was not widely read until 1898, it is best, if we wish to understand Darwin's part in the Reform Movement, to look first at *T'ien-yen lun's* influence second-hand, as it appeared through the writings of Liang Ch'i-ch'ao, and others of like spirit, in the new journals of reform.

The first issue of the greatest of these, the *Shih-wu pao,* was published on August 9, 1896. Its format, throughout all fifty-six issues, published every ten days barring New Year's holidays and other occasional interruptions, consisted of one or two leading essays, important edicts and memorials, translations from the foreign press (English, French, Japanese, and occasionally Russian and German) and one or two educational features, the first a biography of George Washington. *T'ien-yen lun* was not actually mentioned until the twenty-third issue (April 12, 1897), but vaguely evolutionary thinking was there from the beginning.

In the first issue, in his introduction to "Pien-fa t'ung i" (A general discussion of reform), the long series of essays that would first make him famous, Liang Ch'i-ch'ao clearly stated the main point of the entire Reform Movement: Change was necessary for survival. As Yen Fu had before him, however, he made his case not just in pragmatic, political terms, which one might think should have been sufficient, but in cosmic terms, in terms of natural law. "Why must our institutions change?" he asked. "Because of all things between Heaven and earth there is nothing that does not change." Our very world was formed through change: "Dawn and darkness changed to make days. Heat and cold changed to make years. In the beginning, flaming liquid, molten then freezing, through countless changes formed our globe. Seaweed, snails, giant plants and giant birds, flying fish and flying reptiles, marsupials and vertebrates, this kind living, that kind dying, one after another constantly changing—thus was our world formed. Purple blood and red blood flowing through the body, inhaling oxygen, exhaling carbon dioxide, one after another, over and over, a thousand changes in a day—thus was man formed. Without change, Heaven, earth, and man would in the same hour all cease to exist."[20]

This was almost evolutionary; it was naturalistic; it was far more than natural metaphor. It was a new and powerful type of argument that bore the aura of Western science. And with it Liang Ch'i-ch'ao, like Yen Fu before him, sent his readers, and himself, enthusiastically off towards "scientism."[21] But just as Yen Fu had in "On the Race of World Change" and "Whence Strength?", Liang Ch'i-ch'ao, in this *his* first step, also unwittingly let man enter nature's order in a semi-supernatural position. "Natural change," he said, "is the way of Heaven. But whether it be change for the better or the worse depends on the way of man."[22]

IT'S UP TO US. Change was natural, change was vital; but if "things" were to "get anywhere," if they were to progress, they needed man's help. Indeed, without man's help, change could have a positively deleterious effect, especially on human institutions: "If we rely on fate, and let things change as they will, then things will get steadily worse. If we shape up, pull ourselves together, think hard, so that we understand change, things will steadily improve."[23]

This does not sound like a belief in inevitable progress. But Liang Ch'i-ch'ao did claim to believe in inevitable progress. It was just that he was also convinced that progress was not inevitable for all. He mentioned Timothy Richard's translation of *The 19th Cen-*

tury as proof of the remarkable recent progress of the West;[24] he introduced K'ang Yu-wei's Three-Ages doctrine as proof that "the destiny of the world is to move from chaos to tranquility";[25] he even wrote that, "with the whole world now open, the ten thousand nations steadily move upwards, impelled by forces of circumstance which cannot be held back."[26] But the "ten thousand nations" still meant the world in general. When it came to nations in particular, Liang Ch'i-ch'ao was painfully aware of the fact that, far too often, one nation's rise was another's fall. Citing the constantly cited examples of India, Africa, and poor Poland, he made up his own proto-Darwinian law (which did not survive): "The strong flourish; the weak are destroyed."[27]

The "world's" progress was assured, but China's was not. Liang Ch'i-ch'ao's point, however, was that China's progress was up to China. Change was inevitable; in general change was good; but change would only be good for China if China herself changed herself for the good. This was the dictum that first made Liang Ch'i-ch'ao famous, a dictum that distinctly suffers in translation: "If we change we will be changed. If we do not change we will be changed. If changing we are changed, the power of change will be in our hands, and we will be able to preserve our country, our race, and our religion. If not changing we are changed, then we give up the power of change to others, who will harness us and drive us like beasts of burden."[28]

Liang Ch'i-ch'ao bristled with rage against pessimists and fatalists, who, bewailing China's plight, could only moan, "It is Heaven's will; it is our country's fate; there is nothing we can do."[29] "Never!" he kept insisting, "It is up to us." "It's-up-to-us-ism" was perfectly Confucian. Remembering Confucius's dictum that, if aught goes wrong, "the noble man seeks for the fault within himself,"[30] Liang Ch'i-ch'ao quoted Mencius's famous line, "A country must attack itself before others will attack it," in the original passage preceded by "A man must disgrace himself before others can disgrace him."[31] But Liang Ch'i-ch'ao shared Mencius's willingness to accept responsibility for weakness only because he believed, as

Mencius did, that men and countries have it within themselves to be strong. Men could turn wickedness to goodness. Countries could "turn weakness to strength."[32]

"It's-up-to-us-ism" would prove an indomitable faith. None of the coming washes of determinism would overwhelm it. It would persist in the thought of Liang Ch'i-ch'ao, K'ang Yu-wei, and Yen Fu. It would appear in the thought of Sun Yat-sen. It would appear for us most powerfully in the still cryptic, however unoriginal, dictum of Mao Tse-tung, "If [the people of China] take their destiny into their own hands . . . there is no difficulty in the world which they cannot overcome."[33]

There would be grave logical difficulty, of course, not easily overcome, in maintaining that one should take one's destiny into one's own hands in the service of destiny. But there were also logical difficulties in fitting "It's-up-to-us-ism" into an all-embracing scheme of evolution by natural selection. It was some while before anyone tried to do that consciously, but unconsciously the invincible, if unrecognized, Chinese faith in voluntarism from the very beginning helped shape Chinese understanding of evolution and natural selection, and made it more difficult to understand what Darwin himself had said.

Liang Ch'i-ch'ao, however, as he began to move into the Darwinian minefield of mysteries, was realistic enough to realize that "It's up to us" did not necessarily equal "We can do it." In the first twenty-two issues of the *Shih-wu pao,* therefore, before Darwin really bolstered the arguments of either "pessimists and fatalists" or progressive optimists, the question of whether or not China could "do it" was *the* question.

IMPERIALISTS, BARBARIANS, AND RACISTS. Although the modern word *imperialism* was not yet in use, the Chinese were aware of the phenomenon. Indeed, the manager of the *Shih-wu pao,* Wang K'ang-nien, had already coolly analyzed it in proto-Social Darwinian terms: "All nations established on this globe rely on war for their survival. Those who toughen themselves through war all prosper; those who take their ease and refrain from war all perish.

Such warfare has three weapons: religion, to seize a country's people; soldiers, to seize its land; and commerce, to seize its wealth."[34] The Reformers, at least, were quite convinced that the Western nations, attacking with all three weapons, meant to swallow up China, whole or in pieces, as they had swallowed up India and Africa and (as the Chinese always added) Poland.

In order to convince the complacent that this was so, however, Liang Ch'i-ch'ao and his confederates loved nothing better than to quote foreigners. Thus, they triumphantly recorded from a Japanese journal a Russian discussion of English foreign policy, which proclaimed that the policy of the English "is, in a word, to wrap up the profits of the entire world for themselves. This is the Englishman's ambition. The English are extremely contemptuous of other peoples. They say that others are barbarians, who can hope to develop only under British rule. Thus they do not take the conquest of another country as immoral or as anything their religion would decry."[35]

When Englishmen, and others, called the Chinese "barbarians," they sowed the seeds of the greatest national loss of self-confidence the world has ever known. No people had ever been more sure of themselves than the Chinese, and so no people were ever made more painfully unsure of themselves. When outlandish Western upstarts first called Chinese "barbarians," the Chinese must have greeted their remarks with snorts of derision. But, all too soon, after Opium Wars and treaty ports, and missionary criticism, the humor of the situation disappeared, and Chinese self-confidence began to fall. And for many it would fall low indeed before it would come up again. Some, near despair, would in time come to berate their "barbaric" countrymen even more mercilessly than did the Westerners. Even the strongest, even the most self-confident must have known moments when they were forced to wonder, "*Are* we barbarians? *Are* we as good as the Westerners?" Even the strongest were thrown on the defensive.

Liang Ch'i-ch'ao knew such moments, and was often on the defensive, although he tried his best—usually—to argue that the Chinese were as good as the Westerners. He first parried the charge of barbarism, for example, by treating it as a worldwide phenomenon, which, as the world had already entered K'ang Yu-wei's "Age of Increasing Tranquility," he saw as an ill wind now on the decline. He readily admitted that barbaric practices existed in China—foot-binding was his most commonly lamented example—but he comforted himself with the fact that African and Indian "head-warping" and Victorian corseting were just about as bad. Oddly enough, the only two peoples he specifically excused from the ranks of those guilty of barbarous beautification, were "*our Chinese* Manchus and Mongols," the barbarians *par excellence,* in traditional Chinese terms.[36] Needless to say, the Manchus at least were soon to be restored to their former status.

When it came to the more general question of whether Chinese were basically as good as Westerners, Liang Ch'i-ch'ao again appealed first to an assumed universal humanity. Employing a rather curious classical allusion, he appealed to such humanity in expressing his moral outrage at the degree of foreign dominance already obvious in China: "Alas, both peoples alike have round heads and square feet, and yet we Chinese must in everything politely bow and raise clasped hands and take orders from others. How can one keep from sighing?"[37]

He was not sighing just at injustice, however. Despite the similar cut of their feet, Liang Ch'i-ch'ao claimed to see in Westerners and Chinese a qualitative difference in performance. "When China carries out new government policies," he wrote, "if Westerners are used, things usually succeed; if Chinese are used, things usually fail. Why is this so? Because Westerners are enlightened and Chinese are ignorant."[38] But ignorance was not stupidity. "There is little disparity," he went on, "in people's intelligence and morality. If one were to insist that Westerners are all wise and Chinese all stupid, or that Westerners are all saintly and Chinese all degenerate, even a five-foot [*sic*] child would know this is nonsense."[39] Lest any doubt it, and the fact that he felt a need for such proof proved, alas, that some did, he quoted the devils themselves: "I once heard a Westerner say that, when Chinese study in the West, their basic ability and intelligence show themselves in no way inferior to those of the Westerners. There has been a steady succession of Chinese who, after a few years' study, have honorably come out first in their examinations."[40]

Thus, there was no difference in intelligence, only in education. Chinese education, said Liang Ch'i-ch'ao, confined the intellect; it stifled creative thought. Similarly there was no difference in morality, despite the alleged difference in the moral performance of Chinese and Western government officials. Chinese officials are not innately venal, he protested: "The system makes them so."[41] That is why it had to be changed.

Absolute confidence in Liang Ch'i-ch'ao's professed faith in common humanity and racial equality, however, was already being subtly undermined by an ever-increasing number of racial slurs appearing in the *Shih-wu pao,* from a variety of sources, against "black men, brown men, and red men." Liang Ch'i-ch'ao himself was of little help, when he matter-of-factly echoed K'ang Yu-wei's disdainful argument that such peoples were (naturally) treated as beasts of burden "because they *could not* group"—the same reason given for animals themselves being treated as beasts of burden.[42] The situation worsened, however, when "madman" Chang Ping-lin made a dubiously glorious second entrance into the *Shih-wu pao* with the statement: "The races of civilized nations are equal in alone having blood corpuscles larger than those of beasts and foreheads of a steeper angle than those of savages and aborigines." He went on to ridicule certain opponents as being "not even up to the red men or the blacks."[43]

These sentiments did not reflect any age-old Chinese prejudice. They undoubtedly reflected to considerable degree an unconscious acceptance of Western prejudice and, indeed, of current Western scientific opinion. Nevertheless, however psychologically satisfying it may have been for Chinese to look down on peoples in even worse straits than themselves as members of lesser breeds, any admission of racial inequality, in the face of Western threat and criticism, made the stature of the Chinese race itself uncertain. Darwinian concepts, when they spread, would make the stature even less certain.

Doubts were already deepening. Besides Liang Ch'i-ch'ao's

major efforts, the first issues of the *Shih-wu pao* contained a host of minor items also designed to bolster Chinese racial self-confidence. The editors were quick to pick up any allegations, especially foreign ones, that foreigners were not as special as they were made out to be. Thus they happily translated a Japanese article, for example, "On the Sneakiness of the Russian Temperament," which announced that "the Russians are by nature dull and sluggish. They are not as clever as the French, nor have they any such air of refinement. Their avarice and churlishness cannot be equaled even by the Germans. . . . In a word, there is a vast multitude of Russians that has not the slightest idea what civilization is."[44]

An even more encouraging account was found in another Japanese article entitled "A Doctor Discusses the Brain." The doctor, a Westerner most likely slight in stature, proclaimed as scientific fact that "anyone who has large bones and muscles will necessarily have a small brain. . . . Men with extremely tall bodies and extremely good brains are very rarely seen. Bismarck was an exception."[45]

Besides whittling Westerners down to size, Liang Ch'i-ch'ao and his fellows sought to build Chinese up. Somewhat pathetically, however, they often felt obliged to rely on Western authorities to do so, such as the anonymous Englishman whom someone discovered in a British journal arguing that Britain should be good to China and protect her, because someday, like it or not, China would be "the most influential nation in the world." Once China had reformed, he said, "she will steadily advance. When England shall have declined, China will be in her ascendency. And will she not be strong! Will she not be strong! Macaulay once said that it will undoubtedly be New Zealanders who wander through the ruins of London, but I feel that the people who trample down our English capital will probably be Chinese. . . . For if China really adopts Western institutions, she will become the strongest and most invincible of nations. The Mongol race will again be able to lead its armies against the West, as Attila did of old."[46]

This "yellow-peril" argument, even before it was so named, at first seemed very flattering to the Chinese, even though "the

Mongols" had, of course, been the peril of China as well as the West. It offered a vision of hope and power (and not a little revenge) that was authenticated by Western experts—who ought to know. At the same time, it clearly bore the Reformers' basic message: Future glory could only come after *Western-style* reform. Most important, however, the anonymous Englishman's argument implied a cyclical theory of history, very resonant with the Chinese tradition, which almost turned China's weakness into an asset. If peoples rose and fell, China's present weakness could be seen not as a sign of her imminent demise, but as a natural prelude to ascendency, perfectly timed to coincide with her enemies' inevitable decline. Unfortunately, the Darwinian picture of species disappearing forever from the globe soon made it difficult to be complacent with such a theory. Still, the faith that China would one day *again* be the greatest nation on the earth led many to try to rationalize such cycles with evolution, and many more, even without their becoming Taoists, to see in weakness hidden strength.

The editors of the *Shih-wu pao,* however, were not yet attempting to present any one consistent theory. In an effort perhaps to shame their countrymen into action, they printed the opinion of another Englishman who was *not* impressed by the yellow-peril argument: "China is the invalid of the Orient. She has long been half dead. . . . People have said that China will be the number one power on earth. Whom do they think they're fooling? Heaven? China has never fooled me. . . . Lord Wolseley [?] has repeatedly said that Chinese troops are so many, and their generals so brave, that he is deeply afraid that the Western nations will be annihilated by the yellow race. But it is already as clear as a candle in a cave that we can lie back on our pillows without fear."[47]

Such a quotation was supposed to inspire shame, indignation, and the determination to show such scornful foreigners that China could be fearsome. But if, instead, it inspired gloom, the editors could always offer, as dubious consolation, the protestations of yet another Englishman, who said that, although China was indeed one of the "four invalids of the world" (the others being Turkey, Persia, and Morocco), she would probably survive even so because, "with three hundred million people, the Western nations are all unwilling to be bothered ruling her."[48]

Even the foreigner above who was so disdainful of Chinese power and so sure that England could rest assured had snidely remarked that, although China was already quite at the mercy of the Western powers, since the two greatest of those powers, England and Russia, did not yet *want* to partition China, China's sovereignty might yet be preserved. It was "still not too late."[49]

That last assurance, however, was all Liang Ch'i-ch'ao needed to insure his cry for action. China had a chance. It was up to her. Time was short, but she could do it if she would. It was at this point that Liang Ch'i-ch'ao brought in Darwin's doctrine to help show her how.

EVOLUTION TO THE RESCUE

Although Yen Fu had been listed on the last page of the eighteenth issue of the *Shih-wu pao* as a contributor of a hundred "dollars"[50] (the eighteenth issue, published on February 22, 1897, was also the issue in which the journal added the English name, *The Chinese Progress*), it was not until the twenty-third issue, of April 12, 1897, that he was formally introduced, in a footnote, as the translator of *T'ien-yen lun,* which was likewise there introduced for the first time (a rather humble entry for such an influential book). In the same issue, Yen Fu's article "In Righteous Refutation of Han Yü," apparently thought more radical than *T'ien-yen lun* because of its espousal of democracy, was reprinted under a pen name. The article caused trouble—it made Chang Chih-tung so angry he ordered T'u Jen-shou to write a rebuttal, which was published (probably quite reluctantly) in the thirtieth issue, two months later. Apparently it almost got Yen Fu into serious trouble.[51] The book, however, as introduced by Liang Ch'i-ch'ao, did not.

ASCENDENT DESCENDANTS. The footnote in which Liang Ch'i-ch'ao first introduced Yen Fu, *T'ien-yen lun,* and the theory of evolution came in the sixth installment of a major tract, "On Schools."

That installment was on women's education; the section in which evolution made its entrance was on "pre-natal education" (*t'ai chiao*)! Liang Ch'i-ch'ao suggested that, if men were truly aware of evolution, they would make very sure that, by the time their children entered the world, they should already have had nine months in a class womb. Liang Ch'i-ch'ao, given only the barest sketch, though a very poetic one, of the role of heredity as it was then understood in evolution, leapt at once to a belief in the inheritability of acquired characteristics—because that held out hope for the speedy improvement of the Chinese people.

In a pregnant paraphrase that verged on plagiarism of a passage from Yen Fu in *T'ien-yen lun,* Liang Ch'i-ch'ao stated that

> Western scientists who have investigated the principle behind the successive progression of animal species and the races of man say that in all organic creatures there is that which dies and that which does not die. . . . That which does not die leaves the mother and goes with her child, over and over in unbroken succession. This is called transmitting the seed [*ch'uan chung*—The word *chung,* besides meaning seed, means also race, kind or species] . . . . Yet, after the meeting of the sexes, there is the possibility of influence and change that can cause the species to steadily improve. From apes and monkeys there is an advance to man. From wild savages and lowly breeds there is an advance to civilized and noble races. The process began with the infinitesimal, but the final result shall be tremendous.[52]

Here was the clearest statement to date (clearer than the fact) that man had both descended and ascended from the ape. But here also, alas, was the clear assumption that the difference between man and ape was parallel to the difference between civilized man and savage, between "noble race" and "lowly" (*kuei-chung, chien-chung*). The potentially fearsome prospect of racial inequality, however, was mitigated for Liang Ch'i-ch'ao by his second assumption: that men could "themselves improve their race."[53] Liang Ch'i-ch'ao had missed Huxley's scorn for people who believed that "the pigeons . . . [could] be their own Sir John Sebright,"[54] but he did not, lest we exaggerate, yet have any fanatical eugenicist's gleam in his eye. He was not advocating any weeding out of the unfit. He was actually still quite in harmony with Huxley. For he advocated a policy, the aim of which, in Huxley's words, would be "the fitting of as many as possible to survive."[55]

Liang Ch'i-ch'ao's means of cultivating fit citizens for a fit nation, was, of course, education, his solution for almost everything, but how he became convinced that education could begin in the womb remains unclear. He cited two early Han works that advocated pre-natal education, and concluded that "the ancients valued it so strongly, there must be something in it." The belief had now, alas, long been out of fashion in China, but Westerners were recently becoming more and more excited by it. Indeed, "those who study races are convinced that pre-natal education is of primary importance" in racial improvement. He did, however, confess that he found it hard to convince people of the fact. "Most," he said, "claim it is about as practical as plowing to save the starving."[56]

But what did he mean by pre-natal education? In keeping with a very pervasive early Chinese prejudice that virtue is somehow catching, the Han works he cited had held that, if a woman with child was especially virtuous before her delivery, if her behavior was in all things proper, and if she was exposed to the beneficial influence of the proper behavior of others, that is, if she heard no evil, saw no evil, and spoke no evil, but only its opposite, then her own virtue would "rub off" on her child, who would be born with a head start on the road to perfection. Liang Ch'i-ch'ao's thinking seemed to be on a different plane. "All countries," he said, "that wish to have strong soldiers insure that all their women engage in calisthenics, for they believe that only thus will the sons they bear be full in body and strong of muscle."[57]

Although calisthenics was the only specific example of pre-natal education he gave, he obviously had more in mind than the nation's pulling itself up by having its pregnant women do pull-ups. Physical strength was a bare beginning. He wanted "to make the race progress" from "graft to loyalty, from selfishness to public-spiritedness, from disunity to solidarity, from stupidity to wisdom, and from barbarism to civilization."[58] He wanted to change, and believed it was possible to change, his countrymen's very nature—if not necessarily their human nature, at least their racial nature, just as Mao Tse-tung, in his more optimistic moments, would later believe it was possible to change their class nature. Neither man believed, of course, that all this was possible in only nine months of pre-natal training, but Liang Ch'i-ch'ao, at least, believed that the process began there, and he seemed to believe that somehow the effects of both pre-natal and post-natal education could be inherited, so that in several generations' time a race's nature could improve.

Liang Ch'i-ch'ao's first acknowledged quotation from Yen Fu was to the effect that a man's "thought, intelligence, physique, and habits" were fashioned by a combination of both heredity and education, that is, environment, and that both factors should thus be taken into account by those who would preserve the race.[59] Pre-natal education was just the (seemingly) most obvious place where the forces of education and heredity came together. The distinction between pre-natal and post-natal education, however, was fuzzy at best. Women's education in general was "the first step towards the preservation of the race," simply because half of the population were women and the other half were largely raised by women.[60] A woman's education, therefore, would rub off on her offspring both before and after birth. It was the belief in an overall accumulative effect of such education, however, and in the consequent betterment, not just repeatedly of individuals within each generation, but once and for all of the race as a whole after passing through several generations, that implied a belief in the inheritability of acquired characteristics.

There was nothing strange in Liang Ch'i-ch'ao's believing in such a notion. Darwin himself, in both *The Origin of Species* and *The Descent of Man* had attributed "great weight . . . to the inherited effects of use and disuse,"[61] and he had said in the latter that it was "probable" that resultant "modifications would become hereditary, if the same habits of life were followed during many generations."[62] Indeed he had dreamt up a theory of "pangenesis" to explain how this could be.[63] In Darwin's time, and even in

1897, the true science of genetics did not really yet exist. But Liang Ch'i-ch'ao had precious little idea even of what Darwin had said about inheritance and evolution. All he saw of importance at this point in the fact that inheritance was somehow part of evolution was that heredity was therefore not a trap. One was not doomed to be forever like one's ancestors. The race could yet improve. There was, therefore, hope. He did not realize what a terribly complicated picture he was stepping into. He suggested that animal species and the races of man were part of one long progression, at the pinnacle of which would be truly civilized man; but, although individual species and races appeared thus to be different stages, he could not bear to believe that they were cast-off stages, doomed by heredity to remain such forever, immutable. And yet, if "races" could improve, could "lower species"? Could apes become ape-men?

Liang Ch'i-ch'ao did not yet ask such questions. All he said was that the Chinese race needed to improve, although he characteristically added, in his letter to Yen Fu, that "it is not only the yellow race that ought to seek improvement; even the white race should seek to improve."[64] In stating the reason why the Chinese race in particular and all races in general needed to improve, however, he invented a Darwinesque law that was highly un-Darwinian, but which proved to have most powerful survival value among Chinese Social Darwinists. He stated that "one must cause one's race [or species] to progress before one can preserve it."[65]

PRESERVATION OR SURVIVAL. In the natural world (in opposition to what other?), "progress" is not necessary for survival. From the point of view of a species, not even change is absolutely necessary, for if a species changes it is a good question as to just what it is that survives. Gerald Durrell has described the tuatara, a New Zealand lizard-like creature, which is really not a lizard but a "beak head," as a "genuine, living, breathing prehistoric monster," the "last survivor" of a "once widely spread group," rampant over Asia, Africa, North America, and Europe "some two hundred million years ago." "To have come down through all those years unchanged," Durrell writes, "surely makes the tuatara the conservative to end all conservatives,"[66] not to mention one of the fittest creatures of all time.

The tuatara, of course, was actually fit for only one corner of New Zealand. And, as Durrell knew full well, it no more warranted a pat on the back for being conservative than any other species does for being progressive. In the unwitting (one presumes) fashion of a Taoist Sage, the tuatara had simply followed its nature, as had its uncles and its aunts now long extinct (lest one conclude that Taoist non-action be the secret of survival). The tuatara survived, that is, was preserved, because it was one of the "favored races." It *chanced* to live in splendid isolation, free from undue competition. It lived a charmed life. Nonetheless, it is perfectly true that its "survival" is recognizable solely because it did not change. Had it "progressed" very far, we would not say the original species still survived.

Soon some would suggest that China's survival, and indeed her splendor, were also due simply to isolation, and to the lack of any competition to speak of. And they would suggest that now, with the advent of the West, a new competitor, a new species, had been loosed into the environment, and that a struggle for existence of unprecedented gravity was about to ensue.

Darwin, as Yen Fu explained in *T'ien-yen lun,* had claimed that such changes in environment and competition, as much as simple overpopulation, did set the stage for the struggle for existence and natural selection, but he had offered no such hopeful law as Liang Ch'i-ch'ao's, that by making one's race progress one could triumph. He had not said the giraffe had the power, given the will, to stretch its neck to meet new conditions of competition. But, even if he had, the question would still remain: Did the long-necked giraffe really preserve the race of his short-necked ancestors?

Admittedly, Darwin often sounded as if he meant that species change in order to survive, but what he proved was that species *are changed* by the nature of those who *do* survive. It is only after successive segments of a species survive, thanks to happy accidents of environment and genetic combination, that "modified descendants" are suddenly seen to have become so recognizably modified as to be modified almost beyond recognition, at which point biologists announce the existence of a new species, or a sub-species, if not, necessarily, a super-species. But what has happened to the species that wanted to be preserved? Is a race that has so progressed still the race it was? Or were the conservatives right in thinking that, if China wanted to survive, if she wanted to preserve what she was, she should resist change? How far, after all, could China change and still be China?

A tremendous question was in the making as to just what survival meant, and just what it was that one wanted to survive. Twenty years later, when conservatives were desperately entreating their countrymen to "preserve the national quintessence," Lu Hsun would answer, "Ask not whether something is of the national quintessence; ask whether or not it has the power to preserve us."[67] But he did not say who "we" would be without "our quintessence." He himself, as we shall see, had enough of a Darwinian faith to believe that it did not matter what was lost, for whatever survived would be better. But, for the conservatives who did not have that faith, it sounded as if he was saying that China must sell its soul to save it.

Not unnaturally, it was Yen Fu who first saw deeply into this dilemma of survival and preservation. He did so back at the beginning, when in 1897 he analyzed K'ang Yu-wei's movement to "preserve Confucianism" (*pao chiao*), that initially radical movement which first drove conservatives to try to preserve Confucianism from its preservers, but which in the end became the spiritual predecessor of the conservative movement to preserve the national quintessence. Yen Fu was against any active attempt to preserve Confucianism, because he felt, as he told Liang Ch'i-ch'ao, that, "if you preserve the religion and it progresses, then it will no longer be the original religion that you sought to preserve anyway."[68] It was thus better to let it take care of itself, to sit back, in short, and see if it was fit.

When Liang Ch'i-ch'ao read this, he pounded the table in admiration. But he did not apply the same logic to the other two entities K'ang Yu-wei sought to preserve, the nation and the race. To preserve one's race, he insisted, one must make it progress. He did not seem frightened by the possibility that, if the race thereby survived, it might no longer be the original race. He did not yet foresee that the preservative changes he thought necessary for survival might themselves put China's identity at stake. But even if he had, he was not just going to sit back and see if China was fit. He was not going to let China enter the lists of natural selection without putting up a struggle.

But did these logical problems matter? Liang Ch'i-ch'ao's arguments made sense as common sense. If he translated his law, that a race must be made to progress in order for it to survive, into the statement that China must reform in order to preserve her sovereignty, as of course he did, who could doubt it? His argument made probable sense as political science; it just did not make sense as natural science. The trouble was that he claimed to have derived it from natural science. Indeed he claimed it was a natural law. He claimed, alas, the backing of a law that did not exist. Did it matter? Yes. For his illogic was catching and, though not incurable, tenacious. And it spread to both his friends and his opponents.

THE CH'ÜN AGAIN. Be that as it may, in the twenty-sixth issue, Liang Ch'i-ch'ao again mentioned Yen Fu and *T'ien-yen lun,* this time not in a footnote, but in the text, at the very beginning of the introduction to a promised work of grand proportions: "Shuo ch'ün" (On the group). In the very first paragraph, Liang Ch'i-ch'ao brought together K'ang Yu-wei, Yen Fu, and the rather wildly eclectic young thinker and future martyr of the Reform Movement, T'an Ssu-t'ung, in order to pool their wisdom and scholarship in an effort to explain the significance of that all-important word, the *ch'ün.*

Liang Ch'i-ch'ao opened in perfect Confucian master-disciple style: "I asked my teacher, K'ang Yu-wei, the way of ordering the world, and my teacher said 'Take the group as your body and take change as your action (*i ch'ün wei t'i, i pien wei yung*). If these two principles are established, the world can be ordered for ten million years.'" This theory proved so deep (or abstruse), however, that Liang Ch'i-ch'ao despaired of expounding it, until he saw Yen

Fu's *T'ien-yen lun* and T'an Ssu-t'ung's *Jen hsueh* (The study of universal love). Then, he said, "lamenting the fact that so few men of patriotic spirit had as yet been able to hear Mr. K'ang's theories or see the great works of these two gentlemen," he decided to take it upon himself to attempt in a treatise of ten chapters and a hundred and twenty sections to proclaim his master's doctrine and its substantiation by these two exciting new books.[69]

He finished only one of the promised six score sections. But that, together with the introduction, which was all that ever got into the *Shih-wu pao,* was quite enough to prove how readily, indeed how almost instinctively, Liang Ch'i-ch'ao accepted the notion that the *ch'ün* was at once the *raison d'être* and the *sine qua non* of human existence. Of course the question "What is my group?" threatened to be as unsettling as the question "Who is my neighbor?" But sweeping it aside, at least for the moment, Liang Ch'i-ch'ao could immediately agree with Yen Fu and K'ang Yu-wei that what mattered most in human life was the life of the group, not the life of the individual.

That Liang Ch'i-ch'ao should so bring together K'ang Yu-wei's formula for ordering the world and Yen Fu's introduction of the Darwinian theory of evolution was immensely fitting—not because their thinking was actually in harmony, but because K'ang Yu-wei's formula, couched as it was in the popular *t'i-yung* (body and action) dichotomy, could stand as a perfect descriptive motto for most of the Chinese Social Darwinism that was to follow: Take the *ch'ün* as the basic unit in the evolutionary struggle for existence, and take change as the *ch'ün*'s basic means of survival. The human struggle for existence was not to be described by the cry "Every man for himself." It was "Every group for itself," for every man's fate was bound to his group's.

This part of the formula, however simplistic, could easily be justified by Darwinian quotations. True, Darwin had said that new species emerge, and evolution takes place, thanks primarily to the struggle for existence *within* given species (although struggle *between* species might exacerbate that struggle). He had said, indeed, that "the struggle almost invariably will be most severe between *individuals* of the same species." But he had hastily added that "in the case of varieties of the same species the struggle will generally be almost equally severe."[70] If tribes and races of men could be called varieties of the human species, therefore, *The Descent of Man,* which Yen Fu had probably read,[71] although Liang Ch'i-ch'ao had not, could provide more than enough examples of life-or-death struggles between groups.

The second part of the formula, however, was, as we have already seen, on shakier Darwinian ground, for by "change" Liang Ch'i-ch'ao, K'ang Yu-wei, and Yen Fu all meant conscious, active change. They believed in self-change, self-cultivation, self-strengthening, self-improvement, self-reliance. They thus assumed that, in the struggle that was evolution, human beings had the power to change their "corporate self" and make that self survive.

Now Darwin himself was unclear at this point. Will power was not part of his original scheme. In *The Origin of Species,* there were no species that consciously modified themselves. But in *The Descent of Man,* Darwin in several places showed himself tempted to agree with Alfred Russel Wallace that, with the arrival of man, evolution became "a new contest":

> Mr. Wallace, in an admirable paper . . . argues that man, after he had partially acquired those intellectual and moral faculties which distinguish him from the lower animals, would have been but little liable to bodily modifications through natural selection or any other means. For man is enabled through his mental faculties "to keep with an unchanged body in harmony with the changing universe." He has great power of adapting his habits to new conditions of life.[72]

With animals, Darwin went on, the situation was very different: "When they migrate into a colder climate, they must become clothed with thicker fur or have their constitutions altered. If they fail to be thus modified they will cease to exist."[73] On the one hand, "great power of adapting," on the other, the necessity of "being modified"—there was a world of difference, almost beyond analogy, between the active and the passive voice. Wallace's vision, which Darwin seemed to accept, almost had evolution creating man only to set him free, above "the changing universe," again in the super-natural.

Darwin, of course, had precious little place for the super-natural, at least by the time he wrote *The Descent of Man,* and yet, in the very chapter in which he included the above passages, he himself came perilously close to placing man there. He wanted to prove that man's intellectual and moral faculties were both products of evolution, and indeed of natural selection, but, in so doing, he acknowledged that man's "instinct of sympathy, which was originally acquired as part of the social instincts," and which was now "the noblest part of our nature," had developed to such a point that man had no choice but to "bear the undoubtedly bad effects of the weak surviving and propagating their kind."[74] But, if the fit and unfit all survived, then what had happened? And what would happen? Had evolution turned against itself? Had nature selected from itself by natural law a creature that could outlaw natural selection? And, if it had, then what would be the "undoubtedly bad effects"? Would morality make man regress, or keep him down (shades of Nietzsche), or simply stop his evolution? If it was really, as Huxley said, that "after the manner of successful persons, civilized man would gladly kick down the ladder by which he has climbed,"[75] would he, then, climb no more?

Darwin, in 1871, was very close to the position that Huxley would take twenty years later, that for better or for worse—and Huxley thought for better—"the ethical process is in opposition to the principle of the cosmic process" which gave it birth. Darwin never put it quite so baldly, and so was spared, perhaps, the headache of wondering how such a thing could be. Huxley was not spared, but concluded in the end that, if such an antagonism was "logically absurd, I am sorry for logic."[76]

Liang Ch'i-ch'ao was singularly unperturbed by the logical problems of human freedom and evolution. He had obviously not yet thought of them. In bringing things back to earth, however (where he had been all along), he was still momentarily on rather good Darwinian ground. For Darwin had suggested that natural selection was in danger of being thwarted only *within civilized nations.*

The struggle for survival *between* nations, civilized or barbarous, was still very much in full swing. And the general trend was all too clear: "At the present day civilized nations are everywhere supplanting barbarous nations"[77] (to Liang Ch'i-ch'ao a self-evident fact). In that kind of struggle, Darwin had indeed made just the point Liang Ch'i-ch'ao wanted to make: "Selfish and contentious people will not cohere, and without coherence nothing can be effected."[78]

Liang Ch'i-ch'ao had not read that sentence. He had come to the same conclusion quite independently. The conclusion, after all, was simply "Solidarity is strength" again, and like K'ang Yu-wei he could find confirmation of that fact, if he needed it, in Hsun Tzu. But Yen Fu now told him there was further confirmation in Darwin, and what made Darwin's confirmation exciting was that it was from a Westerner, a member of the race most able to make "coherence" work, and one who had scientifically found proof of *ch'ün* strength in the cosmic process itself.

Liang Ch'i-ch'ao had read Yen Fu's statement in *T'ien-yen lun* that "those good at grouping survive; those not good at grouping perish."[79] He now expanded the idea to read, "Groups ordered by the arts of grouping will succeed; groups ordered by the arts of individualism will be defeated." The "arts of individualism" (*tu shu*) he explained thus: "Everyone knows of himself and not of the world," which statement he followed with a long list of parallel phrases designed to prove that everyone in China was out for himself, so that "with four hundred million people there are four hundred million countries."[80] Among those out for themselves, however, there were many *groups:* villages, clans, families, and occupations. If the cosmic process meant a struggle between groups, who could blame them for thinking first of themselves? Moreover, the very success of these groups seemed to belie the idea that every Chinese was out solely for himself, although it suggested that many Chinese might already be torn by conflicting group loyalties. But Liang Ch'i-ch'ao blithely side-stepped this chance for a confrontation with another host of Darwinian dilemmas, and quite arbitrarily decreed that the "national group" (*kuo ch'ün*) was the group for which all others should cohere. Self-interest, he implied, on any other level was not in the group's interest.

Curiously, however, at the very end of his argument, Liang Ch'i-ch'ao almost sabotaged the sanctity of the "national group" that he had just established. Again he swept away the primary importance of the nation by waving the banner of the *Ta t'ung.* He had seemed to be building up to the rather pessimistic conclusion that Chinese society, with each element out for itself, could not hope to survive by simply imitating the outward actions of Western peoples, who were truly united. For China would then be only "a sheep in tiger's clothing." Some basic spiritual change would have to be effected. But then he suddenly thrust the whole gloomy prospect away from him and said: "But I have heard that there are national groups, and there is the group of all under Heaven (*t'ien-hsia ch'ün*). Western administration for national groups is unsurpassed, but they are not yet able to apply it to the universal group." Then he quoted the *I ching* and the *Spring and Autumn Annals* (that is, the commentary thereto) on world unity, and finished with the "Li yun" paean to the *Ta t'ung.* "Is not this the group of all under Heaven?" he asked, "Is not this the group of all under Heaven?"[81]

Here, surely, is an especially obvious example of the *Ta t'ung* providing an emotional escape from the depressing fact of China's ignominious position in the ranks of international competition. But, to be fair, we must admit that it is not only that. The *Ta t'ung* was an ancient Chinese ideal, approximated, at least, in the Chinese Empire, in China's own world order. The supposed Darwinian picture, of a perpetual succession of warring states, was not. No Chinese could really accept unending struggle as the *Tao.* That is why "evolution" was so quickly fused with "progress." Struggle was only good because it was moving man towards the end of struggle. Therefore, the national group, and international struggle, were of prime importance only "at the present stage."

A Fighting Stage. The first in the *Shih-wu pao* consciously to use a stage theory as a philosophical analogy for militant nationalism was Mai Meng-hua, who deftly managed to employ in that cause the Three-Age theory of K'ang Yu-wei that was supposed to herald supra-national harmony. He did so in an essay entitled "Tsun hsia p'ien" (Respect the righteous fighter), a call for a national revival of a chivalric, martial spirit.

The traditional Chinese *hsia* that he celebrated was a figure somewhat like a Western knight, though rarely seen in shining armor, and no member of any officially recognized order, rank, or social class.[82] The Chinese *hsia* might or might not be errant. He was more likely to fight on foot than on horseback, and he could just as well be the master of some karate type of self-defense as a master swordsman. But, like his Western counterpart, he was supposed to right wrongs, exterminate oppressors, and rescue damsels and other innocents in distress. He was a fighter, but one who ideally used might for right.

The kind of *hsia* that Mai Meng-hua admired, however, was not the damsel-saving variety, but the kind who would "sacrifice his family to save his country."[83] The wrongs Mai Meng-hua wanted righted were wrongs against the nation: "How deeply do these Westerners insult us. They cut away our territory, force from us treaties, forbid our raising tariffs, interfere with our officials, drive out our laborers, control our railroads, seize our mines, demand our ports."[84] To avenge these affronts, he cried, China must raise up a host of fierce patriots, men of *hsia*-like spirit, who would dare "seek life through death, survival through extinction, possession through loss, safety through danger."[85]

He admitted that such a spirit might seem at variance with the *Ta t'ung:* "Some may object that the meaning of the *Spring and Autumn Annals* is that in the Age of Great Peace all countries, far and near, large and small, will be as one. In the reign of Great Unity one will love a neighbor's country as one's own." But in that august day there would be equality, Mai Meng-hua protested: "With the West so contemptuous of our China, humiliating us in a hundred different ways, how, in the name of reason, can you call this equality? Now a *hsia* would save the weak and root out the strong, making fair the unfairness of the world. . . . Even in the

Age of Great Peace there must still be retribution. How much more so in our world today, a world in the Age of Disorder." If one does not fight in *our* world, one follows India and Turkey.[86]

Mai Meng-hua said nothing of Darwinian evolution, but he accepted K'angian evolution only with the Darwinesque proviso that in the Age of Disorder one still had to struggle to survive, or rather one had to struggle so that one's country could survive, for countries would be all-important until they were all equal, if not in size at least in sovereign rights. At least two of his ideas, however, were distinctly un-Darwinian. The very idea of either men or countries having any "rights" at all was almost impossible to derive from a biological theory of evolution. Mai Meng-hua declared that the *hsia* ideal of equalizing the unequal was just "what advocates of international law meant by equal rights,"[87] but such rights could not be natural. "Natural rights" could only be granted by the super-natural. Even legal rights were artificial, for in Huxley's terms they were (presumptuous?) man-made restrictions on the cosmic process. By the same token, however, the *hsia* themselves were unnatural, or anti-natural. "Saving the weak and rooting out the strong," they were super-busybodies interfering with the process of natural selection.

Nevertheless, the *hsia* did remain an ideal for many Chinese Social Darwinists. Without worrying that a man of *hsia*-like spirit might interfere with natural selection, they believed that such a man both could and should help influence its outcome. Here was the belief in voluntarism again, for the *hsia* was the epitome of man turning will power into action to effect his "destiny." Clothing voluntarism in the garb if not the armor of the *hsia,* however, had one perhaps unlooked-for result. Mai Meng-hua had written his essay to promote reform. He wanted to inspire a national martial spirit to stand against foreign aggression. But, in the long run, his stage-supported doctrine of the worth of the righteous fighter probably did less to aid reform than to set the stage for revolution.

THE SPREADING OF THE WORD—AND RACISM. The *Shih-wu pao,* however, still maintained that the Chinese government could change itself. Its most direct political campaign was to impress upon the government the frightening urgency of reform. And this it did through its news stories as well as its essays and editorials.

Most of the news items gathered from foreign publications were concerned with the relative strength of the powers and their relative threats to China. Every encroachment, real or imagined, was repeatedly discussed. The Trans-Siberian Railroad, for example, was a source of constant worry. Every rumored change in the military or economic strength of foreign countries was immediately reported, as of course was any news of Chinese growth. Occasionally, there was encouraging news. Someone reported that, after only a few years' practice, grape-growers in North China had produced a Chinese wine that "according to certain foreigners" was already "much better than French champagne."[88] More often, however, there were gloomy prognostications such as that on the very next page: "The Russians are going to swallow up all Asia."[89] Only rarely were there any items as seemingly free of vital significance as the piece of "rare news" that "the English Prince of Wales is good at making boots."[90]

As the *Shih-wu pao's* campaign progressed, the vocabulary of both Darwinian and K'angian evolution slowly spread. The *Ta t'ung* soon turned up in the name of a translating-publishing house, started by the Reformers to spread Western knowledge,[91] and in the name of a school in Yokohama, started by overseas Chinese to foster patriotism among their children. The founders of the school had asked the advice of Sun Yat-sen, who oddly enough recommended that Liang Ch'i-ch'ao, whom he had not yet met, be made its principal, and that the school be named The Chinese-Western School (Chung hsi hsueh-hsiao). He also recommended, however, that the founders seek the further advice of K'ang Yu-wei, whom he had also not yet met, but whom he at that point still respected as a patriot and scholar. K'ang Yu-wei, however, rather high-handedly deciding that Liang Ch'i-ch'ao belonged at the *Shih-wu pao,* instructed the founders to take Hsu Ch'in, another of his disciples, instead. He also decided that Sun Yat-sen's words, *Chinese-Western,* were not eloquent enough, and substituted his own choice for a name, The Ta T'ung School (Ta t'ung hsueh-hsiao), which he wrote in his own hand on a wooden plaque to be hung over the new school's gate.[92]

Darwinian vocabulary did not immediately have any such startling success, but it did become more common; at least Darwinian metaphor did. Darwin's name, during this period, was almost never mentioned. Liang Ch'i-ch'ao embroidered his accounts of evolutionary change "from the world of giant flora and fauna to the world of human beings" with new bits of Western archeological knowledge that traced human evolution "from the age of stone knives, to the age of bronze knives, to the age of iron knives, to the age of today."[93] And evolutionary concepts began to show up in the writings of other contributors to *Shih-wu pao* besides Liang Ch'i-ch'ao, and even in news stories.

"Survival of the fittest" was the most commonly quoted Darwinian expression, although there was as yet no one accepted formula for the term. A Japanese, commenting on the abnormal rate of economic growth in Japan after the Sino-Japanese War, predicted that a moment of truth was coming for the enterprises that had mushroomed at the time: "The false shall be weeded out and the true left. The superior, as they say, will survive, and the inferior will perish."[94] A Manchu, appealing to the same law, implored his kindred Manchu bannermen to pull themselves together before it was too late: "The mind of Heaven is constant. If Burma, Korea, the Ryukyus, and Vietnam have all gone under, it is not because Heaven helps the violent. England, Russia, France, and Japan have flourished simply because the strong survive and the weak do not. Germany is a small country, but her people through self-struggle have stood up. Japan was a weak country, but, fearing subjugation and striving for survival, she has suddenly become strong. . . . Who says that human determination cannot win out over Heaven? My brothers, wait not on fate with complacency. Do not throw yourselves away trusting to good fortune."[95]

This Manchu plea for bannermen self-strengthening, confused as it was over a constant Heaven that could be "bested" by man, was issued out of fear of the foreigners, not of the Han Chinese. Perhaps, however, to some minds it struck a note of further foreboding. For if the Western precept "Only the strong survive" was frightening to the Chinese, it should have been twice as frightening to the Manchus.

Actually, the growing Darwinian influence in the *Shih-wu pao* is nowhere more clearly seen than in the ever-swelling current of Darwinesque racism that eventually helped pull Chinese and Manchu apart. Such racism, as we have seen, developed largely in response to Western racial insults against China. Liang Ch'i-ch'ao said that Westerners consciously conducted vilification campaigns against other peoples and governments as a first step towards taking them over. They called others savages to inspire "virtuous men" to "rescue" them from their savagery. Labeling such insults ploys, however, did not take the sting out of them. Complaining that Chinese had recently been subjected to a veritable barrage of vile slander, Liang Ch'i-ch'ao said: "We have translated a lot of such stuff in the *Shih-wu pao,* but there is much, much more that we dare not translate, like [one piece] . . . that said, 'The Chinese people have not only already died, they are stinking rotten.' Such talk is really unbearable."[96]

His answer was to shout, "The Westerners are wrong! China is no India or Turkey."[97] But in so shouting, only ten months after his universal-humanity arguments, he himself chose to deliver a barrage of racial insults which we, in turn, almost "dare not translate." "India's failure to rise," he wrote, "is due to limitations of the race. All black, red, and brown peoples are in the microorganisms of their blood and the slope of their brains quite inferior to white men. Only the yellows and the whites are not far removed from one another. Hence anything whites can do, yellows can do also."[98]

But, not content with the Chinese being only one of the top two races, Liang Ch'i-ch'ao gave in to visions of even greater grandeur. He dreamt of China's beating the white men at their own game. He became one of the first to voice the hope that China would one day be the greatest of colonialist powers. But this dream was even more obviously held at the expense of the "inferior races." The white men, he said, were strong because of their colonialism. But the world was too big for them. There were vast expanses of the globe that the white men were unable to touch. They were not up to the task of holding dominion over all the earth. Why? Because (and only a Chinese could have said this with such condescension) "there are only so many white men, and they are certainly not enough to monopolize the world's wealth." What then? The answer was obvious: "In all the earth there are only five races of human beings. The white race is as described. The red race is about to expire. The brown and black races are stupid and lazy. They are not able to order their lives. They are not willing to work hard. They multiply, but they are always as before. They are like walking corpses, and there is no talent to be had among them. Thus, to cultivate the virgin wastelands and to clear up the entire globe, unless it be done by us yellow men, there is no way."[99]

It will come as no surprise to learn that these sentiments were expressed in an article entitled "China Will Be Strong." Liang Ch'i-ch'ao and all too many of his confederates accepted the doctrine of racial inequality to disassociate themselves forever from all races subjugated by the white man. ("Third World Solidarity" was a long way away.) They insisted that Chinese were not to be classed with Africans or Indians, red or brown. Chinese were in a class by themselves. In the words of an ancient epithet, Chinese were "descendants of the gods."[100] China was a huge land, with a perfect climate, innumerable kinds of produce, and a people of exceptional intelligence:[101] "Of the ten thousand nations, where is there an equal?"[102] Liang Ch'i-ch'ao, forgetting for a moment his oft-repeated warnings, insisted that "there is no way for China to perish; she must be strong."[103]

The dreadful discrepancy between China's natural superiority, so fervently believed in, her "natural right" to pre-eminence, and her inferior position in present reality created a pressing need for a theoretical explanation. The search for one, however, led, in only two short steps, to a point whence one could find light in Social Darwinism.

The first step was taken by Mai Meng-hua in a significant essay on "The Significance of the People." By "the people" he meant the people politically, not racially; he meant the people as opposed to the government. Racially, he quite agreed that the Chinese people were practically peerless. Politically, however, the Chinese people had been kept down. "Why is the West strong?" he asked, "Because its people are knowledgeable. Why is China weak? Because her people are ignorant."[104]

This was a view strongly shared by Liang Ch'i-ch'ao and Yen Fu. Indeed Mai Meng-hua's next line of argument was very close to Yen Fu's attack on the Sages. He held that an authoritarian system had stifled the intelligence of the people. He was not as harsh on the Sages as Yen Fu was. "The Sages did not," he said, "actually wish to make the people ignorant. The people became ignorant because they were ruled in an ignorant way, so that in time the wise lost their wisdom, and the ignorant became happy in their ignorance."[105] In the end, he put more blame on the people themselves for this state of affairs than on their rulers. "When," he asked, "did the rulers ever completely seize the people's power? It was the people that completely gave up their power, entrusting it to their rulers saying, 'We have no rights, we have no rights.'"[106] It was their acquiescense that led "a brilliant race to became as base and sickly as the blacks."[107]

But this was an unnatural result, an aberration from China's proper destiny. The proper way, if only men had realized it, had been pointed out by Confucius, in his doctrine of the Three Ages, a doctrine of progression towards popular liberation and self-rule. In the Age of Disorder, there had been no order whatsoever. To escape that, "in the Age of Ascending Peace, a ruler was to rule the people. The ignorant masses were to receive order by obeying a single wise man." But that was not the end ideal. From there men were supposed to move on to the Age of Great Peace, in which "the ruler would still be revered, but affairs would be run by the people."[108] It was this that the Chinese people in their apathy and self-contentment had failed to do. It was this that they now must do if their country was to survive. They must realize the truth of the seventeenth-century patriot Ku

Yen-wu's words, "The empire's fate is the responsibility of even the lowliest commoner."[109]

WAVES OF THE FUTURE. Mai Meng-hua again fell back on K'ang Yu-wei's Three Ages, this time to suggest that Chinese society should be evolving towards democracy. Liang Ch'i-ch'ao was to go one step further and say that the Westerners' theory of evolution proved the same thing. He was to fuse his master's Three-Age theory with the theory of evolution he had learned from Yen Fu, even though in so doing he was publicly, however politely, to take issue with Yen Fu over the future course of evolution.

The article in which he did this, "On the Law of Evolution from Monarchy to Democracy," was probably the most important piece he ever wrote for the *Shih-wu pao,* the most provocative politically, and the most exciting and influential intellectually, for it offered his first attempt at what for China was a radically new evolutionary theory of history. K'ang Yu-wei himself had as yet provided only the barest sketch of his Three Ages. And he was much more concerned with the evolution of human society beyond the nation-state into One World than with the evolution of institutions within the nation-state. Liang Ch'i-ch'ao, however, for the first time, specifically and quite elaborately identified the political institutions proper to each stage of history in the life of a single country.

The Three Ages were seen as a progression from government by many rulers to government by one ruler to government by the people. But each stage was further divided into two, so that, in all, a society would move through six distinct phases of political development: "The Age of Many Rulers is divided into the Age of Tribal Chieftains and the Age of Feudalism and Hereditary Aristocracy; the Age of One Ruler is divided into the Age of Monarchy and the Age of Constitutional Monarchy [lit. "Joint Rule by Monarch and People"]; the Age of Popular Rule is divided into the Age with a President and the Age Without a President."[110] Not only would history show the Emperor to the door; it would leave the door open for the anarchists.

This political evolution was not just an architect's dream. It was supposedly grounded in and parallel to the evolution of all things: "All things on earth progress from the simple to the complex, from the rude to the refined, from the evil to the good. There are fixed steps; there are fixed periods, like the levels of stone in geology. And their order cannot be confounded."[111] Liang Ch'i-ch'ao, taking his inspiration, and a good bit of his language, from *T'ien-yen lun,* saw in evolution an inevitable progression. He saw in it one road that all nations must follow, clearly marked by the milestones of his elaborated Three-Age theory, compressed into the life of each and every nation. At least, that is what he seemed to see: "The destiny of the world is to enter Great Peace. This is not something the West can have to itself; it is not something China can avoid. I know that before a hundred years are gone the five continents shall all embrace democracy. Our China will not be able to stand alone unchanged."[112]

There was no mention of nations falling by the wayside. The darker races, whom Liang Ch'i-ch'ao had earlier placed so precariously close to extinction, were in this argument conspicuously absent, although, as inhabitants of the five continents, it seemed by implication that they too ought to reach the promised land. Evolution was apparently for all, in which case every people's future was secure. At least China's was. The uglier aspects of evolution were here totally ignored. It became a doctrine of hope, indeed of assurance.

Before he could firmly establish such a point, however, Liang Ch'i-ch'ao had to refute a more pessimistic argument that came, oddly enough, from Yen Fu. "Yen Fu says," wrote Liang Ch'i-ch'ao, "that 'evolution begins with the embryo and ends with the complete body. For the West to have the democracy it has today, there must have existed at the time of the Hsia and Shang dynasties a seed, a starting point. A country that has nothing but monarchy cannot move from monarchy to democracy though it evolve for ten billion years.'"[113]

Actually, of course, Yen Fu was not as pessimistic as he sounded. He too was sorely tempted to believe in fixed-stage theories, those both of Spencer and later Jenks, and here he was simply arguing that one could not, having discovered the Western evolutionary road to democracy, just declare that the East should or even would turn democratic too. You had to grow into democracy naturally. If your ground had no seeds you had to plant them. At face value, his statement raised a host of disturbing questions: Was evolution not universal? Were different countries, sprung from different seeds, set forever on different courses? Was democracy Western and only Western? Yen Fu obviously did not think so. China could become democratic if men of spirit would plant the seeds, if men who willed it would break the ground. He did not realize how confused he was. He did not see that his planting was unnatural and that he was making evolution a matter of will, but he did believe that, given such planting, China would evolve towards democracy. If he was more pessimistic than Liang Ch'i-ch'ao, it was only in the amount of time he thought such evolution would require.

But Liang Ch'i-ch'ao chose to take Yen Fu's statement at its gloomiest. He said it meant that the West had the embryo of democracy and that the East did not—and then he denied it. The proof, he said, was to be seen in the history of Japan, which had lived under the same imperial house for two thousand years, but which was now as close to democracy as Germany and Great Britain.[114] So all countries *were* evolving along one road. The only difference in their histories was one of timing. Certain countries were further along than others because countries passed over different sections of the road at different speeds. All tarried along the way, but at different places. Hence the main difference between China and the West was that China had spent a short time with a system of many rulers and a long time with the system of only one, while the West had done just the opposite. But both would end up at true democracy. It was "as the Buddhists say: There are both slow and sudden ways to get through the same gate to enlightenment."[115] And China really was not that far behind. Only France and America had entered (barely) the Age of Democracy. China, Russia, England, and Japan were all still in the

Age of Monarchy.[116] (The reds, blacks, and browns were back where China was before the Hsia dynasty).[117]

This was precisely the kind of argument Marxists would later make when attempting to prove that the pattern of history for all nations was basically the same, and that China's history was not an aberration. Liang Ch'i-ch'ao was the first to familiarize Chinese with such thinking. Indeed, in this same article, he familiarized his readers with the term *feng-chien,* later the standard Marxist translation for *feudal,* and one of the most widely used (and abused) terms of abuse for the Chinese tradition. In his scheme of history, Liang Ch'i-ch'ao had already identified a feudal stage, although he also characterized it as being a slave society.[118] But his greatest bit of groundbreaking for the Marxists was simply his insistence on the inevitability of the course of history, which at this point is what evolution seemed to him to mean.

The word Liang Ch'i-ch'ao used for such a "course" was *yun,* a word intrinsically mysterious, and very difficult to translate. It was also often used by K'ang Yu-wei and Yen Fu, but never defined. It could mean "fate." It could mean "destiny." It could mean "trend" with a capital *T.* Liang Ch'i-ch'ao now claimed to be able to see its wake in history, but, far more than that, he claimed to be able to see its course in the future. From the wake of evolution seen in human history, in the "progression" of ages, labeled *post facto,* he claimed to be able to predict fixed ages yet to come. He could see where nature was going and where it was taking us. He took no heed of Hegel's warning about the owl of Minerva (to be fair, he had not heard it).

This conception of the *yun,* which would resonate so well with Marxism, was subtly different from the one he had offered just a short time earlier in his article "China Will Be Strong." There he had expounded at some length his "master's [K'ang Yu-wei's] doctrine of the *yun* of the earth," as strange as only a doctrine of K'ang Yu-wei's could be. K'ang Yu-wei's *yun* was some sort of mystically marvelous civilizing and strengthening ("fittening") force that wandered over the globe throughout human history making nations flourish. It arose in the K'un-lun Mountains and first breathed life (so to speak) into India. Thence it went on to Persia, Babylon, Egypt, across the Mediterranean to Greece, along the sea to Rome, then up the Atlantic coast to Spain, Portugal, France, and finally England. For the last thousand years it had been concentrated in Europe. But in the last hundred it had split in two. One branch was moving east sparking the rise of Russia; the other had gone across the Atlantic to stimulate America. But from Russia and America the two prongs of the *yun* were converging at last on China: "In ten years time the two *yun* will meet, and the strength of China will be first in all the world!"[119]

This was written twenty years before the Russian Revolution, twenty-two before the May Fourth Movement. Like perhaps too many of K'ang Yu-wei's strange visions, it seems strangely prophetic.

But, prophetic or not, K'ang Yu-wei's concept of the *yun* was, and is, metaphysically mystifying. His *yun* was already almost Hegelian in spirit. At least it was a spirit of some sort, passing torch-like from nation to nation. And yet, as Liang Ch'i-ch'ao described it, it was not explicitly a progressive spirit, for perhaps it was only on one nation's decline that the *yun* bestowed the mantle of world leadership on another. At any rate, whereas K'ang Yu-wei's first *yun* was a spirit moving through the world of nations, Liang Ch'i-ch'ao's new *yun* was the course, seemingly predestined, through which the world of nations moved—both singly and collectively.

The collective movement of nations was the ingenious way in which Liang Ch'i-ch'ao returned to K'ang Yu-wei's original Three Ages. First, each nation separately was to run the course: from oligarchy to democracy, from "tribal chieftains" to "no president." At least each nation that survived was to run such a course. Liang Ch'i-ch'ao did not so reverse himself as to give up his newfound belief in the survival of the fittest or the possible nonsurvival of the unfit. His doctrine of a universal *yun* for all nations probably meant only that, if a nation survives, it must evolve in one way (which doctrine immediately became *in order to* survive one must evolve in such a way). Democracy was the prize for the survivors. But political evolution, through Three Ages, to democracy *within* a nation was evolution parallel to a greater evolution towards democracy *among* nations. The world with national units as its actors was itself evolving from an age of many rulers, to an age of one, to an age of none.

The trouble was, the overpoweringly ominous truth was that, although individual countries had run their proper course and were nearing the threshold of the glorious third age of their evolution, the world as a whole was frightfully backward. It was still at the beginning, so that Liang Ch'i-ch'ao, at the end of his essay, could only cry, "Alas, the five continents and ten thousand countries are nothing but one great world of tribal chieftains." The prophecy of the *Ta t'ung* and the *Book of Changes'* auspicious omen of a herd of dragons without a leader "probably speak of a world yet a hundred thousand years away."[120]

That is why the struggle for existence was still so brutal, and national strength so vital. If evolution went from many rulers to one to none, would the world on its way to One World have to bow for a time to one conqueror? Were the tribal nations finally entering a fight to the finish? The vision that soothed the reader's mind at the start of Liang Ch'i-ch'ao's essay, of all nations evolving peacefully together towards the great society, even though at different speeds, faded before a picture in which democracy within a nation meant only strength, the strength necessary for that nation to have even a chance of survival in an unavoidable, bloody contest.

Liang Ch'i-ch'ao, like Yen Fu, abandoned his readers in the uncomfortable ground between two visions, one optimistic, one pessimistic, visions half-connected but still conflicting. Moreover, he soon abandoned them literally as well. Believing that democracy was the way to strength and that education was the way to democracy, he decided to try his hand at classroom teaching, and thus accepted the invitation of Huang Tsun-hsien, Hsiung Hsi-ling, and T'an Ssu-t'ung to be chief lecturer at the Shih-wu hsueh-t'ang (Academy of Current Affairs) that they had founded at Changsha.

He could have reached more students through the *Shih-wu pao,* of course, than through any academy, and he did keep sending in essays on occasion, but he had been having trouble with his colleague Wang K'ang-nien, and even more with Chang Chih-tung, and so, in October or November 1897, he left the journal and left Shanghai. Somewhat in a huff, he went off to Hunan.

FOUR

Darwin in Hunan

Hunan, much to the horror of its thitherto typical gentry-intellectuals, had quite unexpectedly become a center for radical reform, or at least for wished-for radical reform. A new Governor, eager for innovation, Ch'en Pao-chen, had been given control of the province in 1895 during the furor after the Sino-Japanese War, and he thereafter managed to attract a rather remarkable group of hotheaded young scholars to Hunan's capital city of Changsha.[1]

The most famous of these, T'ang Ts'ai-ch'ang, and Liang Ch'i-ch'ao's friend, T'an Ssu-t'ung, were Hunanese themselves but, to make things even better, or worse in the eyes of conservatives, they were joined by Liang Ch'i-ch'ao's patron, Huang Tsun-hsien, who happily came to Hunan as Salt Commissioner. Their efforts, eventually bolstered by Liang Ch'i-ch'ao's own, made Hunan the very center of the Reform Movement until K'ang Yu-wei, capitalizing on the renewed spirit of national alarm following Germany's seizure of Kiaochow, and Russia's of Port Arthur, organized the Pao kuo hui (Preserve the Nation Society) in Peking in April 1898. Their efforts, as Liang Ch'i-ch'ao later pointed out, did much to help create the climate that made the resulting "Hundred Days Reform" possible; they also did much to inspire the opposition that doomed the Hundred Days to failure.[2]

NEW KNOWLEDGE AND IGNORANCE

The Hunan reformers published their ideas in two major journals, the *Hsiang hsueh hsin pao* (The Hunan study news), a magazine published every ten days for some twenty-three issues beginning in April 1897, and the *Hsiang pao* (Hunan news), a daily published for only three months starting in February or March 1898. It is in these two journals that one can first discern the gradual proliferation of Darwinian thinking.

At the same time, however, one can also discern, more readily than in the *Shih-wu pao,* and certainly more so than in the writings of Yen Fu, how far away the average Chinese intellectual was from having any scientific appreciation of Darwinism. The *Hsiang hsueh hsin pao* and the *Hsiang pao* give evidence of a tremendous new interest in Western "science," still at that time translated, perfectly aptly, by the ancient Confucian term, *ko-chih* (the extension of knowledge through the investigation of things), but they also give evidence of China's rather awesome ignorance of the world at large, both the world of nature and the world of other men.

Many of the magazine's readers from the Hunan hinterland who eagerly wrote in, in modern fashion, to have their questions answered were still having obvious difficulty believing that the world was round and revolving about the sun. "If the earth turned over," wrote one, "would not people be hanging upside down? If the earth spun, would not east and west regularly exchange positions?"[3] The editors, busily scraping up every scrap of Western knowledge they could get their hands on—many gleaned, undoubtedly, from John Fryer's Kiangnan Arsenal translations—with patient eagerness passed on their understanding of gravity, the atmosphere, and the orbiting of the earth. As questions got more difficult, they were often hard-pressed for answers (small wonder), but in areas such as earthquakes and eclipses they had excellent success at discrediting ignorant notions of the past.[4]

Paradoxically, however, "ignorant notions of the past" may have been less an obstacle to modern scientific ideas in China than they had been in the West, precisely because they were based on a lower level of scientific knowledge. In China, at the end of the nineteenth century, there was no *scientific* opposition to Western scientific ideas. There was only the opposition of classicists, who could be affronted by new scientific ideas, but nowhere nearly as much as could rival scientists or priests.

As Joseph Needham has long since proved, it would be a gross insult, in view of China's collective past experience, to take literally, in terms of a collective scientific mind, Mao Tse-tung's pronouncement that the mind of China was a beautiful blank sheet, waiting to be written upon.[5] Nonetheless, much of what certain Chinese had once known (for example, that the world was round) had by many been forgotten, and much more remained untaught. Therefore the minds of many of China's younger intellectuals *were* quite blank when it came to science. They had neither scientific prejudices nor any defenses against them. Their rather profound ignorance of Western science gave them both an astounding thirst for it, and an exceptional willingness to believe it—whatever it was.

Most of the questions asked the editors of the Hunan reform journals, therefore, however challenging scientifically, were not at all challenging rhetorically. Very rarely did anyone ask, "How do you know?" or "Who says so?" Rather there was a very trusting, naive, acceptance of the explanations given by the Hunan reformers' set of very recent experts. "Western scientists say" came to ring with an authority that was positively deafening to ears as yet unavoidably uncritical. And that ring was to prove of tremendous aid both to Social Darwinism and Marxism.

THE RACES OF THE WORLD. This combination of ignorance, curiosity, and uncritical faith in Western opinion was unfortunately brought to bear on a new Chinese interest in foreign peoples. In the *Hsiang hsueh hsin pao,* one can see evidence of a genuine curiosity about peoples in far-off lands such as China had not seen for several hundred years. But much of the material available to satisfy that curiosity only furthered racial prejudice.

There were two major pieces on foreign peoples published in the *Hsiang hsueh hsin pao,* "A Study of the Races of the World" and "A Study of the Similarities and Differences of the Customs of the Five Continents."[6] Each was a lengthy compilation of quotations, taken from what were considered the best sources available, arranged according to country and continent, and followed by comments from the compiler. The sources chosen, however, revealed how desperate the Chinese were for knowledge. Alongside some fairly recent items from the Kiangnan Arsenal translations, and an original work by Timothy Richard, were a Chinese version of the Book of Genesis, and Hsu Chi-yü's *Ying-huan chih lueh* (A brief description of lands overseas), one of modern China's first attempts at a Western geography, printed fifty years before. The compilers often took issue with various individual statements quoted, but they gave their poor readers no consistent help in evaluating the relative reliability of their sources. They were probably unsure themselves.

They were quite sure, however, about what they thought they were doing in making such studies. Their unmistakable new curiosity for other lands was only in part the curiosity that still leads millions to read such magazines as *The National Geographic;* it also had a deadly serious purpose. As Liang Ch'i-ch'ao told the *Hsiang hsueh hsin pao's* readers, the purpose of studying the history and cultures of other peoples was "to understand the basis of strength and weakness, survival and extinction. . . . The Tai-tsung Emperor of the T'ang said, 'Take the past as a mirror and you can understand success and failure.' I say, how much more effective if we take the past of East and West together as our mirror,"[7] and the past of all in between. "Success and failure," however, now meant "survival and extinction"—on an ethnic level. World history now meant the history of racial struggle for survival, and those who had eyes were meant to seek in that history the secrets of such survival.

Darwinism was dictating new terms for historical study. If Liang Ch'i-ch'ao failed this time to acknowledge Darwin as his source of inspiration and authority, "P'ing-p'i-tzu,"[8] author of "A Study of the Races of the World," did not. He introduced his study with a summary of Yen Fu's earliest introduction of Darwin and *The*

Origin of Species in "Whence Strength?" making it perfectly clear that Darwin had both inspired him to look at races and had told him what to see.

Perhaps it would be more accurate to say that Darwin had frightened him into looking at races. His first reaction, on reading Yen Fu's essay, he said, had been to cry out in dismay, "Today what we really must worry about is our race, it is our race."[9] For races, Darwin told him, could become extinct. "P'ing-p'i-tzu" primed his readers with Yen Fu's description of "nature red in tooth and claw," making sure they understood that the struggle for existence was actually much more subtly treacherous than any simple tooth-and-claw struggle could ever be (a coincidence of metaphor),[10] for even innocuous changes of habitat and climate could bring destruction. His proof was news of newly discovered remains of prehistoric birds and beasts, those monsters which in the West had struck dread in the hearts mostly of theologians, but which in China frightened everyone, as proof of extinction, which conceivably could overtake even the great and ancient race of the Chinese.

Not surprisingly, however, "P'ing-p'i-tzu's" conclusion on surveying the races of the world was that extinction would not overtake the Chinese. He expressed this conviction, before he let his readers see the evidence, in a series of parallel phrases that approached doggerel verse. The world had four races, he said, yellow, white, red, and black: "Yellow and white are wise; red and black are ignorant. Yellow and white are masters; red and black are slaves. Yellow and white are tight-knit groups; red and black are dispersed."[11]

The evidence, such as he gave, did seem to prove his point. The *Wan kuo shih chi* (A history of the ten thousand nations) announced that the blacks were descendants of Noah's son Ham.[12] But, despite such noble lineage (marred only by the curse of Canaan), their homeland, Africa, according to the *Ying-huan chih lueh,* was "in its climate and people the very worst of the four continents."[13] Australia, apparently, was not much better. The *Wan kuo shih chi* said that the Australian natives "in appearance are like apes, even more ugly than the orang-utans one sees in Malaya, and their skills and intelligence are no better either," to which "P'ing-p'i-tzu" kindly added, "If you give them wine, they get drunk on one drink. They sleep in the mud like wallowing pigs. The males work their females like beasts and kill them in fits of anger. Actually they are not the equals of orang-utans."[14]

The most damning evidence, however, and our first example of native, eye-witness, racial prejudice, came from a quoted report of Hsueh Fu-ch'eng, a respected "foreign expert" who had served from 1890 to 1894 as China's Minister to England, France, Italy, and Belgium. Hsueh Fu-ch'eng had taken the southern route west to Europe, but, if he was diplomatic once he got there, he definitely was not in his appraisal of the peoples he had met along the way. "I went to Saigon, Singapore, and Ceylon," he wrote, "and the natives were ugly and primitive, no different from deer and swine. Indeed, the various peoples of Vietnam, Burma, India, Malaya, and Arabia are all black-faced, squat, and coarse-looking. They are as far from the elegant refinement of the Chinese people and the surpassingly healthy whiteness of the Europeans as earth is from heaven."[15]

In a line-up of the world's races, the Chinese wanted to stand with the whites, not with the "oppressed peoples of the world" or with the "victims of Western imperialism," for they did not want to be victims of Western imperialism. If it could be argued that they had already been victimized enough to owe a measure of sympathy to other victims, "P'ing-p'i-tzu," and many others, had a ready reply: "Heaven does not pity Indians, and I have no time to pity them either" (either Indian Indians or American).[16] He was too busy worrying about China.

DOES YELLOW REALLY EQUAL WHITE? "P'ing-p'i-tzu's" worry somewhat belied the confidence expressed in his protestations of Chinese and European equality, but he protested too much. Too much of the rest of the information in his study of world races, like so much of that marshaled in the *Shih-wu pao,* seemed gathered for no purpose other than to dispel doubts that the Chinese really were the Europeans' equals. The story of Noah was false. The

Chinese were the first people on earth, and all others were descended from them.[17] Europe had been opened up "only a little later" than China; her people had moved to Europe from Asia, and Europeans were thus *almost* a match for the Chinese—but no other peoples came close.[18] As Chinese and Europeans were from one source, there was obviously "no reason for them to harm one another."[19] Chinese and Europeans were both spreading into Southeast Asia.[20] In Singapore and Honolulu, their races were winning out over all others.[21]

But these arguments did no good. The assertion that yellow and white were better than red, black, and brown could never prove that yellow equaled white. So what if yellow and white were superior? Were they *equally* superior? If not, what good would it do to establish that China was number two, if number two meant number-one slave?

"P'ing-p'i-tzu's" evidence was largely gathered in vain. Even if one accepted it as proof of the "equal superiority" of the Chinese and the Europeans, that equality did not prove that there would be no ultimate contest between the two or that the result of such a contest would be a draw. Despite his quoting of the pious statement that yellow and white had no need to fight, being—*ab origine*—from one house, "P'ing-p'i-tzu" elsewhere seemed convinced that they would fight, indeed that they must. For Darwin told him so. "In far antiquity," "P'ing-p'i-tzu" explained, "human beings fought with beasts and won, because they grouped together and the beasts did not. In mid-antiquity the Han (Chinese) race fought aboriginal races and won, because it was intelligent, and the aboriginal races were stupid. When such races were controlled, racial conflict in Asia ceased. But now, suddenly, the races of Europe have begun to compete with the races of Asia, and if the Asians show the slightest weakness . . . ,"[22] the result would be unspeakable. After centuries of struggle, the Chinese had emerged as the fittest in their world. The trouble was that they had thought they were the fittest in *the* world. But they were only the fittest in the East. The world championships had yet to be held.

The question of why the Chinese and the Europeans had won in their respective divisions, however, now became vital to the question of who would win in the end. Were Chinese and Europeans superior in innate ability or only in performance? Was superior performance the result of superior potential or superior "drive," "will," or "spirit"? What in Heaven's name was superior spirit? Was it something born, developed, or what? Unfortunately, there was no answer to these questions that was unequivocably encouraging to China's cause.

Hsueh Fu-ch'eng blamed racial inequality on climate. The "lower" races he had seen and despised on his travels were laid low, he said, by the heat. It was not just that heat made men lazy, or drained them of energy. In Hsueh Fu-ch'eng's Neo-Confucian terminology, heat drained men of something far more important: top-grade *ch'i,* the vital "ether" of which all men were made. "Below the equator," he wrote, "as there is only heat and no cold, the best kind of ether evaporates, so that men are left incapable of great endeavor, either physical or mental." Whether this was a congenital defect handed down from the days of the forming of the race, or whether it was something that happened to each individual born under the tropical sun was not made clear, but it certainly sounded as if whatever was wrong with the southern races' make-up was too serious to be solved quickly by air conditioning or migration north: "From the tropics on down, although people abound, they are of poor quality. Only in the temperate zone [home, needless to say, of the yellows and the whites] . . . could the best etherial stuff be congealed."[23]

This theory, if one ignored, or was ignorant of, the temperate zone in the southern hemisphere, served well enough to set apart the yellows and the whites from most other races (the North American Indians, of course, were still a problem). But it definitely could not give Chinese the absolute confidence in their ultimate survival they so desired. Indeed, its implications could only add to Chinese insecurity. For, if races were really unequally endowed because of climate, with the vital *ch'i* the determiner of racial excellence, then how could one be sure that the white men, the British for example (despite their fog—or because of it?), had not received an even better allotment of the stuff than the Chinese? And if they had, then what hope remained? In a world of supposedly inevitable racial conflict, a theory of fixed, unequal, racial endowments could be of no comfort unless one was sure one's race was the very best endowed of all. And the Chinese were *not* sure.

"P'ing-p'i-tzu," therefore, unconsciously, we may surmise, unable to bear the potential hopelessness of even a seemingly favorable theory of predestination, rejected Hsueh Fu-ch'eng's analysis. Climate did not determine racial performance, he protested. The same race in the same place could both rise and fall. Look at the Indians, glorious in their past, now groveling before the British.[24] The people of south China used to be as barbarous as the present Malayans, but sage men led them out of barbarism. South China had a perfectly noxious climate, but the Chinese now thrive there, just as Chinese and Europeans do in the even hotter climes of Southeast Asia.[25] Races differ. Races rise and fall. Races perish, because they are barbarous. But if they do, "it is not because their races are infertile, or their intelligence weak, or their weapons unsharp. They perish in the end because they have not the government, philosophy, and learning to preserve themselves."[26] It is a matter "not of Heaven, but of human endeavor."[27]

Most of the evidence "P'ing-p'i-tzu" gathered in his study of world races was clearly gathered to prove Chinese superiority, to make it seem natural and assured. Moreover, his descriptions of the darker races were so derogatory as to make their plight seem virtually hopeless, as if their destruction was also natural and assured. But, in the end, he shied away from accepting any idea that the racial rank list was already determined by nature. At least he settled for some kind of "spirit," the spirit behind human endeavor, as being the determining factor, rather than any natural but unequal physical endowment (he failed to ask, of course, whether such "spirit" itself might not be unequally endowed). For the potential power of human endeavor offered an unfailing wellspring of hope. If human endeavor made the difference, then China still had a chance, even, Heaven forfend, if one was forced to admit that the white man seemed for the moment to have the upper hand.

STRUGGLE TO SURVIVE

"P'ing-p'i-tzu's" announced espousal of the naturalist Darwin did not lead him to believe that human affairs were all naturally determined. The "struggle for existence" was not a process to be watched with noble resignation, content in the conviction that the best man (or race) would win, but a call to action, an exhortation to struggle. Indeed, in one of Yen Fu's translations of the phrase *cheng tzu-ts'un,* the first Darwinian slogan widely used in China, and that used by "P'ing-p'i-tzu," it was an imperative: "Struggle for your own existence. Struggle for your own survival."

If this seems a mistranslation, it was Darwin's fault. For there was a tremendous problem inherent in his own use of the word *struggle,* which he had overlooked. *Struggle,* in any language, implies determination; it implies an intensity of will, even if that will is called instinctive. But Darwin, at least in *The Origin of Species,* had never dealt with will, or determination, as a determining factor in the struggle for survival.[28] He had not admitted such a factor, and he had certainly not accounted for such a factor. Instead he had blithely described a world of struggle in which the will to struggle, the will to live, was apparently a constant. Life went not to those who "tried harder," but to those "most fit." The fittest elephants, for example, to relieve the overworked giraffes, were not those who stretched their noses by "sheer determination," but those who *had* long noses, who *happened* to have long noses. When all was said and done, in a truly Darwinian world, survival was a matter of luck.

Luck, however, was not something in which the Chinese of the late nineteenth century were willing to trust. That is why men like "P'ing-p'i-tzu" chose to hear Darwin's message as an imperative, however frightening, rather than as a declarative statement of the inevitable. Ironically, of course, Hsueh Fu-ch'eng's climatic racial determinism, which "P'ing-p'i-tzu" consequently rejected, was actually more Darwinian than "P'ing-p'i-tzu's" would-be Darwinian determinationism. Hsueh Fu-ch'eng's system, despite the exoticness of its Neo-Confucian metaphysics, was at least one in which the most "favored races," favored by climate—and what could be more natural—were those preserved "in the struggle for life." By natural selection of a sort, races in cool climes ended up with the fittest *ch'i*. Perhaps this should have appealed to "P'ing-p'i-tzu." But he could not be sure that Chinese *ch'i* was best. Thus, in the end, he left racial struggle an open contest, and contented himself with exhorting his countrymen, in "Darwin's" words, to "struggle for their [collective] survival" and "to leave fit progeny (*yi yi chung*)."[29]

THERE OUGHT TO BE A LAW. "P'ing-p'i-tzu's" exhortation was soberly accepted by almost all the Reformers. And yet there were a few, of the most optimistically idealistic, who still bridled at the injunction to struggle (however amenable they might be to the injunction to be fruitful). Taking struggle to mean warfare, they said they wanted none of it. They were for "peaceful coexistence," and believed, despite the dourest Darwinian pronouncements, that it was attainable. For they had tapped the progressive stream of nineteenth-century thought before it ran into the evolutionary stream and were hence too excited by the idea of progress to believe in the "naturalness" of world conflict, or at least to believe in its continuing naturalness. They were not, therefore, simply against struggle because it was bad. They thought it was backward, primitive, unbefitting, and unnecessary. They were somehow so convinced a new age was at hand that they believed China could survive without struggle, by putting her trust in international law.

It may seem amazing that, at the close of the nineteenth century, there were still Chinese who believed China could put her trust in the Westerners' sense of fair play. But the fact is that many were impressed with Western law and order, even as seen in the treaty ports, even, indeed, as seen in the "unequal treaties." The harried officials that over the decades had been forced to conclude those treaties had consistently argued that at least, as the Westerners were remarkable sticklers for law, legal treaties would check them from absolutely lawless aggression. Now, however, when Chinese heard of Europeans trying to outlaw war among themselves, and when they saw evidence of the Europeans' great advances in other fields, in technology, government, education, and science, some found it possible to believe, with a dose of wishful thinking, that Europeans really would be able to outlaw warfare. They could control themselves. And the whole world, now pulled together, would be able to outlaw imperialism—just in time to save China. When faced with the most pessimistic Darwinian doctrine of natural struggle, therefore, the most optimistic of the Reformers could reply, in effect, that man had progressed so far that he could indeed outlaw the law of the jungle.

That such an attempt would create a legal crisis of cosmic proportions went unappreciated. Yen Fu, of course, had seen Huxley's recognition of the crisis posed by the laws of men standing in seeming opposition to the laws of nature, and some of the leaders of the Hunan reformers, although attributing Huxley's views to Spencer, also recognized such opposition. But they did so without recognizing the crisis. They seemed to think man's going against nature was perfectly natural.

This was apparent in an official reply of the Nan hsueh-hui (the Southern Study Society) to a letter from a local independent study society requesting admission to the Nan hsueh-hui as a branch member. Lauding the newcomers for recognizing the virtues of solidarity, the leaders of the central office quite clearly set "Spencerian" solidarity in opposition to Darwinian struggle. "Darwin," they wrote, "clearly understands the principles behind the life cycles of living things, but, when it comes to human ethics and good government, then Spencer is obviously the real authority. Heaven's aim is to give birth to living things and keep the seasons in their courses. But man's job is to supersede the work of Heaven. The struggle of animals is simply the struggle for life. But the struggle of man should be the struggle for right and reason. Animals by nature cannot group, and hence only the tough, the ruthless, the violent, and the fierce are able to leave descendants. But humans by nature can group. That is why they are able to order themselves and their society with ritual and government. As Kung Ting-an [Kung Tzu-chen (1792–1841)] has said, 'Heaven [nature] teaches acquiescence; man teaches opposition.' [A Chinese precursor of Huxley!] Animals only acquiesce; they never oppose. Hence their growing up and growing old, their living and dying, all are left in the hands of nature. Man alone can add opposition to acquiescence. Man alone can improve on nature."[30]

This passage was one of the first specific rebukes, however mild, that Darwin received in China, and yet its author, or authors, in telling man to stand against the flow of nature and thus improve upon it, did not in actuality launch a full-scale Huxleian attack against the cosmic process and Social Darwinism. They did not altogether repudiate "the gladiatorial theory of existence."[31] What they were talking about was a curbing of natural selfishness for the common good, but what they wanted was Chinese, and particularly Hunanese solidarity as strength for the gladiatorial contest waiting outside. And they themselves would be the first to fall back on Darwin for documentation of that grim contest's natural existence. This passage thus offers a perfect example of what was to become the most common Chinese position towards Darwinian struggle: "None within to win without."

The author's position, then, was not the same as that of the starry-eyed advocates of international law. The only similarity was in their lack of concern over the philosophical problems inherent in the idea of a natural creature even partially outlawing the law of the jungle. The advocates of international law, who really did want to "repudiate the gladiatorial theory of existence," and said that such a thing was possible thanks to progress, had no fear, for example, that, by artificially preventing the struggle for existence, they might effectively legislate an end to progress. "No struggle," in rather traditional Chinese fashion, seemed the obvious, fitting, and quite sufficient end towards which any respectable progress should lead. Like the Chamberses, whose ideas, through Fryer's *Tso chih ch'u yen,* seem here to have had their effect, they seemed to believe that such an end could gradually be reached by gentlemen's agreement. They refused to accept the alleged Darwinian postulate that no progress was possible without struggle.

The whole idea of international law as a national salvation had come to China largely thanks to another of John Fryer's translations (done with the help of Wang Chen-sheng), *Kung fa tsung lun,* a translation of Edmund Robertson's article "International Law" from the ninth edition of the *Encyclopaedia Britannica.* Two other influential works were W.A.P. Martin's translations of Bluntschli's *International Law, Kung fa hui t'ung,* and of *Wheaton's International Law, Wan kuo kung fa.*[32] The last was recommended by Liang Ch'i-ch'ao; the first two were selected by the *Hsiang pao* as the best of a list of six such works then available.[33]

There was an immediate association of the Western vision of One World of nations united under international law with K'ang Yu-wei's, that is "Confucius's," vision of the *Ta t'ung.* And of course many hastened to point out that the Chinese had had the idea first. The Western idea went back only as far as "Hugo," that is, Hugo Grotius,[34] whereas the Chinese plan had already been revealed in the *Spring and Autumn Annals.* But China, alas, had lost the way and her head start, so that the Westerners with their international conferences in international law now seemed closer to the goal. Western laws, however, offered China a great chance: "If we can acquaint all of China's officials with international law, if they can all gain a basic knowledge of Western statutes, then in future Chinese-Western disputes, if all consistently appeal to such laws in their discussions, they should be able to make the foreigners relent, and end hostilities."[35] If only Chinese "knew their rights," they would be able to make the Westerners control themselves.

Even Liang Ch'i-ch'ao subscribed to this idea, although in a most unstarry-eyed fashion. When the Russians seized Port Arthur and demanded that it be leased to them, Mai Meng-hua, Liang Ch'i-ch'ao, and others jointly memorialized the Throne urging that, as China quite obviously had no hope of *fighting* Russia, she should refuse the Russians' demands on grounds of international law, and appeal for support from other "law-abiding" nations.[36]

This seemed an appeal to international law only as a last resort, but other appeals were much more optimistic. The word *equality* took on a magic ring. Some felt that, as soon as the whole world awoke to the concept that "for strong nations to enslave weak nations is the most flagrant violation of the principle of equality,"[37] then such enslavement would be outlawed and would cease. P'i Chia-yu could envision the day when, because of the principle of equality, Russia, England, Germany, France, and Japan would all be forced to return their leased territories, their spheres of influence, their railroads, and their treaty ports: "The tribes of Southeast Asia will again submit to our protection. . . . The despicable doctrine of other nations' carving up the Chinese melon will be broken. The disastrous warfare that now racks the five continents will end. The word shall lack nothing from the Great Peace of the *Ch'un ch'iu* or the Great Unity of the 'Li yun'."[38]

All nations would be equal under the law, and, at least as "T'an-t'an-tzu" saw it, so would all races.[39] But not everyone was sure that all races would live to enjoy this great age of equality under international law. K'ang Yu-wei himself, as we shall see, felt that only the yellows and the whites could be certain of reaching the *Ta t'ung*. And almost everyone at least admitted that international law was not for "pirates, brigands, and savages." "To distinguish ourselves from savages" thus became a doubly important imperative.[40]

But there was another problem facing those who advocated trusting to law rather than struggle: How soon would such a law-abiding age arrive? P'i Chia-yu ended his description above with the rather discouraging question, "Might this not occur in a few hundred years?"[41] "T'an-t'an-tzu" said the day would arrive "after the twentieth century," but whether this meant in or really after was unclear.[42]

The most optimistic, however, were not worried. The day was close enough. Of course, this did not mean that China could necessarily just sit back and wait for the Westerners to control themselves. One of those advocating that China seek refuge in international law, I Nai, offered the most radical reform program yet suggested. Finding support in the very Taoist phrases Yen Fu had so reviled in "Whence Strength?" I Nai argued that China should seek strength in weakness, by so bowing to Western ways as to become acceptable in Western eyes: "If we want to stand independently in the world, if we want the great conferences to give us equal treatment, then we must change our calendar and our mode of dress. All our institutions must follow the West. We must enter the international league of nations, and we must obey international law."[43]

Such a proposal had every right to be called Darwinian, if one took it to mean that China could survive by "fitting in" to the Western world; but, as Darwin's message was then understood, it was un-Darwinian, for it claimed that China's survival could be won without struggle. And that contention was soundly voted down.

DON'T COUNT ON IT. When we look at the *Hsiang hsueh hsin pao* and the *Hsiang pao* as a whole, we find an overwhelming consensus that in *these* times struggle is the only path to survival. And, over and over again, the authority cited is Charles Darwin. Few indeed denied that a better age was coming, but it was obviously not yet here. The present world was "a world of jealousy and strife, of [races and countries] devouring one another."[44] Some said that "the strong eat the weak now as ever."[45] Others said that was only the way of the first of the Three Ages. But they hastily added that that was the age we were in.

International law could not be trusted because the foreigners could not be trusted. For one thing, they were racists: "Westerners are very strict in differentiating races. . . . They are white and we are yellow. That is why they so despise us. We can wait ten thousand years and they will never look on us as equals, never grant us equal rights."[46] Even K'ang Yu-wei, the most fervent believer in the *Ta t'ung,* had no illusions about the fair-mindedness of the West: "Ever since the Japanese disgraced us, the Westerners have belittled us and treated us as barbarians. . . . Before, they called us a half-civilized country. Now, they class us with the servile blacks of Africa. Before, they hated us for our pride and self-esteem. Now, they call us stupid, deaf, and blind. Their international law promises equal protection for all, but it is only for 'civilized nations,' not for 'barbarians'." China had been spared so far, he said, because the Westerners had been concentrating on the partitioning of Africa, but now that task was done.[47]

And the Westerners were coming on with no holds barred: "The Westerners' forcing us to buy opium is a plot to extinguish our race."[48] "The Westerners use trade to topple other countries."[49] The missionaries were agents of subversion.[50] The treaties were legalizations of aggressiveness to date. That last fact most dampened Chinese hopes for international law: "We are defeated in war; we are defeated in treaties. . . . because they are strong and we are weak. They make a science of the struggle for survival."[51] So China, most insisted, must truly *win* equality before it could demand it: "We must reach their level before we can be equal."[52] "If we want to enter a world under international law we must first resolutely make ourselves strong." "P'ing-p'i-tzu" made it clear: "Mencius says 'strength is good.' Darwin says to 'struggle to survive.'"[53]

The final argument was made by T'ang T'sai-ch'ang, who in two short years was to become the first of the "reformers" to lose his life in armed struggle against the Ch'ing. T'ang T'sai-ch'ang voted for struggle, with Darwin's backing, in, of all things, a dedication piece he wrote for a Study Society of International Law. He said that Darwin said to struggle, that this meant that *ch'ün* and *ch'ün* must struggle, that Western study of society, race, biology, and evolution was all directed towards that aim, and that commerce, missionary work, colonization, scientific discovery, and technological invention were all for the sake of "survival in the great war of human beings."[54] This was natural; this was the way things were. The dream of men like Hugo and Grotius [*sic*][55] was a great dream. It would one day come true, but, for the present, China must hold to the words of Pi Yung-nien: "The starting point towards no struggle is the struggle for survival."[56]

That one must fight for peace, kill to end killing, and struggle to stop struggle was a view that would become axiomatic for revolutionaries. It is fitting, therefore, that we should first find its Chinese Darwinian expression in the writing of T'ang Ts'ai-ch'ang, the first famous reformer to turn to revolution. But T'ang Ts'ai-ch'ang himself had not yet thought of revolutionary struggle as a necessary prerequisite to any successful struggle against imperialism. He and the other reformers still wanted the government to struggle by changing itself.

Most of the rest of the material in the *Hsiang hsueh hsin pao* and the *Hsiang pao* consisted of practical suggestions to the government and to all loyal and patriotic intellectuals as to what China could actually do to struggle for her survival. Some of the suggestions were admittedly radical. Many, probably most, of the Reformers felt the government should at least begin to turn itself into a democracy, because the democracies seemed at present to be "the fittest." The most common suggestions however, were much less shocking: China needed railroads to survive;[57] China needed commerce to survive;[58] China needed legal reform to survive;[59] and—over and over again—China needed talented men to survive, and hence new schools and new education.[60]

As to "Darwin's" suggestion, that China needed an improved race, that was a tall order, and there were few concrete proposals as to how it might be accomplished (although, as we shall see, there were some). All most people could think of was the good first step of abolishing foot-binding. After that could come calisthenics, especially for women, and finally, when women were healthy, a grand effort at out-multiplying the Westerners ("The Westerners are multiplying every day and we do not increase at all. . . . Is it not sad?")![61]

Survival and Religion

There was only one more thing China needed to survive, in the eyes of the Reformers' leader, K'ang Yu-wei, and that was a religion. In his eyes, of course, China had a religion, Confucianism, but it needed to be reformed, institutionally, so that it would resemble a Western religion, and so be able to compete with Western religions.[62] For there was a struggle for survival among religions as well as nations, and that struggle affected nations. As Ho Lai-pao said, "There has never been a country that has survived after its religion has perished."[63]

T'ang Ts'ai-ch'ang raised the stakes even higher: "If Confucianism perishes, the yellow race has not a chance for survival."[64] Thence arose the slogan "Preserve the religion to preserve the nation," with its possible extension, "Preserve the religion to preserve the race." To be preserved, the religion had to be reformed, to be spread by Western-style missionaries,[65] and, in T'ang Ts'ai-ch'ang's opinion, to be subjected to a process of artificial selection, as recommended by the "evolutionists" (*t'ien-yen-chia*), so that it could be strengthened before it was naturally selected out.[66]

WESTERNIZE IT AND US. Who would have imagined it? Darwin was brought into China's first controversy over "religion" (not, of course, into China's first religious controversy),[67] as a champion of religion. And yet, as such, he finally provoked the ire of conservative defenders of Confucius. They were infuriated by K'ang Yu-wei's call to "religionize" Confucianism, and nauseated by I Nai's even wilder call to Westernize "the religion" and the race.

I Nai's argument for complete Westernization was actually a Taoist-Darwinian program for survival. The fittest was not the strongest, but the most adaptable. The fittest was La Fontaine's reed. I Nai did not tell China to rely on the east wind's overcoming the west wind. He said China should give in to the west wind and adapt itself to it, so that it would not be blown over. I Nai's final program for China's self-adaptation to the new world environment, however, was far more radical than any of the later twentieth-century programs for "complete Westernization." For he not only wanted to adopt the Western calendar, Western clothes, and Western laws. He not only wanted to "lower the imperial eminence to protect that eminence," that is, to institute constitutional monarchy.[68] He wanted China, for her very survival, to "integrate her religion" and "mix her race."[69] He wanted to modify Confucianism and the Chinese.

Staunch conservatives like Yeh Te-hui loathed I Nai and found "integrating the religion" ridiculous. Christianity and Confucianism were to co-exist in China. The Emperor was to pronounce a special edict allowing anyone who wished to to become a Christian. Truly knowledgeable Confucians were to "repair" Confucianism, by discarding faulty doctrines and borrowing new ones, to make up for what was lacking. Confucianism and Christianity would complement each other, would influence each other, would become as one.[70]

If such compromising of the faith seemed blasphemous to the orthodox, "mixing of the race" seemed positively disgusting. Mixing races meant "intermarriage between yellows and whites." I Nai said that China must "mix her race to preserve her race." He realized, of course, that many would ask, aghast, "Would you have our pure and fair descendants die out, while that sheep-stench race increases?"[71] But he had a ready answer:

> These are the frightened words of stupid scholars, the snide remarks of little minds. Good men who know the times do not speak thus. . . . There is no better way to preserve our race than for our princesses to be married to the aristocracy of Europe and for our princes to marry princesses of Europe. Indeed, there should be an order encouraging anyone who so wishes, be he gentleman or commoner, to wed his daughter to a Westerner or take a Western wife. The most valuable thing in international marriages will be to choose Westerners of great intelligence. . . . This is what is meant by "using the power of love to draw out one's country's destiny," and by "prolonging a noble race through transformation." A union of two of the same kind produces few offspring. Marriages between people with the same surname have been forbidden throughout the ages. Westerners, too, say the children of consanguineous couples are often epileptics. That is why China has forbidden marriages between cousins. But, if yellows would mate with whites, the children they would bear would necessarily be big and strong, healthy, good-looking, and intelligent.[72]

Chinese stock would be improved, and improved it could survive.

One did not have to be "conservative" to find this idea ridiculous. The patent absurdity of the white race, or any other race, being able to work any appreciable change on a population that already encompassed a quarter of humanity—barring measures that would have made child's play of the Draconian—should have been quite sufficient to keep most people from taking the suggestion seriously. But also, I Nai's suggestion was simply too insulting. That some *were* willing to take it seriously is therefore far more startling than Yeh Te-hui's refusal to do so. The fact that men of high intellect would even contemplate the horrible prospect that Chinese blood might have to be mixed with the blood of white men to survive proved how serious a blow had been dealt Chinese self-respect. It also proved how frightening a "Darwinian world" could seem.

When Yeh Te-hui, a Chinese whose self-respect was quite invincible, first heard of I Nai's proposal, however, he did not even bother to get angry. He said that, when I Nai's article first came out, "the scholars of Changsha rose in a body to attack it." When some of the more worried rushed to tell Yeh Te-hui of I Nai's call to mix the race, he quickly calmed their fears: "I told them . . . that 'ducks and chickens may roost together, but they will not mate. Likewise dogs and swine may walk one road, but they will not pair up. Human nature being more exalted than that of birds and beasts, I Nai's works need not be feared.' Hearing that, they laughed and took their leave. I may be said to have totally disarmed the enemy with a joke."[73]

When the Emperor began to listen to K'ang Yu-wei and Liang Ch'i-ch'ao, however, the humor of the situation, for Yeh Te-hui, vanished completely. Ignoring I Nai, he savagely attacked, in essays and memorials, the doctrines of "mixing the race" and "merging the religion" as parts of a pernicious K'ang-Liang plot to turn Chinese into Westerners. K'ang Yu-wei and Liang Ch'i-ch'ao, he said, were caught up in the insane delusion that by selling both body and soul they could somehow achieve "self-strengthening." Not unreasonably, Yeh Te-hui saw the slogan "Mix the race to preserve the race" as a logical absurdity.

Yeh Te-hui's blaming this idea on "K'ang-Liang" was probably not unjust. When the *Ta t'ung shu* was finally published in 1935, it revealed that K'ang Yu-wei had indeed harbored hopes for the whitening of the Chinese people that far exceeded I Nai's. In a section entitled "Abolish class boundaries and equalize races," he had written:

> Yellow and white possess the world: the silver race is spreading across the globe, but the golden race has the largest population. The white race is admittedly fittest in strength, but the yellow race, with its

> numbers and its intelligence, will never be extinguished, if only it will merge and be transformed. I have seen some of our countrymen who have long lived in England or Australia, or who in China have been especially careful in their diet, nurturing themselves with the best of Western methods, whose "complexions seem dyed with cinnabar," just like the Europeans. Anyone who daily dines on broiled beef, half raw and still dripping with red blood, will look, after several months, as if his face were rouged. If the members of the yellow race will only get out more in the sun and absorb the sunshine, if they will live in areas open to the wind, and breathe good air, if they will move southerners a little to the north and mountain people nearer to the rivers, then, after two or three generations of interbreeding, in no more than a hundred years, they will gradually get whiter. Racial interbreeding will naturally lead to racial amalgamations and, before the *Ta t'ung* is accomplished, the yellows will already all be whites.[74]

This was carrying the admonition "If you can't lick 'em, join 'em" too far: "Can this talk of mixing our race to preserve our race—as if out of several hundred million Chinese children there is not one fit to transmit his seed—be anything else," cried Yeh Te-hui, "than the deranged yapping of a mad dog?"[75] The idea might be mad, but together with the idea of integrating the religion it was more pernicious than even the doctrines of institutional reform, equality, and democracy, for which K'ang Yu-wei and Liang Ch'i-ch'ao were more commonly attacked.[76]

HERESY. Just how much and for how long Liang Ch'i-ch'ao ever shared his teacher's hopes for the whitening of China is unclear. Perhaps, in telling his students that the wave of the future was coming from the West, he did, as Yeh Te-hui claimed, inspire some of them "to wish to throw away the purity of body they received from their parents and merge with the whites."[77] But he has left no writings that come close to those of K'ang Yu-wei or I Nai on the subject. Before, during, and after his Hunan teaching, he wrote essays protesting that racially the Chinese were every bit the equals of the whites, so that, if he ever did believe in the necessity of mixing the race to preserve it, it must have been only briefly.

Liang Ch'i-ch'ao was, however, "guilty" of professing another doctrine, which in Yeh Te-hui's eyes lay even closer to the heart of the K'ang-Liang "heresy," the doctrine, as Yeh Te-hui put it, "that human nature will not be good until the human race has improved."[78] Liang Ch'i-ch'ao had indeed said this, and he had directly related the idea to the Western theory of evolution, even though he gave credit to Confucius for having first divined it. In an argument sublimely unsullied by any textual backing, he declared that Confucius had said (or meant) that human nature would progress, or evolve, from the evil to the both good and evil to the good, in neat congruence with society's progression through the ages of Disorder and Ascending Peace to that of Great Peace. The theory of evolution had substantiated Confucius's teaching. Westerners were promoting studies towards the improvement of the race, and, when such studies bore fruit, Confucius's prophecy would be accomplished.[79]

Yeh Te-hui, totally unmoved by Liang's Confucian "scholarship," seems to have been one of the first to realize that, so interpreted, the theory of evolution aimed a potential death blow at the very heart of at least the orthodox Mencian strain of the Confucian faith. The fact that Liang Ch'i-ch'ao chose to propound his "Confucian" theory of the evolving goodness of human nature in an essay expressly intended to reinterpret Mencius only suggests that he had probably come to the same conclusion even sooner. For Mencius had, in fact, never even hinted that human nature might be evolving. The Mencian argument was that human nature was—*already*—basically good, that is, that human nature already had within it that which it takes to be good, the full potential to be good. This potential, which was all that really mattered, was complete in every man[80]—and the whole Mencian-Confucian moral imperative depended on it. Yeh Te-hui could not believe that the human race had been struggling under a hopeless handicap, that the Confucian dream of peace on earth had hence been a mirage from Confucius's day to this, and that it would remain one until man had changed his very self. He believed with Confucius that goodness had always been at hand,[81] and that it was still at hand, if only men would grasp it. Goodness was at hand, and order was at hand. There was no waiting on evolution.

We can see that, at least on this one issue, evolution apparently challenged the Confucian canon much as it challenged the Bible. It is thus far from surprising that there were Confucians who rose to their canon's defense.

Yeh Te-hui was no divine. As a "defender of the faith" he was largely self-appointed, but he brought against this evolutionary notion of the imperfection of human nature the closest Chinese equivalent to the *odium theologicum* with which men like Bishop Wilberforce had first met evolution in the West. Indeed, in magnificent coincidence, Yeh Te-hui turned on Liang Ch'i-ch'ao with the same rhetorical question Bishop Wilberforce had asked of Huxley: "Liang teaches that human nature will be good only after the species has been improved. I wonder from what seed Mr. Liang is descended? I wonder what creature's nature he has?"[82]

In associating nature with ancestry in his insult, Yeh Te-hui at last gives evidence that the evolutionary (and revolutionary) notion of man's having animal ancestors could, after all, offend Confucian minds (or emotions) as it could Christian. Any Confucian, as we have said, could readily admit that man was related to the lowest of the ten thousand things "in principle" (just as any Christian could admit that man was related to all creatures through the hand that made them), but it took considerable philosophical sang-froid to admit that man and beast could be related in blood. Yeh Te-hui was obviously afraid that such an admission would deny the distinction between man and beast, and lead men to beastly behavior.[83] Mencius, after all, had said that, although the difference between human nature and animal nature was very slight, it nevertheless made all the difference in the world. The good man preserved that difference; the weak man let it go.[84] Liang Ch'i-ch'ao, said Yeh Te-hui, had lost it: "Mencius says that to live in leisure without any doctrine is to be almost an animal. A man like Mr. Liang lives in leisure with a doctrine, but he is still almost an animal."[85] And as for K'ang Yu-wei, "his lack of human nature is such that he is not even a vegetable's equal. How can he be fit company for bird or beast?"[86]

Mencius himself would not have been as upset as Yeh Te-hui

was at the thought of man's having animal ancestors, for Mencius had admitted from the beginning that man was a mixed bag, with animal appetites as well as a humane heart. What mattered to Mencius was man's make-up, not how man was made. The unique spark of goodness in man's nature was a mystery with or without evolution, but it existed, and no number of weird skeletons in man's closet could deny it. For the proof lay not in man's forbears but in his being.

Evolution would not have destroyed Mencius's faith in his good-natured man, nor should it have destroyed his faith in the completeness of man, or rather in man's complete potential for completeness, for, strictly speaking, every "creature" that ever evolved was complete in itself. If something else evolved from it, it was something else.

Yeh Te-hui's antagonism to the idea of evolution was not, therefore, altogether rational. But he had no way of knowing enough about evolution to be able to make a truly rational assessment of its seeming challenge to his Confucianism. His only informant was Liang Ch'i-ch'ao, whose own understanding of evolution was, quite understandably, sophomoric at best. We cannot, therefore, dismiss Yeh Te-hui's antagonism to evolution, as he heard of it, simply as part and parcel of his political aversion to Liang Ch'i-ch'ao and his reformist doctrines. Evolution, as Liang Ch'i-ch'ao presented it, was hard to take.

Consider, as one example, this bit of "evidence" of man's animal ancestry, quoted by Yeh Te-hui from one of Liang Ch'i-ch'ao's comments on a student's paper:

> Many of the ancient emperors portrayed in the Han period ancestral temple of Wu Liang have human heads and serpents' bodies or animal bodies with human faces. This is proof of the veracity of the ancient legends. The *Shan hai ching* (Book of mountains and seas) definitely does not speak nonsense. Tung Chung-shu quoted an ancient saying that "men must realize that they are exalted above all creatures." From this we can see that in ancient times the difference between man and beast was not at all great. For who today does not know that he is nobler than the animals? Who would need to be told so by the Sages?

> The Westerners' ancient books contain repeated prohibitions against men committing intercourse with animals. This can only mean that many must have done so.[87]

What might have been a shrewd argument in an investigation of the social customs of Sodom and Gomorrah was a less-than-helpful gambit for convincing skeptics of the truth of evolution. When such an argument was seen beside the injunction to mix the race to save it, Chinese purity seemed assailed at once by past and future—and all in the name of Darwin. Small wonder, then, that a proud Chinese like Yeh Te-hui could growl in scornful anger, "If such a breed wants to act like the men of the *Book of the Mountains and Seas,* then let them, but why must they drag the entire yellow race into bestiality?"[88] Bishop Wilberforce would have shouted, "Amen!"

Ironically, however, in the religious controversy over Confucianism, in which Bishop Wilberforce would most definitely not have agreed with Yeh Te-hui, Yeh Te-hui himself fell prey to the infectious quality of Social Darwinian rhetoric, even though he usually deplored the modernisms with which the Hunan reformers, so he thought, were sullying the Emperor's Chinese.[89] He recognized a struggle among the world's religions. But he was convinced that "Confucianism, being the religion most widely in harmony with the human heart and the laws of Heaven, must one day flourish in all the civilized countries of both East and West. . . . Heaven, foreseeing the present situation, prepared for it in the Warring-States period. Thus, the more strongly men held to heterodox religions, the more strongly others protected our Sage's holy teaching. The Westerners speak of the 'struggle for survival.' That is of course a fixed law."[90]

Yeh Te-hui was convinced that Confucianism would win in that struggle because it was best. Momentarily forgetting his own quarrel with Western science, he reveled in the fact that, in Western countries, "the scientists, more often than not, are at loggerheads with the religions [Protestantism and Catholicism]. This will prove their religions' undoing; it will clear the way for Confucius."[91] Thus he could argue against those who sought to preserve Confucianism by

Westernizing it, "There is no need to weep over Confucianism; there is no need to preserve it."[92] It would win out naturally in the end. In religious matters, Yeh Te-hui believed in the survival of the fittest.

So Darwinian language did sneak in to what the Reformers would have called the rhetoric of reaction. Chang Chih-tung, whom Liang Ch'i-ch'ao considered to be, with Yeh Te-hui, one of the two greatest intellectual enemies of the Reform Movement, also spoke occasionally of the struggle for survival. He even attributed European wealth and power to the intense struggle for survival necessary among European nations,[93] although he did not go on to conclude that as a general principle competition was thus a good thing. Chang Chih-tung, of course, no less than Liang Ch'i-ch'ao, wanted China to gird herself for what everyone saw as the inevitable struggle with the West, and, if he had not thought that K'ang Yu-wei and Liang Ch'i-ch'ao threatened to go too far, he certainly would have accepted Liang Ch'i-ch'ao's Darwinian exhortations to make China strong.

Unfortunately, however, there was someone else struggling for survival *within* China, for whom Liang Ch'i-ch'ao's exhortations seemed as pernicious as the weapons of the Westerners. The Empress Dowager was already struggling for the dynasty's survival, as much as for the nation's. And she was struggling for her own survival no less fiercely. The young Kuang-hsu Emperor could accept the K'ang-Liang reforms as being in the best interest of both his country and his throne, but she, his adoptive mother, seeing the characters of revolution on the wall, could not. Thus, with an intuitive grasp of the struggle for survival that would have done credit to a Mongolian tiger, the "Old Buddha," at the eleventh hour, lashed out—and the Reform Movement was crushed. She who was least fit to save China won one last battle.

And lost the war. By crushing the Reformers she virtually insured the eventual triumph of the Revolutionaries. But she had at least won supreme imperial power for herself for the last miserable decade of her life.

She had also put an end to the first Chinese movement in which

Charles Darwin had taken part. She had not turned on K'ang Yu-wei and Liang Ch'i-ch'ao in order to stamp out the Darwinian heresy. She had not done so to placate angry Confucians like Yeh Te-hui, although knowledge of their anger must have added to her confidence as she measured the chances for success in her coup. Most probably, she did not even remember Darwin's name. And yet that name, because of her actions, would come back in a far more frightening form to haunt her to the end of her days.

Darwin's loudest champion, Liang Ch'i-ch'ao, fled with Darwin's banner to Japan. But once there, Darwin's banner was torn in two. The Revolutionaries claimed Darwin as their own. And in their tracts they sent him back to China in their behalf. The Empress Dowager had not wittingly attacked Darwin, but Darwin would attack her. Through the mouths of revolutionaries he would shout, or be forced to shout, "You and your race are unfit. You will perish through natural selection. Evolution will sweep you aside."

FIVE

A Good Book for Bad Times

In the gloomy calm that followed the ending of the Hundred Days Reform, Chinese men of feeling quietly read Yen Fu's *T'ien-yen lun,* his translation of Thomas Huxley's *Evolution and Ethics,* which had finally in 1898, a year of great hope and disappointment, seen publication in an edition that reached all China. Discouraged men found in its pages new cause for hope; complacent men at last found cosmic reason for concern. Some, alas, found reason for despair. Mature men of sophisticated literary and philosophical mind—for whom the book was written—read it through, nodding at very least in admiration for its elegant Chinese; schoolboys, for whom it had expressly *not* been written, read it through, struggling with the difficult richness of its classical prose, impatient to decipher its exciting if somewhat frightening new vision of the world. Two generations found answers in one book to the question then assailing every disheartened patriot's mind, "What do we do now?"

The immediate individual excitement generated by this book has been captured in the memoirs of both generations, most famously by the eventual champion of Chinese liberalism, Hu Shih. But Hu Shih's memoirs are perhaps too famous, for they have fooled many people into thinking that his discovery of *T'ien-yen lun* in 1905 or 1906 marked the beginning of that book's influence.[1] It is thus better to cite here the memoirs of another great intellectual figure ten years Hu Shih's senior, Lu Hsun, who has left an account of discovering the book in its very first year of publication, when he was a student of seventeen in a "foreign-style school" in Nanking:

> As it became popular to read new books, I learned that China had a book called *T'ien-yen lun*. Therefore, when Sunday came, I ran down to the southern section of the city and bought a copy, a lithographed copy on white paper, for exactly 500 *wen*. I opened it and took a look. It was printed in excellent characters. The first lines read: "Huxley sat alone in his house in the south of England; with mountains behind him and fields before. The scenery outside his windows was as clear as if at his fingertips, and he wondered what had been there two thousand years ago, before even Rome's great general, Caesar, had arrived. And he guessed that there had been only wilderness, created by nature."—Oh, so the world has a Huxley, thinking like that in his study, and thinking so freshly. I read on without stopping. I came upon "the struggle for existence" and "natural selection," I came upon Socrates, and Plato, and the stoics.[2]

And he came upon Homer, Hamlet, Kant, and Hume, Shakespeare, Thales, Haeckel, Job, Alexander the Great, Alexander Pope, cynics, Tudors, chimpanzees, biology, logic, nebulae, and nerves,[3]—he came upon whole new worlds.

For Huxley, in arguing that "cosmic nature is no school of virtue, but the headquarters of the enemy of ethical nature," in attacking "the fanatical individualism of our time [that] attempts to apply the analogy of cosmic nature to society," in repudiating, in short, "the gladiatorial theory of existence,"[4] introduced not only Darwin and the theory of evolution, but famous men and ideas from Greek—and Indian—philosophy, from Western science and literature, from Western history and religion. With Yen Fu's help, he opened up for Chinese eyes the whole awaiting field of Western intellectual history. He also, by letting people see natural fields in a new way, opened up the whole vast field of natural history. For most Chinese intellectuals had never looked at nature as naturalists. When men like Lu Hsun looked with Huxley through his windows at English fields, they quite literally saw nature as they never had before—and were excited by it.

T'ien-yen lun, therefore, was exciting in many ways at once. It was not exciting just as a political map or compass for the future, or as something that seemed to make cosmic sense out of international chaos. It was not exciting simply frighteningly as a Darwinian warning of impending doom, or, at very least as a disheartening protestation of the "naturalness" of China's political predicament. Nor was it exciting simply as a Darwinian promise of progress. For those reflective enough to think beyond China, it was deeply exciting scientifically and philosophically.

T'ien-yen lun meant many things to many people, especially as so many conflicting points of view were expressed within it. Huxley, playing his own devil's advocate, gave quite generous expositions of many of his opponents' views—before rebutting them—and Yen Fu, in his commentaries, gave *his* views, as well as those of Spencer. Except, therefore, as the prime source of a handful of Darwinian or Social Darwinian slogans, which, once translated, proved to have a most wildly creative, influential life of their own, it makes no sense to talk of "the influence" of *T'ien-yen lun,* for that influence was actually the many differing influences of many different people—and it is no easy matter to determine whose was the greatest.

If, as Benjamin Schwartz contends, *T'ien-yen lun* was really published not just as a translation of *Evolution and Ethics* but as a running argument between Huxley and Yen Fu, if indeed it really "consists of two works—a paraphrase of Huxley's lectures, and an exposition of Spencer's essential views as against Huxley"[5]—then who won? Whom did Yen Fu's readers believe, Huxley or Spencer, Huxley or Yen Fu? The place to find out, of course, is in the quotations from *T'ien-yen lun* that run rampant throughout the polemical essays of the first two decades of this century, essays on revolution, political, social, and philosophical. But to answer the question well, we must first probe the depths of disagreement between Huxley, Spencer, and Yen Fu, and put off our examination of others' excitement over *T'ien-yen lun* until we have inquired into Yen Fu's own excitement over *Evolution and Ethics.*

SOCIAL DARWINISM

Professor Schwartz, in his excellent chapter on Yen Fu and *Evolution and Ethics,* is both disturbed and intrigued by an "enormous paradox." Having determined that Yen Fu became a "Social Darwinist" largely thanks to the works of Spencer, and that he thus wanted above all else to introduce to China Spencerian Social Darwinism, he is puzzled that Yen Fu should translate Huxley. "Huxley's lectures are decidedly not an exposition of Social Darwinism. They actually represent an attack on Social Darwinism! They must be understood in terms of that widespread revulsion against Spencerian orthodoxy which set in at the end of the century in both Great Britain and the United States. . . . Why, then," wonders Professor Schwartz, "does Yen Fu choose to translate a work so little in tune with his basic message?"[6]

But Huxley's work is really much more in tune with Yen Fu's message than it would appear, and Yen Fu was trying to make it even more so. Professor Schwartz is perfectly aware of the fact that there is much in Huxley's book to excite Yen Fu. He notes the "brief, vivid and almost poetic account of the main tenets of Darwinism," and he notes what we have noted above, Huxley's almost mystical description of nature and his invitation into the new worlds of natural history and Western thought.[7] Yen Fu was the first to be excited by so much of that which, thanks to him, excited readers of *T'ien-yen lun.* But, philosophically, Professor Schwartz still sees Huxley and Yen Fu at loggerheads. He claims that Yen Fu "makes clear his profound commitment to Social Darwinism and to the ethic implicit in Social Darwinism," and concludes that Yen Fu translated Huxley's *Evolution and Ethics* largely as "a foil for the master," that is, for Spencer.[8]

These last two statements can be misleading on several counts. Spencer was not, strictly speaking, Yen Fu's master, for Yen Fu was no one's disciple. He never accepted any Westerner—not even Spencer, not even Darwin—as an infallible seer, nor did he take any Western work as gospel. He consequently was never caught in the fundamentalist bind, religious, Marxist, or whatever, of having to accept, make sense of, and defend every sentence of a canon. He chose to translate certain books by certain men to express certain ideas, but there was no one man and no one book that he felt could stand as the one and only guide for China. He had analyzed British power as the product of a combination of ideas that he could introduce only by a package of translations, a synthesis of those Western ideas that he, Yen Fu, thought good for China. He offered a synthesis that was clearly his own, even though he never sat down and wrote one. In fine Chinese tradition, he confined his own writing largely to footnotes, but in the tradition of the Sung philosopher Lu Hsiang-shan, the "Classics," that is, those works he translated from Spencer, Smith, Huxley, Montesquieu, Mill, and Jenks, were in reality *his* footnotes.[9]

Yen Fu was tremendously impressed with Spencer, but only selectively impressed. He never tried to justify all of Spencer; in *T'ien-yen lun* he certainly did not simply defend Spencer from Huxley's attack, nor did he simply attack Huxley with Spencer. Much, perhaps most, of what Huxley said was left unchallenged, and much of Spencer was ignored. He ignored, for example, that part of Spencer that could justify "non-action," that part that was against strong government or strong nation-states. He ignored, indeed, that part of Spencer that most seemed *for* the "fanatical individualism" Huxley was so much against.

We have already seen that Yen Fu did not even come close to being a "fanatical individualist." He was a fanatical "groupist," a *ch'ün*-ist, from start to finish. "Individualism" was good only in that it liberated individual energies for group effort, which energies were only good if channeled or indeed harnessed towards such effort. Yen Fu could thus be impressed by Spencer's vision of the social organism, and at the same time be impressed by Huxley's attack on a kind of uncontrolled individualism that in Yen Fu's eyes could only harm that organism—Yen Fu could feel that both Spencer and Huxley were for the group.

That is why it is misleading to say that Yen Fu was a "Social Darwinist" committed to "the ethic implicit in Social Darwinism," and hence at odds with Huxley. The label "Social Darwinist" encompasses too much to tell us much, and it fools us if we think it means one thing. There is no single ethic implicit in Social Darwinism. If we take "Social Darwinist" as a fit label for someone committed to the ethic of ruthless, uncharitable self-interest that Huxley so abhorred, then Yen Fu was not a Social Darwinist at all, while, if we take it, as I think we must, as a label fit for anyone who attempts to explain society in Darwinian terms, then Huxley, even in *Evolution and Ethics,* was as much a Social Darwinist as Spencer himself. He simply explained society in *different* Darwinian terms. He held that evolution had produced the ethical sense that now enabled man to rise above his former self, to "kick down the ladder by which he has climbed," and to repudiate "the ape and tiger methods of the struggle for existence."[10]

There was already in Huxley a strong proto-Kropotkinist bent, with which our "Spencerian" Yen Fu was in total sympathy. Huxley believed that social morality, the ethical process of human "cooperation," was not only a phenomenon produced by evolution; it was man's only means of further evolution. He held that societies—all societies, whether of men or of ants—"have arisen out of the advantage of co-operation in the struggle for existence."[11] Human societies have been able to arise, however, only thanks to a curbing of man's natural self-assertiveness. Individuals have come to have self-restraining "consciences" thanks to their gradual recognition that their own survival depends on the survival of their group, which can survive only if its members restrain themselves and work together. Huxley saw "an artificial personality, the 'man within' as Adam Smith calls conscience . . . built up besides the natural personality. He is the watchman of society, charged to restrain the anti-social tendencies of the natural man within the limits required by social welfare."[12] Huxley distinguished between an "artificial" and a "natural" personality, but, as the artificial was also the result of natural selection, he logically meant that the artificial was simply a later natural addition to the originally natural. At any rate, he termed "this evolution of the feelings out of which the primitive bonds of human society are so largely forged, into the organized and personified sympathy we call conscience, the ethical process." And he insisted that "it tends to make any human society more efficient in the struggle for existence with the state of nature or with other societies."[13]

Now this ethical process, so close to Kropotkin's later dynamic of "mutual aid," was also incredibly close to Yen Fu's concept (partially borrowed from Spencer!) of "the people's virtue, that *ch'ün*-ing spirit so necessary to the preservation of China. It is ironic, therefore, that at this very point in his translation Yen Fu chose to quibble with Huxley. He even took the opportunity to comment, rather condescendingly, that Huxley's "discussion of social theory is not as tight as Spencer's."[14]

But he was merely quibbling, and at this point quite unfairly. He was sidetracked by a Huxleian aside about parental love as a factor in the earliest formings of society. Yen Fu assumed Huxley meant that man was able to group because he was innately possessed of a sympathetic spirit. He therefore claimed that Huxley had confused cause and effect. Yen Fu wanted no innate factors that were not the product of natural selection. But this comment proves only that Yen Fu had missed Huxley's point, so obvious in the above quotations, that there *had* been an "*evolution* of the feelings out of which the primitive bonds of human society [were] . . . forged," that conscience had indeed been "built up." The real point, however, is that, despite this misunderstanding, Yen Fu was in total agreement with Huxley's main thrust in this section, that the "gradual strengthening of the social bond, though it arrests the struggle for existence inside society, up to a certain point improves the chances of society, as a corporate whole, in the cosmic struggle."[15]

This sentence, which Yen Fu subtly turned into a policy rather than a simple statement of natural fact ("The reason those who seek a cohesive group suppress competition within that group is in order to withstand the natural forces without")[16] was not only a basic "Darwinian" conviction of Yen Fu; it was one of the most pervasive "Darwinian" tenets in the whole Chinese Social Darwinian movement. The tenet "Cease struggle within to win without" was one of Yen Fu's main grounds for opposing revolution. But revolutionaries of all sorts made equally easy use of his doctrine, once they identified themselves as the in-group and their enemies as outsiders, even though they all struggled inside China. For they could also say, as they did, that such internal struggle was a struggle for the internal unity Yen Fu felt so necessary for China's survival.

In as much as Huxley recognized the survival value of solidarity and its corollary of self-restraint, he was, if "original Social Darwinism" means what it means to many, the unhindered glorification of self-interest, a proto-Kropotkinist-Social-Darwinian revisionist—and he was one of the first. But if Huxley was, so was Yen Fu. Yen Fu, most assuredly with a joyous inner shout of "Eureka!" composed, as a comment to Huxley's passage above, his own proto-Kropotkinist formula: "Evolution has brought it about that those who are able to group shall survive; and those who are not able to group shall perish. Those who are good at grouping shall survive—and those who are not good at grouping shall perish. What does it mean to be good at grouping?—To be good at mutual sympathy."[17] Thus, with Huxley's help, did Yen Fu rescue the *ch'ün*-ing virtues from the Darwinian jungle without abandoning Darwin. Perhaps one could say that, for Yen Fu, such Social Darwinian revisionism was Hsun-tzian,[18] but, if it was, it was Hsun-tzian, Kropotkinist, Darwinian, and Huxleian all at once.

CONFUCIAN OPTIMISM

Yen Fu agreed, then, with Huxley's contention that "the instinct of unlimited self-assertion" which "every child born into the world will still bring with him" and his "inheritance (the reality at the bottom of the doctrine of original sin) from the long series of ancestors, human and semi-human, and brutal," was now, despite the fact that it had once been "the condition of victory in the struggle for existence," the "sure agent of the destruction of society, if allowed free play within."[19] Yen Fu also agreed, just as strongly, with Huxley's immediate qualification of that contention: that, if allowed to go too far, the ethical powers that restrained that self-assertion could also lead to the destruction of society; that goodness could be suicidal; that, if the members of a group became too selfless, they might refuse to stand up for even their corporate self against aggressors; that, to *live* in the real world, it was impossible to live absolutely in accord with the Golden Rule, in either its Christian form or its Confucian ("Do not unto others what you would not have them do unto you").[20] For "what," lamented Huxley, "would become of the garden if the gardener treated all weeds and slugs and birds and trespassers as he would like to be treated if he were in their place?"[21]

To Yen Fu's great satisfaction, Huxley thus rejected as a practical philosophy Christianity's most famous maxim (he had already, of course, to Yen Fu's even greater satisfaction, rejected Christianity's theology).[22] He also, again with Yen Fu's blessing, rejected Buddhism, because of its "total renunciation of that self-assertion which is the essence of the cosmic process."[23] Despite the main entreaty of his book, for men to fight as the good fight a civil war within themselves against such self-assertion, he admitted in the end that, without some measure of it, men would not live at all.

Yen Fu agreed with all this. Where he did not agree was with Huxley's mood. For Huxley was obviously saddened, depressed, even angered—though irrationally, if there was only nature to be blamed—by what seemed to him the invincible imperfection of a world in which men were forced by evolution to live, if live they would, by adding only a measure of selflessness to their selfishness. Selflessness and selfishness, these two "necessities of life," he saw as an irresolvable contradiction that he could not just cheerfully accept, in the manner of a Mao Tse-tung, as something "perfectly natural." He did not believe men could strike some perfect balance between the two. Nor did he see any middle path leading, whatever the cost in time, through contradiction to perfection.

He nowhere saw any promise of perfection. "We also know modern speculative optimism," he wrote, "with its perfectibility of the species, reign of peace, and lion and lamb transformation scenes; but one does not hear so much of it as one did forty years ago; indeed I imagine it is to be met with more commonly at the tables of the healthy and wealthy, than in the congregations of the wise."[24] "The prospect of attaining untroubled happiness, or a state which can, even remotely, deserve the title of perfection, appears to me to be as misleading an illusion as ever was dangled before the eyes of poor humanity. And there have been many of them."[25]

Darwin had ceased to cheer Huxley up. "The theory of evolution," he wrote, "encourages no millennial anticipation."[26] His new Darwinian explanation of original sin did nothing to change his basically biblical evaluation of human nature—and he saw no reason to believe evolution was about to change that nature. "Ethical nature may count upon having to reckon with a tenacious and powerful enemy as long as the world lasts," he wrote.[27] He might just as well have said that Satan would torment the world until the Day of Judgment.

Here at last we find something with which Yen Fu most certainly did disagree. Huxley's pessimism and his rebellious dislike of the "cosmic process" went against the grain of Yen Fu's Darwinism, Confucianism, and Taoism all at once. It did indeed offend his "deepest religious proclivities."[28] Yen Fu, like so many Chinese intellectuals throughout history, had come to have both a Confucian and a Taoist side, and now he had a Darwinian side as well, but, far from pulling him apart, these elements so happily reinforced each other in his thinking that all the pieces of the human predicament suddenly fell together in his mind to make "perfectly good" sense out of the contradictions that so needlessly, in his eyes, made Huxley despondent.

Yen Fu did believe in perfectibility. One of the first messages he had heard from Darwin was that evolution was working its way towards (our) perfection. In his very first comment in *T'ien-yen lun* he had written that, "once Darwin appeared, men knew that man was an element in evolution at once evolving and progressing."[29] Yen Fu already believed in the mystical force of the *Tao* as the constant sustainer but also the changer of all things (so readily identifiable with Spencer's "Unknowable"). From there it was an easy leap to a belief that the *Tao* was not just the constant force behind cyclical change but the constant force behind progressive change (*chin-hua*). As Yen Fu translated a line from Huxley in the opening passages of *T'ien-yen lun,* "Although nature ever changes, there is something changeless moving within it. What is it that is changeless? It is called evolution."[30]

Taoism, therefore, helped explain the mysterious force of evolution. But it was Confucianism that helped "socialize" it. The evolving *Tao* was not only pushing our species forward; it was pushing our society forward as well. The Confucians had always protested that men should be able to live in perfect harmony, that perfect order had indeed once reigned under Sage rule, and that it could do so again. Now, the Confucian side of Yen Fu could shout in exultation that Darwin, and Spencer, had proved that the *Tao* was at this very minute irreversibly bearing our society towards that harmony.

It is not surprising that Yen Fu labeled the chapter in which Huxley ridiculed "modern speculative optimism" as "his worst,"[31] nor that he then cited a long Spencerian argument in support of the theory that perfect order could be reached.[32] It is true that the Taoist element in Yen Fu's mind surfaced once again, this time to force him to admit that it would be presumptuous to say that one could literally know the destination of the Unknowable, but he showed his faith, nonetheless, in his own "guarded" conclusion: "What we can know is that the way of the world must progress, and that what comes will be better than what is."[33]

With such a faith, he was not to be dismayed with Huxley's argument that the Golden Rule would not work in our world, or that the contradiction between selfishness and selflessness was necessary. His belief in ultimate perfection made it easy for him to accept such imperfection along the way, and so did the ancient Confucian doctrine of the cosmic correctness of the middle way, of the Golden Mean, which had always acknowledged and accepted the existence of imperfection on its either side.

Yen Fu could casually dismiss Huxley's distress at the "failure" of the Golden Rule with the comment that in neither its Western nor Eastern form was the Golden Rule "sociology's final formula for perfect peace. The formula for perfect peace is this: 'Everyman's freedom shall be bound by the freedom of others.' If one uses this formula, the above error will be corrected."[34] But he was able to believe that this ideal was a workable substitute thanks largely to his Confucian faith that reasonable men could indeed walk the middle path. Everything could, like the little bear's porridge, be just right. One could be perfectly free without being too free. One could be self-interested without being too self-interested. A Western slogan again happily fit the Confucian spirit. Of course a measure of selfishness was necessary, a measure of self-assertiveness. Of course one had, to some degree, to pursue one's self-interest. But all that was perfectly all right, as long as one pursued "what the Westerners call . . . 'enlightened self-interest.' Enlightened self-interest does not go against morality."[35] Enlightened self-interest was just enough. It was the middle path.

Similarly, Yen Fu was not disheartened by Huxley's Darwinian proof of original sin. For Yen Fu had again the protection of two thousand years of this-worldly Confucian optimism. Huxley, of course, had had the protection of two thousand years of other-worldly Judeo-Christian optimism, but that was based on a belief in a God Huxley though Darwin had disproved. Even worse, poor Huxley, unlike so many of his simple-minded contemporaries, for some reason found it impossible to secularize that Judeo-Christian optimism and see in evolution the promise of a Heaven on earth. Consequently, he was caught with what in effect was a biblical evaluation of human nature, long after he had lost his biblical faith in that nature's possible redemption.

For Yen Fu the problem was different. Good Confucians, of all persuasions, had always believed in the possibility of the *Tao* prevailing on earth. And they had believed—whatever they thought of man's original nature—in the perfectibility, again on earth, if not of man at least of men. No confrontation with "the reality at the bottom of the doctrine of original sin" could shake that belief, at least not for a Hsun-tzian Confucian, for even Hsun Tzu had believed—indeed just as strongly as Mencius—that everyman could be a Sage.[36] Every man could so better himself or so get the better of himself that he could best his originally evil nature. Original sin was no cause for despair.

As far as the classic Hsun-tzian–Mencian argument over original goodness was concerned, Yen Fu was certainly more of a Hsun-tzian than a Mencian. It is true, he said that Hsun Tzu had "gone too far"[37] in labeling our original nature evil, but his quarrel was with Hsun Tzu's logical right to make such a value judgment, not with his basic description of human nature. He seemed to believe that the concepts "good" and "evil" were themselves not original and hence could not be fairly applied to what was.[38] Still, he seemed to accept without qualms the idea that men's earliest instincts were selfish. Our *social* instincts, those occasional sparks of selflessness which Mencius took as proof of our original and latent goodness, Yen Fu would have said had developed slightly later, as a result of natural selection. The theory of evolution, therefore, seemed to resolve or at least to end the controversy. Hsun Tzu may have been closer to the truth in the beginning, but even he had noted that what distinguished man from beast was man's ability to *ch'ün* or "socialize." That ability was proof of the evolving Mencian nature, destined to triumph in the end.

Such Darwinian Confucianism, however, as we have seen in the case of Yeh Te-hui, was anathema to any orthodox Mencian Confucian, and not just because it denied the orthodox doctrine that human nature was good *ab origine.* Yen Fu's subtle, if unconscious, transfer of Confucian optimism—from belief in the perfectibility of the individual to belief in the perfectibility of the species, from belief in the ever-present possibility of social order to belief in the ultimate inevitability of social order, from belief in the potentially good nature of men to belief in the eventually good nature of man—virtually destroyed the original promise of the Confucian faith. For the assured future triumph of the species logically ruled out any individual triumphs in the present, except perhaps for a handful of "supermen," or "advanced" human beings. Most men were not yet good-natured enough to be good.

Characteristically, however, Yen Fu overlooked this point. He was so obsessed with China's predicament that he could not see the individual human predicament as an individual predicament. Consequently, he could not appreciate the pathos in Huxley's implied question, "How should we live as sinners without promise of redemption?" But he could appreciate the spirit of the answer. He could deeply admire Huxley's determination to struggle for a better world even as he chided, and perhaps pitied, him for being unable to see that the world could become perfect. He could see that Huxley had that spirit ascribed to Confucius himself by the mocking guard at Shih-men. He was one "who acts though he knows he cannot succeed."[39] Yen Fu's only quarrel was with the "unnecessary" pessimism that kept Huxley from seeing that he *could* succeed.

Yen Fu was obviously deeply moved by the lines from Tennyson's "Ulysses" with which Huxley closed his argument:

> *It may be that the gulfs will wash us down,*
> *It may be we shall touch the Happy Isles,*
> *. . . . . but something ere the end,*
> *Some work of noble note may yet be done.*[40]

He translated Tennyson's lines into verse that had the ring of the *Book of Odes,* and he let them stand virtually without comment as a fitting end for *T'ien-yen lun:*

> *Hoist sail upon the deep blue sea.*
> *Great wind and waves stretch out forever.*
> *Perhaps we shall sink in the bottomless depths.*
> *Perhaps we shall reach the Immortal's shore.*
> *How can we know*
> *Which it shall be?*
> *O Tempora, O Mores!*
> *I shall give my all,*
> *Without giving up,*
> *For such is man's duty.*

To the poem he added only one last line of prose: "I wish to join with all men of conscience in the world in swearing to this resolve."[41]

It was precisely this resolve to act, to work, for a better world that in the end let Yen Fu join with Huxley—against the objections of Herbert Spencer. Yen Fu and Huxley might still in one sense be worlds aparts in their resolve. In their immediate dreams each certainly had his own world most in mind, Huxley a *better* England, Yen Fu a *stronger* China, Huxley a more humane society, Yen Fu a more united—and, hence more militant—"social organism," but philosophically both still agreed that it was human action that would decide success or failure, or at least that, without human action, human dreams could not come true.

CONFUCIANISM IN ACTION

This shared belief in the necessity for resolute human action finally resolves "the enormous paradox" of the "Social Darwinian" Yen Fu choosing to translate an "anti-Social-Darwinian" book as a Darwinian guide to survival for a beleaguered China. Yen Fu, after all, had chosen to act himself, in writing his essays and translating *T'ien-yen lun,* because he wanted action. He wanted his government to act immediately, to get his people started acting on the long, hard course of action that alone could bring them the physical, intellectual, and moral fitness necessary to survive. Yen Fu wanted action, and Huxley gave a call for action—much more clearly than did Spencer. Yen Fu might see prescriptions in Spencer's descriptions of evolution, but he could not find in Spencer any Tennysonian poems that cried out for "work of noble note."

Huxley appealed to Yen Fu's Confucian conscience, even though he offended Yen Fu's Taoist faith. This may seem paradoxical, but it is fact, made possible by the Taoist-Confucian conflict in Yen Fu's head.

For Huxley, without realizing it, was writing about the very same Taoist-Confucian conflict that so divided Yen Fu's "proclivities." Huxley's whole book was unwittingly a modern recast of the classic Taoist-Confucian argument over the proper relationship between *jen wei* (human action) and *t'ien hsing* (the cosmic process). Indeed the "classic" nature of the argument was one of the reasons Yen Fu was drawn to Huxley's book in the first place. It was also one of the reasons Huxley's book fell so elegantly into classical Chinese, one of the reasons the scholar Wu Ju-lun could claim that Yen Fu's prose had carried Huxley almost to the level of the philosophers of the late Chou.[42] In this argument, Huxley, of course, emphatically took the Confucian position, while Herbert Spencer took the Taoist. Actually, Huxley took the Hsun-tzian–Confucian position. He argued with Hsun Tzu that human action—*jen wei*—consciously and intellectually directed towards inhibiting our naturally selfish selves was the only way to goodness. Hsun Tzu had man struggling against his basic nature. Huxley had him struggling against his beastly nature, against the "original sin" inherited in the instinct to survive. In either case the good fight was with oneself. Man was engaged in a civil war, his intellect against his appetites; and so man's "better half" was also at war with the cosmic process, the general that had trained man's baser, basic self.

There was illogic in Huxley's position, the illogic, fully recognized, of the microcosm pitting itself against the macrocosm whence it came,[43] but even the illogic was Confucian. Confucius himself, although he may not have recognized such cosmic illogic as illogic, nonetheless, felt it and revealed it, in his self-defending cry, "If under Heaven the Way prevailed, I would not be trying to change things."[44] And Hsun Tzu, even more clearly, caught the same illogic in his line, "Heaven and earth give birth to the good man, and the good man orders Heaven and earth."[45] For Confucius, Hsun Tzu, and Huxley, the natural order was not naturally ordered. All three sought to change things.

To this the Taoists, and Herbert Spencer, cried, "No! No!" In their eyes, it was precisely this Confucian kind of human action, directed against the natural process, that brought mankind to grief. Man should follow the doctrine of *wu-wei,* "no action *against* nature," and let nature change things herself. Nature knew best. Nature would change things for the best. Man at *his* best should make sure he was not in the way of the Way.

Now, clearly, Yen Fu was drawn to both positions. He wanted

the comfort of the Taoist side with its claim that the Way *was* prevailing, and he felt the need for the Confucian side, for he felt that China had to do something. He had faith in the *Tao,* which now, thanks to Darwin and Spencer, he identified with progress and evolution, and he took comfort in the faith that the *Tao* was bearing mankind upwards. But he also believed that individuals, men or countries, had to struggle to survive, that they had to act. He believed in the necessity of Huxley's "horticultural process." He believed in Confucian self-cultivation, and he believed in China's self-cultivation. He did not believe in leaving China's garden up to God.

Yen Fu refused to see in these beliefs any conflict. *He* saw no paradox in his having at once both Confucian and Taoist beliefs, or in his being drawn at once to both Huxley and Spencer. For, in his own mind, he had resolved their conflicts, indeed, denied their conflicts. He knew that Huxley *thought* that he and Spencer were in conflict, but he thought that Huxley was wrong to think so, for *he* could see that the "ethical process" and the "cosmic process" were not in conflict. Yen Fu's quarrel with Huxley was not with Huxley's call for ethical action but simply with Huxley's "unfortunate" notion that such action was unnatural, that it went against the *Tao.* Yen Fu applauded Huxley's ethical process. It was just what the *ch'ün* needed to pull itself together and survive. But he refused to say that this ethical process was at odds with the cosmic process, because, as he saw it, the ethical process was simply a part of the cosmic process. Yen Fu's proto-Kropotkinism preserved the good name of the *Tao.* The *ch'ün*-ing instincts were the *product* of evolution, as was man's ability to harness both nature, with a small *n*, and himself. Thus the would-be man of virtue did not have to rebel against the universe, for the universe itself was evolving virtuous men.

Wu Ju-lun, who wrote the opening preface for *T'ien-yen lun,* understood perfectly Yen Fu's position: "Huxley arose and completely changed the earlier theory, claiming that nature [lit. "Heaven"] could not be given free reign, but that man must resist nature. Man must resist nature, using to the full his naturally endowed abilities, and ever renewing his efforts, so that his country might exist forever and his race never suffer decline. This is what is called struggling for victory over nature. And yet man's struggle with nature, and man's victories over nature, are both the results of the working of nature. Thus it is that natural action (*t'ien hsing*) and human endeavor (*jen chih*) both are part and parcel of evolution (*t'ung kuei t'ien-yen*)."[46]

Yen Fu had won—to his own satisfaction—a simple semantic victory over paradox. He had resolved the age-old conflict between *t'ien hsing* and *jen wei.* He had resolved the conflict between Huxley and Spencer. He had simply declared the mystical word *evolution* big enough for all of them.

This declaration enables us at last to understand Yen Fu's usually misinterpreted mistranslation of *Evolution and Ethics* as *T'ien-yen lun,* which only means "The theory of evolution," and leaves out "ethics" altogether. Actually, in an earlier draft of his work, Yen Fu apparently did attempt to put "ethics" into his Chinese title. In the May 12, 1897 issue of the *Shih-wu pao,* Liang Ch'i-ch'ao mentioned his reading of Yen Fu's "Chih-kung t'ien-yen lun."[47] The term *chih-kung* is not a good translation of *ethics,* but it is a revealing one, though very difficult to put back into English. It means something like "good social organization or social planning." It refers to the human effort of ordering, indeed of creating, a good society, but it looks at human effort primarily from the point of view of governments, not of individuals. It looks towards ethical organizing, not ethical behavior. Thus, if taken in the right sense, it could almost be translated as "good government."

Now neither "Good government and evolution," nor "Evolution and ethical organization" catches the full connotative force of Huxley's title, *Evolution and Ethics,* but they both come close to what Yen Fu, with considerable justification, identified as Huxley's subject. Huxley *was* talking about good government and good social organization. He was not simply telling individuals that "cosmic nature is no school of virtue";[48] he was telling governments as well. Moreover, Yen Fu, however much he disagreed with Huxley's contention that cosmic nature was indeed "the headquarters of the enemy of ethical nature,"[49] nevertheless totally agreed that "fanatical individualism," the absolute selfishness that cosmic nature was supposed to teach, was anathema to the *ch'ün.* He was thus, in principle, in favor of Huxley's ethics, as he interpreted them, and would not have excluded "ethics" or *chih-kung* from his title because, as a "Social Darwinist," he thought ethics interfered with evolution. If he left *chih-kung* out for anything other than aesthetic reasons, it was because he did not want to imply by juxtaposing *chih-kung* and *t'ien-yen,* that the two were in opposition. He thus called his translation "The Theory of Evolution" and hoped that in his commentary he could convince his readers that "ethics," despite Huxley's own confusion, were an all important, but perfectly natural, *part* of evolution.

As long as his readers could understand that fact, and see in it proof that Huxley was wrong to be so skeptical of evolution as the path to perfection, then, thought Yen Fu, they could and should listen closely and eagerly to what else Huxley had to say, for "he says much," said Yen Fu, "about self-strengthening and the preservation of one's race."[50] That, we must remember, was Yen Fu's ultimate reason for translating Huxley's work. He wanted his countrymen to hear Huxley's Darwinian call to action, his Darwinian imperative. He wanted them to see what he, Yen Fu, with his optimistic, Taoist, Confucian illogic, saw most clearly in Huxley's work, proof that the *Tao* is prevailing, but also that the *Tao* helps those who help themselves.

There is, then, no "enormous paradox" in Yen Fu's translating *Evolution and Ethics.* Yen Fu translated the best book he could find to instill awe for evolution, to deliver Darwinian warning, and to inspire in response the patriotic spirit (patriotic selflessness supposedly born of enlightened self-interest) that alone could bring Chinese into the active unity that was strength, the unity that alone could preserve China in her struggle for existence.

Slightly transposed into Yen Fu's choice of key, Huxley's book fell well enough in tune with Yen Fu's basic message. And, where there was discord, Yen Fu felt he could resolve it. Yet, in the end, there was one note that surely he did not resolve, a deep, quiet almost ever-present note—Huxley's sadness.

Huxley wept at the *price* of evolution. Even if he could have convinced himself that evolution was still working for our good, he would have remained saddened at the way it worked, at the toll it took. For him, evolution was something "full of wonder, full of beauty, and at the same time full of pain."[51] As a younger man, the wonder and beauty had almost blinded him to the pain, but in the end he found no way to overlook it. It kept him from worshiping evolution. It spoiled his scientist's joy. He had a tragic sense that there was something ineffably wrong with the universe, and he could see no way it could be put right.

Yen Fu was not a man without feeling. When he first explained the process of natural selection, he vented at least one heartfelt "Alas! That living things should produce so many seeds, and that so few should survive."[52] And he cried, "Alas!" again when he told of aboriginal races disappearing from the earth.[53] But somehow his sighs did not stay with him. Unlike Huxley, he seemed able to forget evolution's price, or to accept it. Undoubtedly his Taoism did much to help him. When, later, he annotated Lao Tzu's *Tao te Ching,* he excitedly saw in the famous line, "Heaven and earth are not humane; they use the ten thousand things as straw dogs," the "opening idea of the evolutionary school."[54] Heaven and earth worked naturally. They did not have favorites. They did not "interfere" with their system. They were not "soft-hearted," because they were great and up to great things. Heaven and earth worked at a level "beyond the humane and the inhumane."[55] And yet, without question, Heaven and earth, the *Tao,* and, for Yen Fu, evolution, were good, for they were the force of perfection. Evolution's lack of "humanity" was a case of "benign neglect."

In the end, it was Yen Fu's vision of progress that preserved him, for better or for worse, from Huxley's sadness. He himself was neither brutal nor ruthless. The necessary action he imagined was a gentle cultivation that would aid growth, not a ruthless weeding out that worked through death. He never thought in terms of China's having to help out with nature's dirty work, of China's surviving only over the dead bodies of other nations. And yet, unseen, a certain callousness did creep into Yen Fu's thought. For he had a vision of a coming peace on earth, a peace that could–and would–include China, a peaceful end that justified the brutality of evolution's means. He had a vision of the *Tao,* of evolution triumphant, that could lift his eyes above the myriads of sacrificed straw dogs that lay strewn along the Way.

Part Three

Darwin for Revolution—and Darwin Against It

SIX

The Revolutionary Liang Ch'i-ch'ao

While scholars and school boys read *T'ien-yen lun,* after the failure of the Reform Movement of 1898, the watchword *pien-fa* (reform) began to lose out to the word *ko-ming* (revolution) in the struggle for survival of political ideas. *Ko-ming* came to dominate Chinese rhetoric in the first decade of the twentieth century; it still does. However ill-defined, it was and is a word to be reckoned with.

Until mid-1903, however, the man in China who did most to promote revolution was not Sun Yat-sen or any other revolutionary but Liang Ch'i-ch'ao, the arch-reformer. For one thing, before the famous *Su pao* (The Kiangsu journal) incident of June 1903 and the publication a bit earlier of Tsou Jung's *Ko-ming chün* (The revolutionary army) and the radical journals, *Che-chiang ch'ao* (Tides of Chekiang) and *Kiangsu,* Liang Ch'i-ch'ao had China's reading public virtually to himself. His new journals, *Ch'ing i pao* (Honest criticism, December 1898–December 1901) and its successor, *Hsin min ts'ung pao* (A new people, February 1902–November 1907), published in Japan and smuggled back into China through the treaty ports, were *the* things to read. But Liang Ch'i-ch'ao was a political exile, a "rebel" with a price on his head, angrier than he had ever been in his life, furious at the Manchu concubine, now Empress Dowager, who had ruined his dreams of reform and murdered his comrades. His writing, therefore, in the first years after the Empress Dowager's coup, was more radical and inflammatory than ever before. It was almost revolutionary. Liang Ch'i-ch'ao even began to use the term *ko-ming,* and he used it as a good word. Indeed, even after he became Sun Yat-sen's most eloquent enemy, in the *Min pao* (People's journal)–*Hsin min ts'ung pao* confrontation period, he could still find it within himself to define *ko-ming* as a good word.[1]

Liang Ch'i-ch'ao never really meant by revolution what Sun Yat-sen and "the Revolutionaries" did, but he called for much that they and others thought both necessitated and justified their kind of revolution, and to that extent he helped their cause. And yet, far more important than his contribution to the Revolutionaries' Revolution of 1911, great, ironically, as that was, was his contribution to a broader, deeper kind of revolution. Indeed, his greatest contribution was perhaps simply to the broadening of the term *revolution* itself. For Liang Ch'i-ch'ao was one of the very first in China to carry that word beyond the political realm into the social, cultural, intellectual, philosophical, even the psychological. More than a decade before the May Fourth Movement began, Liang Ch'i-ch'ao was already speaking of the need for a "literary revolution" and for a "writing-system revolution." Without claiming to have coined all such terms himself, he publicized the words, "academic revolution," religious revolution," "moral revolution." He called for revolutions in customs, in poetry, in production, in historiography, in economics, and in music.[2] He actually cried out for a "revolution of the spirit."[3]

More than any other individual, Liang Ch'i-ch'ao set in motion intellectual currents that would lead not only to the May Fourth Movement, but to the "Great Cultural Revolution" of Mao Tse-tung. More than any other individual, Liang Ch'i-ch'ao launched an attack on China's traditional society and culture that did indeed lead to revolutionary change.

For simplicity's sake, we can say that Liang Ch'i-ch'ao, with all his many mentioned revolutions, was for three kinds: political revolution, cultural revolution, and "spiritual" revolution. What he really had in mind, of course, was changing and complex but, for our purposes at this point, one fact is all-important. In the cause of all three revolutions, Liang Ch'i-ch'ao, more than any other individual, called over and over again on the powerful authority of Charles Darwin.

RACIAL REVOLUTION

Liang Ch'i-ch'ao was soon to gain fame as the foremost intellectual opponent of the anti-Manchu revolution of Sun Yat-sen. But in the last days of 1898, he published in the very first issue of the *Ch'ing i pao,* his first journal printed in exile, an article more devastatingly anti-Manchu than anything yet written by any of Sun Yat-sen's Revolutionaries. Entitled "Reform must begin with the elimination of the barrier between Manchu and Han," it was significantly the last important article he included under the famous title with which he had opened the *Shih-wu-pao* two years before, "A general discussion of reform," for in it he unwittingly furnished Sun Yat-sen's Revolutionaries with a powerful new rationale for revolution.

Liang Ch'i-ch'ao's argument was that, for the good of China, the traditional laws granting Manchus special privileges and powers in government should be abolished. He based his argument, however, on a threatening new use of Darwin's "natural laws" of the struggle for existence and the survival of the fittest:

> The myriad years from the time when first there were living things upon the earth until today. . . . can be described in a single word—racial strife. In the beginning animals struggled with animals, then men struggled with animals, and finally men struggled with men. First savages struggled with savages, then the civilized struggled with savages, and finally the civilized struggled with the civilized, in endless struggle stretching back throughout all time. Alas, this is the law of the struggle for survival. Not even a Sage can do anything about it. . . . According to the law of the survival of the fittest, members of an inferior race must be devoured by a superior one. Day by day, month by month, they will slowly be eaten away until there are no more of them left, and their race no longer lives upon the earth.[4]

On the surface, this was nothing new. Both Liang Ch'i-ch'ao and Yen Fu had been trying to frighten their country into action with

Darwinian descriptions of a national world of ruthless, racial conflict ever since the Sino-Japanese War three years before. But here, for the first time, the races in question were not the yellow versus the white, not the Chinese versus the evil Europeans, but the Chinese versus the Manchus, and Liang Ch'i-ch'ao left no doubt whatever as to which race he—and nature—were for: "Manchu and Chinese—it takes no expert to establish which is the superior race and which the inferior."[5]

Thus Liang Ch'i-ch'ao, not Sun Yat-sen, first turned inwards with Social Darwinism's slogans, to aim them at the Manchus. Liang Ch'i-ch'ao first raised the specter of a potential Darwinian contest not only without but within the Middle Kingdom. Poor Liang Ch'i-ch'ao, who was to spend most of his life arguing passionately for gradual, peaceful change, loosed into China her first pseudo-scientific rationale for internal strife, a rationale that would subtly justify, in the eyes of many, the revolutions and civil wars of Republicans, warlords, Nationalists, and Communists alike, wars, which together with imperialist wars, also "justified" by Darwin, would ravage China for the next fifty years.

Liang Ch'i-ch'ao was not calling for a war against the Manchus. He was tempted to join forces with Sun Yat-sen's Revolutionaries, but for various reasons he never did.[6] Even in this angry article, he refused to call for war. What he wanted was for the Manchus to step back, if not down, of their own will, for their own good. But younger and/or hotter heads would easily jump to the conclusion that a better solution, with perfect Darwinian blessing, would be to force them to step back, or wipe them out.

The trouble was that Liang Ch'i-ch'ao's reasons for wanting the Manchus to step down seemed to the Revolutionaries the perfect reasons for pushing them aside—and they were Darwinian reasons. Liang Ch'i-ch'ao had not for a second forgotten China's life-and-death struggle with the white races. He was anti-Manchu because the Manchus were hindering and imperiling China in that struggle. The Manchus were keeping down the Han Chinese, upon whom hung the fate of the entire yellow race.[7] Liang Ch'i-ch'ao explained the embarrassing and unnatural situation of an inferior race like the Manchus ruling a race like the Han as the result of an unfortunate seventeenth-century victory of brawn over brain. Such victories, he said, were not uncommon in the struggle for existence, but they were always only temporary. In the end, final victory would go to the intelligent. That is why the Manchus were totally unfit to compete against the whites: "Today's Manchus have lost their former fierceness, without having changed their stupid nature."[8] Their only hope was to throw in their lot with the Han Chinese, who alone could save them from the Europeans. But the Han Chinese could only save them if the Manchus got off their backs, if the Manchus abolished their special privileges and allowed the Han Chinese, with their superior wisdom, to organize a national reform and a national defense.

If the Manchus continued to insist on lording it over the Chinese, they were doomed, for they would invite a revolution like the American or the French.[9] Even if the Europeans conquered China before such a revolution could take place, the Chinese, with their intelligence and vast numbers, would survive to win another day, but the Manchus would not: "The Manchus, being few, stupid, and weak, even if they were to escape the vengence of the Han, would be forever beasts of burden for the white man."[10] As things stood now, Liang Ch'i-ch'ao concluded, with the Empress Dowager's reassertion of Manchu despotism, the Manchus were "choosing the perfect road to their own destruction."[11]

Thus Liang Ch'i-ch'ao, without calling for revolution, threatened the Manchus with revolution. Without calling for Han vengeance, he made it sound perfectly natural. The Revolutionaries could hardly have done better. And they certainly could never have dreamt up a more damaging epithet for their chosen enemies than Liang Ch'i-ch'ao's Darwinian label of *lieh chung* (inferior race or inferior breed). Henceforth anti-Manchu racial slurs, like Liang Ch'i-ch'ao's own ("few, stupid, and weak"), could be paraded as scientific descriptions. Worst of all, the label *lieh chung* proclaimed that, with or without revolution, the Manchus' days of glory were virtually over. Darwin had stated all too clearly what happened to *lieh chung*. One way or another, the Manchus were about to be weeded out by evolution.

Liang Ch'i-ch'ao himself did not keep up a steady campaign of anti-Manchu insults. Indeed, he soon spent much of his time attacking the idea of a revolution based on a desire for racial vengeance. Nevertheless, his low opinion of the Manchus continued to slip out on occasion in his articles, and he continued to allow others to insult the Manchus in his journals.

The following poem by Chiang Chih-yu, for example, proves how thin the line was at this time between Reformers and Revolutionaries. In 1903, Chiang Chih-yu's poem was gleefully included by Tsou Jung in his inflammatory pamphlet, "A Revolutionary Army," but it was first published on July 26, 1901, in Liang Ch'i-ch'ao's *Ch'ing i pao*!

Nu-ts'ai Hao (How good to be a slave)

Slavery's just tremendous.
Slavery's just stupendous.
Diplomacy and politics? Poo poo, such things offend us.
We do our best to sleep through any worries others send us.
The ancients have a saying that proclaims the lofty notion:
"Three cheers for loyalty and filial devotion!"
(and let not any peasant dare to raise undue commotion.)
The Manchus came across the wall full ten score years ago,
And thus we've been their slaves, you see, quite long enough to know
That all our wealth is rightly theirs, as is each stream and hill.
How natural, then, that they should calmly divvy it up at will.
But lo! White foreigners cross the waves.
And they too vie for us as slaves.
Fine.
If they want a coal mine, I'll dig day and night.
If they start a company, I'll be their boy.
If they need an army, I'll fight their fight.
If they want a privilege, I'll find a ploy.

For after all—to serve white masters after masters yellow
Is easy for a race that finds the taste of slavery mellow.
So what's this talk of bloody "revolution"?
This "freedom" stuff and "equal wealth"? Boy, that's sure some locution.
You want to lose your bloody heads for anything so un-Confucian?
Not I, Not I.
No, slavery's fine.
Slavery's grand.
For a slave's at home in any land.
I'll be damned if I'll take a racial stand.[12]

DEMOCRATIC REVOLUTION

If Liang Ch'i-ch'ao in the end shied away himself from taking the racial stand he so clearly helped set up for the followers of Sun Yat-sen, he did not shy away from lambasting his countrymen for nurturing a "slave mentality" (*nu-li hsing*). Over and over again, he echoed Chiang Chih-yu's hatred and scorn for the servile spirit, inherited from the past, and bad at the best of times, but now unspeakably tragic, as it blinded the Chinese to an unprecedented danger:

> Our country's ignorant masses, four hundred million of them, have for thousands of years been kept under control by a people-ravaging government [*min-tsei cheng-t'i*], so that now they are like blind fish born in a black cavern, who come out into the ocean and still cannot see. . . . They do not know that the words *popular sovereignty* [*min-ch'üan*] exist on this earth. If you tell them, "You have your own free rights," they are startled and embarrassed. They cover their ears and beat a retreat. This is what I have kept referring to as a "slave mentality." But people with a slave mentality are not just content to be slaves themselves. They insist on ridiculing those who are not slaves. Alas, to pit such people against the races of Europe in this world of struggle for survival and the survival of the fittest—What hope is there? What hope is there?[13]

The blind fish belonged to Darwin. As Liang Ch'i-ch'ao had already related in an earlier article, a species of blind fish had once been found in a subterranean lake inside an Italian mountain, and Darwin had explained that they had not been blind originally, but had become so because they had no use for their eyes. Now that the lake had been opened by mining to fish that were not blind, however, the blind fish were doomed to extinction, for they had no hope of competing with the non-blind in the struggle for existence. "Alas," Liang Ch'i-ch'ao had commented, "this can be a warning for our four hundred million people."[14]

Liang Ch'i-ch'ao was repeating his Darwinian warning, but he was also making another point—and more clearly than ever before. He was blaming Chinese "blindness," the Chinese "slave mentality," directly on China's system of government, on a "people-ravaging system of government." He was blaming it on despotism, not on Manchu despotism, but on a three-thousand-year tradition of despotism. He made it perfectly clear that he thought that a Han Chinese despot would be just as bad as a Manchu one.[15]

He shied away from an anti-Manchu revolution, but not from the idea, at least, of a political, democratic revolution. He showed himself to be more passionately than ever pro *min-ch'üan,* pro "democracy," "popular sovereignty," "people's rights," "people power." He had been pro *min-ch'üan* long before Sun Yat-sen made the term one of his revolutionary "Three People's Principles" (*san min chu-i*), and he continued to be pro *min-ch'üan* long after.

Where Liang Ch'i-ch'ao differed from Sun Yat-sen was in his idea of the proper way to get to *min-ch'üan,* although the difference was not always clear. Even though Liang Ch'i-ch'ao's would-be first move, and the proposed policy for which he was most known at the time, to restore the Kuang-hsu Emperor to power, seemed the opposite of Sun Yat-sen's aim, to wrest power from the Empress Dowager and deposed Emperor in one and the same fell swoop, Liang Ch'i-ch'ao constantly confused the distinction by savagely attacking monarchy at the same time that he supported the restoration of the "rightful" monarch (Manchu or no Manchu). He frankly admitted, of course, that he wanted to restore the

Emperor as a reforming emperor, who would get China back on the road towards democracy, but he further helped Sun Yat-sen by eventually referring rhetorically to that road as a revolutionary road. He tried to distinguish between a bloody revolution and a non-bloody one, favoring the latter, but that distinction was overlooked by many of his readers.[16]

His attack on monarchy was often Darwinian. His basic argument we have seen before in the *Shih-wu pao:* Monarchy stifled progress, stunted the people, and wasted their energies. "Western evolutionists," he declared, "say that the world progresses through competition." Competition was good for the world. But how could one's country win out in such competition? "To survive one must seek strength and intelligence." And whence came they? They came from "pooling the intelligence of the people" and from "pooling the strength of the people."[17] Monarchy, however, made pooling impossible. The people had no voice. They were not free to pool their wisdom and their strength. As one of Liang Ch'i-ch'ao's colleagues explained things to a student who was puzzled that modern men, despite evolutionary progress, were not up to the Sages of old: "Ever since [China] was united, rulers and prime ministers have conspired to keep the people ignorant. They have obstructed freedom of thought, so that men's minds have been trapped in a circle from which they could not escape."[18]

"Freedom of thought" became a new watchword, as did the word *freedom,* or *liberty* (*tzu-yu*) generally, which was a watchword with profoundly revolutionary consequences.

There has been much discussion among Western scholars of China about *tzu-yu* (probably reintroduced into Chinese from Japanese), as a translation of the Western word *freedom.* Many have suggested that Chinese have refused to warm to certain Western notions of freedom precisely because of the allegedly "bad" connotations of their word for the idea. They have pointed out that *tzu-yu* contains the word *self* (*tzu*), that it literally means something like "self-direction," and that it consequently suggests to "the Chinese mind" selfishness and libertinism. There is some truth in this idea. But it is well worth remembering that Liang

Ch'i-ch'ao, who, as usual, was one of the first to use the new word, used it in the beginning almost exclusively as a good word. Indeed, for several years it seemed his favorite word.

One of his most important writings, begun for the *Ch'ing i pao,* was a long series of short essays published one or two at a time, with individual titles, under the collective title *Tzu-yu shu* (On liberty). Perhaps Liang Ch'i-ch'ao had already seen a manuscript copy of Yen Fu's translation of Mill's *On Liberty,* finished but not yet published by 1899.[19] Perhaps, instead, he had read Mill's book in a Japanese translation, or read just quoted bits and pieces. At any rate, he acknowledged that the inspiration for his title came from one of Mill's sentences: "I have named my book from the words of the Western scholar John Stuart Mill: 'Nothing is more important to the progressive evolution (*chin-hua*) of human groups (*jen ch'ün*) than freedom of thought, freedom of speech, and freedom of publication.'"[20]

So *tzu-yu* was the key to *chin-hua;* freedom was the *sine qua non* of progressive evolution. Progressive evolution was the secret to survival. Despotism, by definition the enemy of freedom, was therefore the enemy of evolution. As such, it was public enemy number one. The monarchy would have to go.

It mattered not at all that monarchy was itself the product of evolution. That was how evolution worked. The fit of today were the fat of tomorrow. And monarchy was the fittest of yesterday.

Liang Ch'i-ch'ao had long since learned, from both K'ang Yu-wei and Yen Fu, to love stage theories of history, and he now brought forth an amazing array of such theories, some borrowed some invented, to prove that democracy was the wave of the future. One of these, which showed both his new love of the word *freedom* and his now constant reliance on "Darwinian" dynamics, established a four-stage scheme of social evolution. Human societies progressed from an "age of barbaric freedom" to an "age of aristocratic autocracy" to an "age of complete autocracy," to an "age of civilized freedom," in which "the whole people would rise up and take back their political power." These stages were the same for "every country and every race," everyone of which would "progress through them in accordance with a fixed natural law." The only differences in the evolution of different peoples lay in the length of time each dwelt in each stage, and what determined that time was the "severity of competition" faced by each people. The fiercer the competition, the faster the evolution.[21]

It should be obvious that, just as the competition Liang Ch'i-ch'ao was talking about was the competition of "peoples" with other "peoples," so the freedom he spoke of often meant freedom for "the people," not freedom for the individual. He wanted his *ch'ün* to be free, but he did not want his people to be free of their *ch'ün*. Here we do see an early qualification of the "goodness" of the word *tzu-yu*. "Barbaric freedom," in which all men were free to do as they would, or at least as they could, was *not* a good thing. Indeed, despotism was good, in its day, precisely because it put an end to such freedom, because it forced individuals, and small *ch'ün*, into a great *ch'ün*. The ideal was "civilized freedom," but that was only good because people were only to be as free as they could be without infringing on the freedom of others. "Truly free citizens" had to bow to three things: reason, the laws established by the people, and the will of the majority. For "freedom without control," wrote Liang Ch'i-ch'ao, "is the robber of the people. Freedom with control is the people's treasure."[22]

The above was still a very "liberal" view of the necessary limitations to individual liberty. A slightly less liberal view, however, at least in rhetoric, showed up in Liang Ch'i-ch'ao's chapter "On Liberty" in his famous work, *Hsin min shuo* (Towards a new people). There he wrote, "Freedom means freedom for the group, not freedom for the individual. In the age of barbarism, individual freedom triumphed, and the group's freedom was lost. In the age of civilization, the group's freedom will be strong, and individual freedom will decline."[23] Even less liberal rhetoric showed up later, in the chapter "On Unifying the Group": "Men must not be slaves to other men, but they must be slaves to their group. For, if they are not slaves to their own group, they will assuredly become slaves to some other."[24]

Thus Liang Ch'i-ch'ao's *ch'ün*-ism tempered his love of liberty.

The greatest of the first generation of China's modern "liberals," he included in his cry for liberty the age-old Chinese prejudice for the primacy of the *ch'ün,* and thus helped voice a modern rationale, later used by both the Nationalist and Communist Parties, for severely limiting people's freedom—in the name of the People's freedom.

Nonetheless, even freedom for the People was a revolutionary cry, and it obviously did the Revolutionaries more good than it did the Reformers, for the Revolutionaries could simply add, "We must free *our* people from the Manchus." Liang Ch'i-ch'ao helped the Revolutionaries even further, however, by using still more revolutionary rhetoric. For a time he almost hopelessly confused the distinction between *reformer* and *revolutionary* by his own favorable use of the word *revolution.*

In December 1902, in an article entitled "Shih ko" (An explanation of *revolution*), Liang Ch'i-ch'ao upbraided his countrymen for being afraid of the term *ko-ming*. He himself quarreled with it as a translation (which he blamed on the Japanese) of the word *revolution,* but he publicly accepted the English word and said those who were afraid of it simply did not know what it meant.[25]

He himself preferred to translate *revolution* as *pien-ko,* which sounded more like "change," but, whatever the translation, he made it clear that sudden, complete, revolutionary change was a good thing and "an inescapable law in the world of evolution." Yet again he described the dynamics of evolution: survival depended on "selection." (The Chinese term *t'ao-t'ai* literally refers, more frighteningly, not to the "selection" of the fit, but to the "elimination" of the unfit.) There were two kinds of selection, natural selection and selection by man: "Selection by man means carefully to seek out the unfit in oneself and change it, to make oneself fit to survive."[26] "Selection [or elimination]," he wrote, "is revolution."[27]

Liang Ch'i-ch'ao's conclusion, therefore, was that "the work of revolution (what the Japanese call *ko-ming* and what I call *pien-ko*) is today our only way of saving China. Seeking survival and strength by any other way would be like grinding a brick to make a mirror or boiling sand to make rice."[28] And failure would be forever: "The most unfavorable thing would be for our people to be sunk forever through the horror of natural selection into the great pit of evolution, from which there is no resurrection, not even in ten thousand eons."[29] Therefore, "if our people want to survive, they must begin by forcefully advocating a great revolution, and by carrying out a great revolution. If our rulers and officials want to align themselves with the people in order that they themselves may survive, they must begin by not fearing this great revolution but by welcoming it."[30]

It would have taken a stout-hearted ruler or official to welcome what Liang Ch'i-ch'ao seemed to have in mind, however, for he described "the work of revolution," undoubtedly to the great delight of Sun Yat-sen and Company, in what was probably the most violent paragraph he ever put in print. In yet another installment of his *Hsin min shuo* entitled "On Progress," in which he voiced dissatisfaction with the eleventh-hour Manchu program for reform, he argued in favor of "destruction" (*p'o-huai*), for "destruction," he said, "was and is the one and only unavoidable law for all nations past and present that have sought progress."[31]

Echoing his first famous formula for the necessity of change, he wrote that, "if we destroy there will be destruction, if we do not destroy there will be destruction. Destruction in the end is unavoidable."[32] His point was the same as his point about change, and about natural selection, and about the one important message of Darwin's theory of evolution. He was convinced that only the active survive and that the passive perish. If one waited to be changed instead of changing oneself, if one waited for natural selection instead of practicing selection by man, if one waited for others to destroy instead of oneself destroying what had to be destroyed, then one waited for death. Unconsciously siding with Huxley, Liang Ch'i-ch'ao protested that to survive one had to take the initiative.

He called on his countrymen to initiate the kind of "conscious destruction" that alone could make possible simultaneous reconstruction.[33] And he made it savagely clear what it was he most wanted to destroy—China's system of government. He fervently prayed such destruction could be bloodless, but, bloodless or not, come it must and come it would:[34]

> We must pulverize our thousands-of-years-old brutal and filthy system of government. We must wrest our millions of tiger-like, wolf-like, locust-like, maggot-like officials from their temple-rat and wall-fox positions of security! For only then will we be able to cleanse our bowels, so that we may ascend the road to progress![35]

This, then, was "the work of revolution," the work of "political revolution." And it was something Liang Ch'i-ch'ao, no less than Sun Yat-sen, was very obviously for. Even in 1906, when Liang Ch'i-ch'ao was voicing his most bitter opposition to Sun Yat-sen, he could truthfully say, "I have held from the beginning in my essays that political revolution is the only means to save the nation."[36] To prove it, he had only to point to the last words of "An Explanation of 'Revolution'": "Alas, it is not just politics that China should revolutionize—but, until we can revolutionize our politics, how, alas, can we speak of anything else?"[37]

HISTORICAL REVOLUTION

Liang Ch'i-ch'ao, of course, could easily speak of much else. Indeed, if he sounded in the above sentence as if he believed that political revolution was a prerequisite for other kinds of progress, he could elsewhere sound as if he believed just the opposite. Whatever the sequence, however, there is no question that he believed in the absolute necessity, sooner or later, of a Great Cultural Revolution.

Liang Ch'i-ch'ao called out in the *Hsin min ts'ung pao* for revolutions in everything from production to poetry, and, though he did not follow up his call with detailed guidelines for cultural revolutions in each of the areas he mentioned, he did point the way in just enough areas to set in motion an unprecedented process of cultural re-evaluation.

Perhaps, in the long run, the most important of the specific cultural revolutions that he led himself—certainly the one in which he most relied on the help of Darwin—was in the study and writing of history. This revolution, more than any other, not only threw open all of China's past to agonizing reappraisal, but seemed in itself to force Chinese into the creation or discovery of a whole new philosophy of life. More specifically, it laid the absolutely necessary groundwork for Chinese Marxism and the Thought of Mao Tse-tung.

It was Darwin, not Kant, Hegel, or Marx, who revolutionized Chinese concepts of history. K'ang Yu-wei, reading *The 19th Century,* and watching it progress before his eyes, may have come to believe in progress before he knew of Darwin, but it was Darwin who "proved" progress, who brought "scientific proof" that the world was "evolving," and, more important, "evolving upwards." It was Darwin, therefore, who proved to K'ang-Yu-wei, Yen Fu, and Liang Ch'i-ch'ao, and through them to at least two generations of their countrymen, that "meaningful human history" had to be the history of human evolution—progressive evolution.

Liang Ch'i-ch'ao declared this "truth" as he defined history in his deeply significant "Hsin shih hsueh" (The new historiography), the second article in the first issue of his journal, *Hsin min ts'ung pao.* By this time, confusion of the two concepts, evolution and progress, was already complete in Liang Ch'i-ch'ao's mind. He now used almost exclusively the term *chin-hua* (progressive transformation) for both; that is, he saw in the two only one. Evolution and progress had become hopelessly synonymous in the Chinese language. The reader must therefore hear both terms, or a new one like *evolutionary progress,* when he listens to Liang Ch'i-ch'ao's statement that "history is the story of the *chin-hua* of the human group [or of human groups—*jen ch'ün*].[38]

The notion that history was the story of human progress was the first of at least five tenets of Liang Ch'i-ch'ao's "New Historiography" that were to become major tenets of the even newer historiography of Mao Tse-tung, and that is no coincidence. For it was certainly thanks to Liang Ch'i-ch'ao, not Karl Marx, that Mao Tse-tung first came to believe: "The world is progressing. The future is bright. No one can change this basic course of history."[39] Liang Ch'i-ch'ao had taught almost every literate member of Mao Tse-tung's generation that this was *the* great fact of life. And, if it was, it was obvious that history should be its story.

Now this idea has become so commonplace in the West since the time of Kant, Hegel, and Darwin, that it is easy to forget how revolutionary an idea it was. Once this interpretation of history was declared a "natural law," it changed, for awhile, the whole form of historical study. Historians no longer delved into the past to see what they could see, but to demonstrate progress. They sought progress and, seeking, they found. In China, the first to do this was probably K'ang Yu-wei's fifteen-year-old daughter, set to the task by her father, the first in China to be accused of distorting history to prove progress.[40] It was Liang Ch'i-chao, however, after K'ang Yu-wei's rather wild Confucian experiment with the Three Ages, who made the first serious attempts to write revisionist, progressionist history.[41]

The attempt to find progress in history had many serious consequences. The first was drastically to reduce the very scope of history—in several ways. History was henceforth to be reserved for the progressive. "Of a hundred things," declared Liang Ch'i-ch'ao, "those that are born, grow, develop and progress belong in the realm of history. Those that do not cannot belong in the realm of history."[42] A new definition of *relevance* thus arose to limit the ranks of historical facts. It was no longer enough for something to happen. To count, a happening had to be part of a story of progress.

It had further to be part of a story of *human* progress. This anthropocentric restriction was hardly revolutionary, but it was slightly odd under the circumstances that Liang Ch'i-ch'ao, first attuned to the "New History" by Darwin's revelation of natural evolution, should so easily rescue man from his seemingly deflated position as a chance figure in the greater history of evolution itself and place him, instead, larger than life, back at the center of the universe. "It is not because we humans are simply partial to our species," he wrote. "Man is the ultimate in evolution,"[43]—or as the materialist Mao Tse-tung would later put it, "Of all things on earth, man is the most precious."[44] Thus, man, however he got there, was still in his pre-Darwinian position at the top of the *scala naturae,* still as he had been in China's first *Book of History,* "the best of the ten thousand things" (*wan wu chih ling*).[45]

But history as human history was only the obvious first step. History must be the people's history—and here again Liang Ch'i-ch'ao put Mao Tse-tung in his debt. Again Liang Ch'i-ch'ao came to postulate the supremacy of the *ch'ün* over the individual, this time because he was dead set on seeing progress. He recognized in the human life cycle a dead end. Individuals lived and died and got nowhere. "Therefore," he wrote, "if you wish to find evidence of progress, you must look at the group. If you make men stand alone, then you can expect no progress, and there can be no history. For human progress means the progress of a people, not the progress of a person."[46] He rather curiously admitted that, "if you look at individuals, men today certainly in no great way surpass the ancients. . . . It is commonly acknowledged that in mental ability the Duke of Chou, Confucius, Plato, and Aristotle could all hold their own against modern man." And yet, he triumphantly went on, there are things Confucius and Aristotle could not understand that today "even a child still wet behind the ears can know."[47] For the child could eat the fruits of the progress of his *ch'ün.* Progress was something social, and social progress was thus what counted.

Now surely Liang Ch'i-ch'ao "had something" in this simple realization. But, when he came close to echoing Kant's view that, "if nature has set man a short term of life (as is, in fact, the case), then (perhaps) nature requires an endless procession of begettings of which one transmits enlightenment to another in order finally to push the genus of human kind to that level of development appropriate to the purpose of nature,"[48] and when, unknowingly, he echoed Hegel's, that "in world history, the individuals we deal with are nations and the wholes that are states,"[49] when, indeed, he himself declared that "history should really pay attention only to the group; if things do not concern the group, be they words however strange or actions however extraordinary, they have no right to be entered into history,"[50] he was not only restricting history, but de-personalizing it, in a way that was most strange for China, and in a way that was not without consequences. The individual was surely devalued. For history was no longer his story.

The people or peoples, the *jen ch'ün,* that Liang Ch'i-ch'ao spoke of, the "individuals" of the new history, were really very close to what Hegel meant by *nations,* and, one might think, far less close to what the Marxist Mao Tse-tung would mean by *the people.* Perhaps Mao Tse-tung's *individuals* should properly have been only classes, but, by including more classes than he probably strictly should have in the ranks of "the Chinese People," Mao Tse-tung showed himself almost as nationalistic in his definition as Liang Ch'i-ch'ao. Both agreed that "the Chinese People" were *the* actors, or *the* actor, on history's stage to watch, applaud, and work for.

In one unfortunate respect directly related to his Darwinism, however, Liang Ch'i-ch'ao was decidedly more "nationalistic" than Mao Tse-tung. For Liang Ch'i-ch'ao was still desperately concerned not only with the Chinese nation, and the Chinese people, but with the Chinese "race." He insisted that history was racial—and thereby made it racist. "What is history?" he asked. "It is simply the story of racial development and racial strife. Aside from races, there is no history."[51]

Worse yet, there was no history for three of the five human races. For there were "historical races" and "non-historical races," the former being those that could unite themselves against others and extend themselves at the expense of others, and the latter being those that could not, those that were losing, or had lost, "their original positions on the stage." Looking at that stage at the height of the "age of imperialism," Liang Ch'i-ch'ao concluded that "the only two races that [could] be considered historical races are the yellow and the white."[52] In one fell swoop he plucked the black, brown, and red races out of history. He eliminated them because, so he thought, they were *being* eliminated.

The only historical drama left was the contest between the whites and the yellows. And the whites had the upper hand. Indeed, Liang Ch'i-ch'ao painfully "admitted" that the Europeans alone had the right to be called a "world history race,"[53] because they alone held the cultural and military power to influence world history. At least the "Teutons" did. Liang Ch'i-ch'ao had somehow convinced himself that, although "knowledgeable people all admit that European civilization is the mother of all civilization for the modern world,"[54] only the "Teutons" had the power and talent to shape that civilization. Somehow, having fallen prey to the worst of intra-European racial prejudices, he declared the "Teutons" (whom he defined as the Germans, Dutch, Swedes, Norwegians, Danes, and English) the cream of the white-race crop, the fittest in white evolution. The Slavs, he said, in a somewhat prophetic aside, might one day supersede them, but, as things now stood, "sovereign rights to ninety percent of the land of the globe belong to the whites, and the so-called whites are really only the Aryans, and the so-called Aryans are really only the Teutons; truly, the Teutons alone are the masters of today's world."[55]

Now it was precisely because the "Teutons" seemed to Liang Ch'i-ch'ao to be so fully in control of history that the study of history seemed to him so important, for, when one sought in history the secret of the "Teutons'" power, one discovered, he said, that it was in their racial consciousness of their history. Liang Ch'i-ch'ao came to the conclusion that only a people conscious of its history could have the national spirit necessary to preserve for itself a place in history. He called for a "new historiography," therefore, specifically to inspire the nationalism that seemed the *sine qua non* of survival.

"Today," he wrote, "if we want to advocate nationalism, so that our four hundred million compatriots may stand firm in this world, where the fit survive and the unfit perish, then every man, woman, and child among us, whether young or old, good or bad, must study our nation's history."[56] "The study of history," he said, "is the broadest and most important field of study. It is a mirror for the people. It is the wellspring of patriotism."[57] The proof lay in the potent historical consciousness of European nations: "Half the credit for the development of nationalism in

Europe today and for the powers' daily advance in civilization belongs to the study of history,"[58] because a true study of history made men conscious not of self but of group. "What is valuable about written history," he wrote, "is its ability to make clear the way in which the members of a group interact, compete, and write, the way they live and grow and advance (*chin-hua*) as one body. It is this revelation that brings to the reader an outpouring of feeling of love and admiration for his *ch'ün.*"[59]

There was also a more scientific purpose of history, "to describe the phenomenon of the progressive evolution of human groups so as to seek out the natural laws thereof"[60]—so that one's group could be taught to follow them. It was in this way that the study of history was to "use past progress to lead to future progress."[61] But love of group was an integral part of that process: "The duty of every historian is, in accordance with the great principle of progressive evolution, to show proof of that principle in the reality of past events, in order to evoke a nationalistic spirit."[62] For the necessity of a nation's having national spirit to be fit was one of the first "natural laws" of evolution that history taught, at least that "good" histories taught. The trouble, of course, thought Liang Ch'i-ch'ao, was that "for thousands of years our China has had no good histories,"[63] no histories that understood evolution or that wrote of the people. Indeed, he said, "it would be no exaggeration to say that, before now, China has had *no* histories."[64] That is why he so fervently cried out for "A New Study of History." That is why he so angrily moaned, "Alas, if there is no historical revolution, our country cannot be saved."[65]

THE REVOLT AGAINST THE ANCESTORS

Liang Ch'i-ch'ao's cry for an historical revolution is our first textual evidence of a cry for cultural revolution leading directly into a cry for "spiritual revolution." The bitterness of his cry against traditional Chinese historiography, however, was itself evidence of a revolutionary change in spirit.

For when Liang Ch'i-ch'ao called all China's historians, from "the Grand Historian," Ssu-ma Ch'ien, on, "badgers from the same mound,"[66] he was insulting figures second in esteem only to the greatest philosophers in China's intellectual tradition. When he said that China's "stacks upon stacks of historical works are all lifeless as effigies in a waxworks,"[67] he was repudiating books deemed second only to the Classics in serious social value. And he was telling his fellow intellecutals that at least half their vaunted education had been worthless—worse than worthless. Of the twenty-one official dynastic histories, written in imitation of Ssu-ma Ch'ien's *Shih chi,* he asked, "How could their slavishness reach such depths? They are like lutes all playing one note. Who can bear to listen? Whenever one reads them the only battle is to keep awake. That is why thought does not advance."[68] But it was not just that "reading them is a waste of brain power." The real trouble was that "China's histories have been instruments not to increase our people's knowledge, but to decrease it." For, "as it has been said, 'to know the past and not know the present' is 'to be sunk in stupid ignorance.' Surely the blame for sinking our people in stupid ignorance must be borne by our historians."[69]

Yen Fu had blamed the Sages; Liang Ch'i-ch'ao now blamed the historians. Between them they denounced almost their entire scholarly tradition. They denounced and renounced some of their most illustrious intellectual ancestors.

This was revolutionary. In a land in which ancestors, intellectual or otherwise, had been revered, if not worshiped, since before the dawn of history, cursing one's ancestors was a mortal sin, a revolting sin. For millennia, "good" Chinese had looked back to their ancestors with reverence, thanks, and a sense of continuing filial obligation. Yen Fu and Liang Ch'i-ch'ao looked back in anger—and with a sense of shame—as they searched for the culprits who had led *their* country to the brink of Darwinian disaster.

ANCESTRAL LOSS OF FACE. Liang Ch'i-ch'ao was beginning to have in Japan the same experience Yen Fu had had two decades before in England. Suddenly brought face to face with the riches of the Western intellectual tradition (even through Japanese translations, and through Yen Fu's own), he began to feel that his own tradition paled by comparison, that Western historians were better than Chinese historians. He began to feel ashamed of the imitators of Ssu-ma Ch'ien, and of Ssu-ma Ch'ien himself, and by implication at least, through the *Spring and Autumn Annals,* even of Confucius. Especially because Western power seemed the result of Western thought (the conclusion Yen Fu and Kuo Sung-t'ao had come to in London in the late 1870s), because Western nationalism seemed the result of Western historiography, Liang Ch'i-ch'ao began to feel ashamed of the whole legacy of Chinese thought.

All patriotic Chinese were being forced by Western power to feel ashamed of China's present, but somehow it was even worse to be forced by that same power to feel ashamed of China's past, for Chinese had always been able to feel confident about their past. That had been one of the amenities that came with being born in the Chung-kuo, the "Middle Kingdom." One inherited "the great tradition," and also the cultural confidence of one's ancestors that came with it. But now, if all that was to be denied, Chinese would suddenly be on their own, deprived of "family support," not only in a crude outside world that paid no heed to their pedigree, but even at home. They would have to make something new of themselves, or be nothing.

The degree to which Chinese intellectuals were beginning to lose confidence in their ancestors and culture can nowhere be better seen than in Liang Ch'i-ch'ao's thitherto unheard of expression of anger and shame at, of all things, his country's name. "Nothing makes me more ashamed than the fact that our nation has no name," he wrote, except for dynasty names, like Han and T'ang, "which defy the principle of respecting the people," or foreign names, like *Chen-tan* or *Chih-na,* which take away "our right to name ourselves," or the names *Chung-kuo* (the Middle Kingdom) or *Chung-hua* (the center of civilization), "which make us guilty of puffed up self-esteem and liable to the ridicule of foreigners." He decided, in the end, that the first two types of names were even worse than the last, and that China "for absolute want of any better way out" should "still use [the name] to which we are most accustomed," *Chung-kuo,* but he did so with none of the age-old confidence of his ancestors. Far from it—he said that China should call itself *Chung-kuo* because, "even though the name is a bit arrogant, it is common practice in today's world for each people to show pride in their country." Moreover, "if our fellow countrymen seriously examine the difference between name and fact, it may yet prove a way to stir up some spirit."[70] The first reason was almost as sad as Chang Chih-tung's final reason why Chinese should study Confucius: because "Western nations consider the preservation of the Classics to be very important."[71] The second merely admitted the sad fact that China no longer *was* the Chung-kuo.

That admission, painful enough in the realm of politics, was devastatingly painful in the realm of culture. Politically, after all, China had been weak many times in history, but culturally never. Cultural self-doubt threatened a deathblow to Chinese self-confidence.

After the Opium War, Chinese had very slowly come to feel that China's armaments were inferior. Later, many had come to feel that China's political institutions were inferior. But now Liang Ch'i-ch'ao and Yen Fu said that China's historians, philosophers, and scholars were inferior, seemingly that China's whole civilization was inferior—*because* China's "ancestors" were inferior. This was a loss of "faith *in* our fathers" probably more traumatic for Chinese than the post-Darwinian loss of "faith *of* our fathers" was for so many nineteenth-century Westerners.

Darwin, of course, was the great discreditor of ancestors for all peoples, indeed for all people. He had said, with a *we* that certainly included both Englishmen and Chinese, that "there can hardly be a doubt that we are descended from barbarians."[72] Worse yet, he had held that "the ancestors of man were, no doubt, inferior in intellect and probably in social disposition to the lowest existing savages."[73] We had, indeed, "ape-like progenitors."[74] Our ancestors were rather rough, and we ourselves, Darwin had added, "are not removed by very many generations" from their "savage state."[75]

It was this last allegation that hurt the Chinese. Most Chinese probably found it easier than most Westerners to accept Darwin's thesis that man was the product of a myriad natural transformations.

Most Chinese would also readily have admitted that their *far distant* ancestors had existed in a savage state. But those far distant ancestors had supposedly been transformed from that state thousands of years ago. Their *recent* ancestors had been "fully civilized" for countless generations. It was this article of faith that Western power and Darwin's theory so irreverently challenged.

Still, Darwin's theory provided a rationale that should also have prevented the loss of that faith from being so traumatic. For Darwin, if he was the great discreditor of ancestors, was also the greatest liberator of descendants, the great absolver of descendants from all responsibility for their ancestors (a logically ludicrous concept with or without Darwin). He himself was not in the least dismayed by the brutish portraits he painted of our forefathers: "Unless we willfully close our eyes, we may, with our present knowledge, approximately recognize our parentage; nor need we feel ashamed of it."[76] On the contrary, "Man may be excused for feeling some pride at having risen, though not through his own exertions [a phrase almost universally ignored] to the very summit of the organic scale."[77]

It is true that Darwin occasionally sounded as if he meant that not "men" but "white men" alone stood on "the very summit of the organic scale," and he also sounded, at least once, as if he found it easier to accept inhuman ancestors than some of his human ones:

> He who has seen a savage in his native land will not feel much shame if forced to acknowledge that the blood of some more humble creature flows in his veins. For my own part, I would as soon be descended from that heroic little monkey who braved his dreaded enemy in order to save the life of his keeper, or from that old baboon, who, descending from the mountains, carried away in triumph his young comrade from a crowd of astonished dogs—as from a savage who delights to torture his enemies, offers up bloody sacrifices, practices infanticide without remorse, treats his wives like slaves, knows no decency, and is haunted by the grossest superstitions.[78]

But the point (although Chinese of Liang Ch'i-ch'ao's generation, intensely sensitive to Western criticisms of "barbarous" Chinese customs, might have thought Darwin was saying he would rather be descended from a monkey than from a Chinese) was that Darwin did admit that he was descended from savages, but said it did not matter, because in actuality he had not *descended* but *ascended* from them: "All civilized nations were once barbarous,"[79] but once civilized, barbarous beginnings were of no account. Indeed, the lowlier the beginnings, the more remarkable the evolution, and the more reason for civilized men to "be excused for feeling some pride at having risen."

As we have already seen, in Robert Mackenzie's *The 19th Century*, late-nineteenth-century Englishmen could feel proud of shameful ancestors, because they were so proud of themselves, so absolutely convinced that they were better than their ancestors, convinced that evolution meant advance, and that they *were* advanced. With that sort of imperturbably self-satisfied Darwinian faith, lambasting one's ancestors could be fun.

But Chinese of Liang Ch'i-ch'ao's generation were *not* self-satisfied. They were not convinced that they were better than their ancestors. Consequently, if their ancestors were shown to be inferior, then so were they.

THE UNFILIALLY FIT. The trouble was that, for thousands of years, Chinese had been told that the greatest disgrace was to be unlike (*pu hsiao*) one's ancestors.[80] The great fear had always been that generations would degenerate. No one had ever told Chinese that their very survival depended on leaving their ancestors in the dust. But that was what Charles Darwin now *seemed* to be saying—that Chinese were in trouble precisely because they were like their ancestors, because they had been too filially pious. They had not changed, and were hence no longer fit.

Men like Liang Ch'i-ch'ao were beginning to believe this was true. That was why they were both ashamed and angry at their ancestors, for they blamed their ancestors for betraying them, and the early promise of Chinese civilization for selling out too early, as Yen Fu had said, to the ideals of stability and peace, for "keeping down" their descendants and stifling Chinese progress.

That is why Kao Hou ("Chien Kung") could write, and Liang

Ch'i-ch'ao publish, the following bitter, revolutionary, Darwinian poem, fittingly entitled, "Pu Hsiao" (Emulate not):

Fathers, I tell you, do not emulate your forebears!
Sons, on your lives, do not emulate your fathers!
Ruthless is the drama of natural selection,
And selection by man but helps it on.
Evolving ever, onwards, upwards,
The present must assuredly best the past.
If only each day we evolve and progress,
We shall yet ascend to civilization.
Still, no organic living thing
Dare scorn the force of national selection.
If the fit are to triumph, the unfit must fall.
That is the unbreakable natural law.
To avoid being unfit, and to rank with the fittest,
It helps to alter one's race's blood.
But our five races have not intermarried,
For we have obeyed our philosophers' words.
Yet Confucianists killed people guilty of incest
Precisely because incest weakens the race.
So today I raise on high this banner,
But who will know my bitter grief?
If I cry, "Emulate not your fathers,"
Men will swarm up from all sides to attack me.[81]

Despite his pen name, Chien Kung ("The swordsman"), Kao Hou feared attack with good reason. For he himself was attacking the most ancient and hallowed of China's moral laws, filial piety, the seeming soul of Chinese civilization from its beginnings, the first, or at least the second, commandment of the *Book of Odes,* and the supposed pillar of "Confucian" virtue, even before Confucious was born. In a charge that was hence for China the ultimate in anti-ancestral Darwinism, Kao Hou was accusing filial piety of being "counter-evolutionary."[82]

It seemed a devastating argument, although Kao Hou quite rightly recognized that it would not immediately win the day. It was indeed soon echoed, in one way or another, in a thousand essays, but for decades it seemed to many to need repeating. Lu Hsun, for one, was still making the same argument in the "May Fourth" year of 1919, a full fifteen years after Kao Hou's poem—which Lu Hsun must have read as a young man—had been published. In an essay as concerned as the poem had been with China's sons and fathers—"What is Required of Us as Fathers Today?"—Lu Hsun wrote:

> The [Confucian] saying "Changing not from one's father's ways throughout the three years' mourning may indeed be called filial" of course is a misguided statement. Indeed it is the root cause of infantile underdevelopment. If in ancient times single-celled organisms had respected such instructions, they would never have dared split in two and reproduce, and the world would never have seen mankind.[83]

And almost another decade later, in a speech at the Whampoa Military Academy (on April 8, 1927), Lu Hsun was *still* making the same argument (this time, exactly a quarter of a century after Liang Ch'i-ch'ao first called for a "literary revolution") in a discussion of "Literature for a Revolutionary Age":

> Biologists tell us, "Men and monkeys are in no way very different. Men and monkeys are cousins." But why did men become men, and monkeys remain monkeys? This was simply because monkeys were unwilling to change—they *liked* to walk on all fours. Perhaps there was once a monkey who stood up and tried to walk on two feet, but the other monkeys must have said, "Our ancestors have always crawled, you are not permitted to stand," and they must have bit him to death. They were not only unwilling to stand up, they were unwilling to learn how to speak—because they were conservative. But men were different. They finally did stand up, and spoke, and so finally were victorious.[85]

Lu Hsun should have known better. A full twenty years before he wrote that paragraph, he had written a long essay entitled "Man's History," at the time one of the most sophisticated accounts (far surpassing anything written by Yen Fu or Liang Ch'i-ch'ao) yet written in Chinese of human evolution and the evolution of evolutionary theory. He was one of the few men of his generation who should have known better, or who could have known better, than to suggest that monkeys had remained monkeys because they were "unwilling to change," that men had become men thanks to the happy "decision" of amoebae to split up.

Perhaps he did know better. His immediate purpose in the paragraph above was to make monkeys of the conservatives of his day, not to explain the intricacies of evolution. Yet, it is strange that Lu Hsun should have knowingly fostered a misconception about evolution, for his own belief in evolution was in deadly earnest. He spoke of himself as one who "believed only in evolution,"[86] and he seemed to look to evolution as the only possible "redeemer" of mankind, of which he at the time had a famously low opinion. He said, "I have believed all along in evolution, and have always felt that the future must be better than the past and that the young must prove better than their elders,"[87] and he said "Hope is in the future."[88] Therefore, although he could joke about evolution, he would not have wanted his listeners to misunderstand it, nor would he have assumed that they, especially at the Whampoa Military Academy, had a sound enough scientific understanding of evolution to be able to laugh at his analogy without being misled.

It seems more likely that under the license of his humor there still was a misconception (in addition to the almost universal misconception of evolution being progress), that even Lu Hsun did not yet see that "spirit," the revolutionary spirit he so hoped to inspire in Chinese patriots, had in no way yet been proven to be the motive force of evolution. At very least Lu Hsun perpetuated, wittingly or unwittingly, the misconception, so ubiquitous in the writings of Yen Fu and Liang Ch'i-ch'ao, that evolution had been willful, that man's existence was a triumph, that men became men because they wanted to. That had been Emerson's old delusion, that ". . . striving to be man the worm/Mounts through all the spires of form."[89] The corollary, of course, which was what counted to Lu Hsun, Yen Fu, and Liang Ch'i-ch'ao, was that in the future it would be those who willed to be fit who would be.

This was not, perhaps, a total misconception. Will cannot totally be ruled out as a motive force of evolution, "the will to live" being one of the mysterious "givens" behind the whole process, at very least the *sine qua non* of the struggle for existence. But even so, evolution was not the result of "the will to live"; it was the result of "favorable" mutations among "creatures" who willed to live, mutations that, as far as Darwin could tell, came not through will but by chance.

In modern times, at least one orthodontist has suggested that the human race is gradually losing its wisdom teeth.[90] But, if this is true, it is certainly not because generations of forward-minded dentists have been pulling them out (the inheritability of de-acquired characteristics), and it is even less because generations of wise parents have simply not willed their wisdom teeth to their children. It would be more accurate, therefore, if one rules out design, to say that man was the product of innumerable reproductive malfunctions than to say that he was the product of ever-ascending aspirations. Evolutionary offspring may indeed have been "unfilial," but not because they wanted to be.

Lu Hsun's argument, therefore, was based on a totally false analogy, for he was arguing that, as man *was* the product of unfilial conduct, progressive men *should* be unfilial, that, as man's very being was the product of evolutionary "revolution," revolution, against family, society, and state, must be a good thing—and always would be. Like Liang Ch'i-ch'ao he was trying to make "revolution" sound "natural":

> Actually, revolution is not at all unusual. Only thanks to revolution has society been able to get better and has mankind been able to advance. That it has been possible to move from the first worm to mankind and from barbarism to civilization is simply because there has not been a single moment without revolution.[91]

This simple but false analogy, this confusion of social revolution with biological evolution, this "naturalization" of revolution, and its consequent legitimization, represents one of the most radical changes of mind in the last two thousand years of Chinese thought, a change that has in physical fact revolutionized, deeply if not completely, the social and political realities of Chinese life. Of course no intellectual idea has been the one and only cause of any of China's revolutions. Legitimization of revolution through "naturalization" was in large part a rationalization. The very fact that men so different as Liang Ch'i-ch'ao, Lu Hsun, and Sun Yat-sen all saw in Darwin's theory the un-Darwinian equation of revolution with evolution suggests that there was "rebellion in their hearts" before there was revolution in their heads. But, if will preceded rationalization, it was, nevertheless, rationalized will which alone had the force to move intellectuals to move the masses. And the masses were moved. They did not spontaneously arise in any of modern China's revolutions; they were led into revolution, by intellectuals who *decided* to revolt.

All of China's revolutionaries have eventually claimed that revolution was "natural" as well as "right," that it was natural and *therefore* right, that it was (strange notion) a "natural duty." But the Communist Revolution has so eclipsed the others, the Republican and the Nationalist, that Mao Tse-tung and Karl Marx are now popularly awarded all the credit for the revolutionary act of legitimizing revolution, Mao Tse-tung, the great Sinifier of Marxism, having proclaimed that, "in the last analysis, all the truths of Marxism can be summed up in one sentence, 'To rebel is justified.'"[92]

Actually, Mao Tse-tung had justified rebellion long before he ever heard of Marx. He had justified his own rebellion against his father, against the blind obligations of filial piety, and he had justified his attempt to take part in the first great Revolution of 1911. Surely many ideas and emotions went into these early justifications, and even more into his eventual justification of Communist revolution; but, before he discovered Marx, by his own admission his head was filled with the Darwinian ideas of Liang Ch'i-ch'ao. He had memorized whole passages from the *Hsin min ts'ung pao,* he had "worshiped K'ang Yu-wei and Liang Ch'i-ch'ao,"[93] he had suggested in his first "muddled" expression of a political opinion that they be made, respectively, Sun Yat-sen's Premier and Minister of Foreign Affairs,[94] and he had taken Liang Ch'i-ch'ao as his literary model.[95]

It is no coincidence, then, that Liang Ch'i-ch'ao's Darwinian arguments for revolution can be seen in the writings of Mao Tse-tung, even in his relatively late theory of "continuous revolution."[96] "Continuous revolution" was deemed meet, right, and necessary, in the last analysis, because "mankind is continuously developing and the natural world is continously developing. They will never stop at a given level."[97] Lu Hsun had said the same thing, long before: "Revolution has no stopping place,"[98] just as "there has never been a moment without revolution." But the *locus classicus* of this idea was in Liang Ch'i-ch'ao's 1902 essay, "An Explanation of Revolution," and in another essay of the same year, entitled, with marvelous ambiguity, "The Theory of the Evolutionary Revolutionary, Kidd":[99] "Selection [lit. "elimination"] by man means revolution, and, as the external world is at no time unchanging, selection by man may at no time stop." Therefore, "Revolution is an inescapable law of the world of evolution"—inescapable, that is, for those who want to win out in the struggle for existence: "'Selection by man' means deeply to examine that in oneself which is unfit and change it to make it fit, so that one may survive."[100]

China's first, and most important, "scientific" justification of revolution, therefore, was "purely" Darwinian, except for the sorry fact that it was based on a mistranslation of one of Darwin's most important sentences. Just who was responsible remains unclear. Perhaps Yen Fu, perhaps a Japanese, perhaps Liang Ch'i-ch'ao himself, mistranslating the translation of a Japanese. At any rate, Liang Ch'i-ch'ao, in his article on Benjamin Kidd, claimed, as the ultimate backing for all his above reasoning, that "Mr. Darwin has said, every living thing, no matter of what kind, must frequently change its form and make it beneficial to itself, for only thus may it survive."[101] But what Darwin had actually said, in his introduction to *The Origin of Species,* was that "any being, if it vary, however slightly in any manner profitable to itself, under the complex and sometimes varying conditions of life, will have a better chance of surviving, and thus be *naturally selected* [Darwin's italics]."[102]

The great mistake was to take "if it vary" to mean "must change its form." Evolution as defined in *The Origin of Species* had contained no imperatives. No species described by Darwin had ever changed its form, "however slightly." No species had changed itself. No species, not even man, had adapted itself to its environment. Man might, as Darwin said, have shown "great powers of adapting his habits"[103] to his habitat, but what that really meant was that man had great powers of adapting his environment to himself. He could dig caves or build igloos. He could float fields in the waters or irrigate the dry lands. He could wear the skins of animals—but he could *not* re-grow a fur coat.

Granted, the proposed new science of eugenics, which Yen Fu had already sketched in *T'ien-yen lun,* might have seemed to hold out such a possibility. By repeated matings of the hairy, man might conceivably regain, for at least a segment of humanity, at least a semblance of the fur of his forefathers. But even that possibility could not make revolution "an inescapable law of evolution," for eugenics had not been the motive force of evolution. Not even man had selected himself, and surely no other creature had ever "examined that in itself which was unfit and changed it to make it fit, in order to survive."

Nonetheless, Liang Ch'i-ch'ao's argument became the evolutionary argument for revolution, an argument that seemed unassailably scientific, not only to Lu Hsun and Mao Tse-tung, but even to Hu Shih, Hu "the Fit," leader of the second generation of Chinese Darwinists, leader of the revolutionary "New Culture Movement," but also leader of the anti-Communist liberal intellectuals most vehemently opposed to the revolution of Mao Tse-tung. Even Hu Shih felt compelled to admit, "I do not condemn revolutions, because I believe that they are necessary steps in the process of evolution."[104] Indeed, he even stated in one essay that revolution —"forced evolution"—was superior to evolution proper, because revolution was "conscious," while evolution was not.[105]

Neither Hu Shih nor any of the others realized, as Wallace had, that this strange difference made almost all the difference in the world, that at very least it pulled the "perfectly natural" ground for revolution out from under it. There was something very strange about one "natural process," revolution, being superior to another natural process, evolution, of which it was a part and out of which it had come. Perhaps this was no stranger than the oak being "superior" to the acorn, but somehow, if revolution was "conscious" or "forced," the difference between it and evolution became something almost beyond analogy. If man was "forcing his evolution," "consciously evolving," or "shaping his destiny," odd concepts all, then something very different was happening from anything that had ever happened before in the evolution that had made him, in the evolution that was the origin of species. If revolutions were consciously forced, then they seemed to be matters of choice, and choice had never before played a part in evolution. Choice had not separated the sheep from the goats. Choice had not kept Lu Hsun's monkeys monkeys. If evolution was perfectly natural, then revolution was something else, and man had no right to say revolution was right because it was natural.

But Lu Hsun's conservative monkeys, or rather his revolting amoebae, have pulled us ahead of our story. The evolutionary argument against filial piety, which Kao Hou had first put into poetry, or at least into verse, did, indeed help the cause of political revolution. The Empress Dowager, mindful of Confucius's comment that "few are they who, respectfully obedient to their parents and elder brothers, yet like to rebel against their superiors,"[106] had every reason to read his poem, if she did, with displeasure. But Kao Hou's cry, "Emulate not your fathers," was of even greater help to Liang Ch'i-ch'ao's cause of cultural revolution. It was a cry that worked towards social revolution, intellectual revolution, philosophical revolution, religious revolution, revolution in the home, and revolution in the mind. It challenged the Chinese family system and Confucianism. It threatened to free Chinese both from their fathers and from the faith of their fathers.

Revolution and Evolution in the Home. The revolt against parental authority and the Chinese family system was a logical necessity, given Kao Hou's belief in the evolutionary benefits of filial impiety. If Darwin had proved that would-be "descendants" had to break from the ways of their parents and ancestors, then spirited youth, like Mao Tse-tung, had at last a scientific excuse, if not a duty, to follow their own inclinations. Mao Tse-tung's rebellion against his family is only one example, but a good one, because it included a rebellion against the marriage his family had arranged for him. He was among the first to take part in the "marriage revolution," so traumatic for his generation and for many that have followed it.

The "marriage revolution" was yet another revolution first justified in China by the "laws" of Charles Darwin. Kao Hou had radically redefined the term "good marriage" by declaring that would-be progenitors of the fit of the future had a Darwinian duty to "mix their blood" (echoes of I Nai). But an even earlier article in Liang Ch'i-ch'ao's *Ch'ing i pao,* "On Free Marriage for Men and Women," had issued a revolutionary appeal in Darwinian language for Western-style freedom of choice in marriage (and for the abolition of concubinage). The article's arguments were actually couched in a mishmash of the *T'ien-yen* vocabulary of Yen Fu and the Three-Age vocabulary of K'ang Yu-wei, but the point was simply that (somehow) marriage was the source of social development, morality, and progress, and that the Western way of marriage was the most modern, the most civilized, and the most progressive way that had yet *evolved.*[107]

The most interesting Darwinian marriage argument, however, was provided by Liang Ch'i-ch'ao himself. He protested that, although admittedly "more than one aspect of Chinese marriage customs should be reformed," the most pressing need was to outlaw early marriage.[108] For Darwin, he was convinced, had proved that "the forbidding of early marriage" was absolutely vital to the survival of the Chinese race.

Liang Ch'i-ch'ao had convinced himself that early marriage was a sign of cultural primitiveness: "The more barbaric a people, the earlier they marry. The more civilized a people, the later they marry." This was because of a law of evolution: "Advanced organisms require a long time to reach maturity." This law was obvious, claimed Liang Ch'i-ch'ao, if one compared human beings with birds and beasts, but, far more important, "It is true," he said, "even within the human race. Inferior peoples mature quickly.

Superior peoples mature late," and "the age at which they marry varies accordingly." Therefore it was common for Indians (Indian Indians) to bear children at age fifteen, while it was common for Europeans ("especially the Teutons") still to be unmarried at age thirty. As a *consequence,* it was the Indians who "age especially quickly," while the Europeans "remain strong, healthy, and vigorous even in old age."[109]

Chinese married later than Indians, but earlier than Europeans. Therefore, as "one need only look at marriage customs to determine the degree to which a people is civilized," China's degree of civilization was clearly worrisome, especially because relatively early marriage was not only a sign of relative cultural backwardness, but also a cause of relative racial unfitness. Worse yet, if early marriage went unchecked, it could lead only to increasing unfitness.

Liang Ch'i-ch'ao was convinced that early marriage was injurious both to the early married and to their early offspring. Adolescent newlyweds, immature of body and certainly of mind, were all too "often apt to drown themselves in the momentary delights of sexual desire, mindless of the [resulting] pain of a lifetime of ill health." Such youthful dissipation, he protested, "at worst will lead to early death, at best to decrepitude by middle age." Early marriage, therefore, amounted either to "suicide" or to an almost criminal "infringement of one's future freedom." Surely in either case, early marriage was a crime against one's *ch'ün,* for, "if everyone in a *ch'ün* should kill himself or infringe upon his freedom, would not the effect on the *ch'ün* be indescribable?" He found it possible, nonetheless, to describe the present condition of his *ch'ün:* "Our Chinese race has no spirit, no courage, no power of heroic determination. The reasons for this are many, but surely early marriage must share the blame."[110]

What really frightened Liang Ch'i-ch'ao was the thought that his race was not only weak but getting weaker, for he had convinced himself that early marriage led not only to degenerate parents but to the biological degeneration of their descendants. Dissipation weakened one's seed, and weakened seed weakened the breed. The proof lay in American and Japanese statistics showing that, "among the people of any country, the vast majority of those who die young and of the crippled, the sickly, the weak, the stupid, and the criminal, are children of parents who married early." Therefore, lamented Liang Ch'i-ch'ao, "if I, because of early marriage, produce a weak son, my son will be weaker than I, and if he, again because of early marriage, produces a weak grandson, then my grandson will be even weaker than my son, and if this goes on with each generation getting weaker and weaker . . . ," then one's family would fall, and one's country would perish.[111]

The fact that "Chinese are world famous for being good at reproduction" brought no comfort, for, after a description of "the ruthless destruction of natural selection" in the biological world, Liang Ch'i-ch'ao proved that numbers alone were no guarantee of survival.[112] Indeed, almost for the first time, he showed himself to be singularly unimpressed with the fact of China's millions:

> It is thanks to early marriage that in numbers the Chinese people are the greatest in the world. But it is also thanks to early marriage that in power they are the weakest. Does a race establish itself on this earth, then, through being many or through being strong? As the proverb puts it, "Several hundred vultures are no match for one eagle." Several tens of thousands of Englishmen have totally subjugated three hundred million Indians. The members of our race living abroad number at least several million, but they live as other men's beasts of burden. There are only ten thousand foreigners living in our country, and they seize our sovereignty. Can a race rely on its numbers?[113]

Frantic reproduction was no way to show one's love for one's fatherland. Grandsons of thirty-year-old grandfathers were better off not had.[114] The Spartans had understood this: "Spartan infants had to pass government inspection, and, if they were not physically or mentally qualified to be citizens of Sparta, they were thrown away without regret in cold alleyways. But Spartans were not brutal or unfeeling. They realized that only thus could their race prove fit to survive in this world."[115]

Thus a Spartan—or, in Chinese terms, a Legalist—element crept into Liang Ch'i-ch'ao's Social Darwinism. Suppressing his Confucian "compassionate heart" (*pu jen jen chih hsin*),[116] he let himself fall back on the classic proto-Social Darwinian justification for "inhumanity," the "necessities of survival," and for the first time sounded like a hardened eugenicist:

> From now on, if we do not eliminate our weak seed and propagate only the strong, there will be no way to continue the sacrifices to our ancestors, already so near the brink of extinction. This is exactly what the Sage [Mencius] was talking about of old, when he said, "Of the three unfilial acts, leaving no posterity is the worst."[117] It is crime enough for a clan or a family to leave no descendants, but what of a country? If we wish our country to have descendants, we must begin by forbidding early marriage.[118]

Happily, Liang Ch'i-ch'ao did not go on to spell out any plans for "eliminating weak seed" that were more Spartan than the forbidding of early marriage. Indeed, he did not even call on the imperial government (perhaps only because he realized the Empress Dowager would not listen to him) to move in on traditional family prerogatives and outlaw early marriage. He suggested, instead, that the Chinese people should not wait for their government to restrain them but should restrain themselves. They should at least take to heart the foresighted wisdom of the ancient Sages, as written down in the *Book of Rites:* "A man weds at thirty, a woman at twenty."[119]

The Chinese Communists, for a time at least, placed the proper marriage ages at thirty and twenty-six. Half a century after Liang Ch'i-ch'ao's suggestion, they have, without forbidding early marriage, successfully "discouraged" it. Their reasons have not been quite the same as Liang Ch'i-ch'ao's. Although they would probably be loath to admit it, their reasons have actually been more Malthusian than eugenic-Social Darwinian. For they have finally admitted to themselves, and the world, that China needs much more production than reproduction.

Nonetheless, the Chinese Communists would certainly have agreed with Liang Ch'i-ch'ao that early marriage was bad for the *ch'ün* and that the *ch'ün*'s interests must be put before either those of the family or of the individual, even though they would definitely not have quoted *Mencius*'s views on filial piety to prove it.

Liang Ch'i-ch'ao's use of Mencius was odd. Many proponents of the "marriage revolution" started their arguments with *attacks* on filial piety. Liang Ch'i-ch'ao ended his by embracing it. But in so doing he subtly redirected it. He actually wrenched it away from the family and aimed it at the people. He appropriated the power of an ancient virtue and put it to new use (a pre-Maoist application of Mao Tse-tung's maxim: "The past for the use of the present" [*ku wei chin yung*]).

Many have pointed out that the Chinese Communists, and the Chinese Nationalists before them, have both in fact consciously tried to redirect the cardinal Confucian virtues of loyalty and filial piety (*chung hsiao*) from emperor and family to chairman, party, and people, although, if they have, they have rarely admitted it, even to themselves. But, decades earlier, Liang Ch'i-ch'ao quite unashamedly sought to rechannel the old Way's power into a new revolutionary nationalism that both Nationalists and Communists would long for.

The seeming sincerity of Liang Ch'i-ch'ao's continued use of even a redefined Confucian virtue, however, suggests that he had not yet wholeheartedly joined in still another revolution his writings were at that very moment doing much to inspire—the revolution against Confucius himself.

CONFUCIUS'S STRUGGLE FOR SURVIVAL

The twentieth-century revolution against Confucius, which has proven of such incalculable consequence, politically, socially, intellectually, psychologically, and emotionally, for modern China, did not begin with the "New Culturists'" efforts in the late 1910s and the 1920s to "tear down Confucius and Sons." Far less did it begin either with the Chinese Communists' early efforts to de-sage the Sage by shelving him, labeled Chou, in a Levinsonesque museum,[120] or with the "Gang of Four's" endeavors to beat the devil out of him (and out of the country) by soundly thrashing his corpse, *ad absurdum et ad nauseam,* alongside that of his most unlikely "disciple," Lin Piao. The revolution against Confucius began in the very first years of this century in the journals of Liang Ch'i-ch'ao. And Charles Darwin was deeply if confusingly involved in it from the beginning.

Kao Hou's Darwinian poem was already in revolt against Confucius in 1904, not openly, but devastatingly nevertheless. For a derisive attack on filial piety could not help but be an attack on the Sage that had made it China's foremost virtue. So too, all of Liang Ch'i-ch'ao's arguments in favor of revolution (whatever he meant by the word) undermined the Sage's sageliness by undermining the second cardinal Confucian virtue of *chung,* loyalty to the king. If the two great virtues, *chung* and *hsiao,* could both be attacked as virtues "outmoded" by Darwinian social evolution, how could Confucius survive as a "model for all time" (*wan shih shih piao*)?

POOR PROTECTION. Ironically, the undermining of Confucius's position as a Sage for all Seasons had begun even earlier, in K'ang Yu-wei's movement to "protect Confucianism" (*pao chiao*). K'ang Yu-wei, at once frightened by Western missionary influence and impressed with "missionary spirit," wanted to protect Confucianism by transforming it into a "bona fide religion," in the Western sense and style. And he wanted to do this primarily for a reason his disciple, Liang Ch'i-ch'ao, at least, acknowledged as Darwinian—to make China fit for its struggle for existence. For K'ang Yu-wei, so said Liang Ch'i-ch'ao, had decided that the fittest nations were the most religious, that, "if the religion is strong, the country is strong."[121]

But the admission that Confucius's religion had first to be itself protected before it could possibly protect China was proof that Confucius was already in trouble. Some who detested Western missionaries seemed to fear that the (ignorant) masses might be fooled into abandoning Confucius for Christianity. K'ang Yu-wei probably shared those fears. But he also seems to have grasped intuitively that the greater danger lay in the possibility that Confucius might be abandoned not *for* anything, but simply because he seemed to have nothing to say to his countrymen in their "death struggle" with Western imperialism.

K'ang Yu-wei had convinced himself that Confucius did have something to say, but to prove it he felt impelled to launch a "Confucian reformation," which Liang Ch'i-ch'ao, who optimistically dubbed his master "Confucianism's Martin Luther"[122] described as a "religious revolution."[123] This "religious revolution," however, designed so hopefully to enhance the authority of Confucius, helped instead to de-sage him.

Two things went wrong. First, to rescue "the true Confucius," whose teachings could still save China, K'ang Yu-wei had first to destroy "the false Confucius," who just happened to be the Confucius the Chinese had revered for almost two thousand years. Chinese had been misled, claimed K'ang Yu-wei, by the forged Classics of the perfidious Liu Hsin, back in the first century A.D. But this was a risky allegation, for, if K'ang Yu-wei were to succeed in destroying the old, false Confucius (admitting that he had been bad for China) without firmly establishing the new, then Confucius would be in trouble indeed—and that, in part, is what happened. But there was a second problem. Even when K'ang Yu-wei did seem successful in establishing a new Confucius, the new Confucius proved self-defeating. For he seemed so much like a Westerner that he might just as well have been one.

It was Liang Ch'i-ch'ao whose help proved most harmful in this regard, for he translated K'ang Yu-wei's new Confucianism into newfangled Western terms. "What K'ang Yu-wei discovered," he wrote, "was the true good of Confucius's teaching: evolutionary-progressivism not conservatism, equalitarianism not autocracy."[124] Indeed, Liang Ch'i-ch'ao even claimed that K'ang Yu-wei's "Confucian" doctrine of the Three Ages, Confucius's newly found belief that "the present surpasses the past and the future will surpass the present," was actually "the very theory of progressive evolution the Westerners, Darwin and Spencer, have proclaimed."[125]

Perhaps Liang Ch'i-ch'ao thought such a revolution really would enhance the sageliness of the Sage. It did seem to, for at least a few of his readers. One concluded that Confucius's doctrine was therefore "the ancestor of the theory of evolution." Better yet, he proudly added, "Darwin realized this principle [evolution] after the ten thousand things were already complete, but Confucius discovered it *before* they were complete. Is it not amazing?"[126] Amazing it was, but, in the long run, no praise could have been more damaging. For to "preserve" Confucius as a proto-Darwinian was to preserve him in formaldehyde. If Confucius was a forerunner, if "the present surpassed the past," if Darwin and Spencer were of the present, and if they spoke, as they did, in more detail, then why not just respectfully shelve Confucius and follow Darwin instead?

It was Darwinism not Marxism, therefore, that first shelved Confucius, that first revoked his epithet, "a model for all time." K'ang Yu-wei's own proto-Darwinian theory of progress had helped, of course. For K'ang Yu-wei had reduced Confucius to the model of a man who believed in a progress that would leave him behind. But it was Darwinism that brought "scientific proof" that Confucius was right in that belief. Of course, the New Confucius did have the *Ta t'ung* vision, which might have brought new life to his epithet. K'ang Yu-wei had given him a new claim to fame as the provider of a blueprint for the future, but the Great-Unity blueprint was so sketchy (K'ang Yu-wei's own elaborately detailed blueprint was not yet published) and so obviously *only* for the future, and for the far distant future at that, that it still left Confucius with nothing to say to the present—as a model for *some* time, perhaps, but not for this. Here again, Darwin hurt the New Confucius. For he reinforced with "science" the stark conclusion that Chinese all too easily came to on their own, that the only "Great Unity" possible in their time of ruthless imperialism was the dread unity that would come with a white man's world conquest.

But Darwin dealt yet another blow to the New Confucius, when he attacked (or seemed to attack) "religion." For K'ang Yu-wei had just decided that the only way to "preserve" Confucius was to establish him as a "religion founder," a *chiao-chu*. Smarting under the taunts of Western missionaries that China had no religion, he insisted that it did, that Confucianism was a religion, that Confucius ranked with, and indeed above, Sakyamuni, Jesus, and Mohammed, that the Chinese, therefore, were not "to be classed with the barbarian peoples of Java, India, and Africa," but were as religious and hence as fit as any race in Europe.[127] At least, they *could* be as fit, if they would only realize that they had a religion, and rally around it.

K'ang Yu-wei never abandoned this belief. Indeed, he forlornly tried to propagate it to the end of his days, but he won few converts. Instead, he lost his best disciple, Liang Ch'i-ch'ao, who, as early as 1897, had secretly admitted to Yen Fu that he shared his doubts about the wisdom of "preserving the religion,"[128] and finally, in 1902, openly broke with K'ang Yu-wei over that very issue. In an emotional essay entitled "'Preserving the Religion' is not the way to Revere Confucius," he declared that he had changed his mind ("Does this mean that my thought has progressed or gone backward? I will let that be decided by the advanced or backward nature of the reader's own thought.")[129] He no longer believed that Confucius was a religion founder or that religion was good for China: "Confucius was a philosopher, a pragmatic statesman, an educator, but not a religious leader." And it was a good thing that he was not. For "Darwin's theory of evolution has arisen as . . . a rival power against religion"[130] and has proven that "religion is not an instrument of help to human progress."[131] Indeed, Darwin's theory, said Liang Ch'i-ch'ao, undoubtedly with immense psychological satisfaction, had proven that "what the Westerners call religion refers to nothing but superstition."[132]

The despised missionaries had scorned China as a backward nation with no religion. Now it was clear that they were the backward ones, benightedly clinging to the superstitions of the past. "Confucius," proclaimed Liang Ch'i-ch'ao, "never called himself the Son of God as Jesus did."[133] He eschewed all talk of gods. He taught only of man. He taught "how a man should be a man, how a people should be a people, how a nation should be a nation."[134] These were the questions that would be ever "more important to study, the more civilization advances." And these were the Confucian concerns that would keep Confucious alive. Confucianism would not perish; the Westerners' religions would: "For as truth becomes clear, superstitions disappear. They will not be countenanced by the civilization of the future. This is the law of evolution."[135]

So Darwin had solved the missionary problem. Evolution would take care of the missionaries. China need fear no "Christian conquest." The "backward" Chinese were really ahead, and always had been. Those who felt they must "protect the religion" could "go to bed with a high pillow and relax."[136] Darwin had solved the missionary problem—but he had also dashed K'ang Yu-wei's hopes for a Confucian church.

Now it may seem that Liang Ch'i-ch'ao, in rescuing Confucianism from the doomed ranks of the world's religions, had murdered his master's New Confucius only to bring back to life the old one. He might, indeed, have done so, if he had stopped with his anti-religion argument. But he did not. He reburied Confucius under mounds of damning praise.

First, he claimed that Confucianism would never perish, which led him to conclude that Confucianism therefore need not be protected. It could (safely) be left to the test of evolution. For "doctrines," he said, "are to protect people; they are not to be protected by people. According to the law of the survival of the fittest, if a doctrine is good, it will necessarily be victorious over foreign doctrines."[137] Confucianism could—and should—be left to the test of evolution. That was Yen Fu's private opinion in 1896. But there was something ominous in the way Liang Ch'i-ch'ao now broadcast it. "Fear not," he seemed to say, "truth will triumph." But why, then, did he add at the end, "I love Confucius, but I love truth more?"[138]

Second, Liang Ch'i-ch'ao redefined the importance of Confucius in a way that further broke Confucius's hold on the future and therefore left him even weaker in his now single-handed struggle for survival. "The reason Confucius is Confucius," wrote Liang Ch'i-ch'ao, "is that his thought was free." Freedom of thought was vitally important to Liang Ch'i-ch'ao. He had written, eleven lines earlier, that "there is more than one reason civilization advances, but the ultimate reason is freedom of thought."[139] But, if Confucious was for freedom of thought, and if China could only "make progress" if it had freedom of thought, would not Confucius himself have wished China to be free from *his* thought? Liang Ch'i-ch'ao made it perfectly clear that it was all too possible for "those who call themselves Confucians [to] go against Confucius's spirit in his name."[140] That had always been true, but never with more serious consequences:

> Perhaps those who speak of "protecting the religion" today have a slightly different idea in mind from people in the past. They want to increase the scope of Confucianism, and therefore they attach to it all the new theories of the modern age, saying, "Confucius already knew of this," and "Confucius once mentioned that." I respect their perseverance, but I regret that, in fact, they are slandering Confucius and obstructing the path of other people's freedom of thought. Confucius was born two thousand years ago. How is it to his detriment that he did not know all the theories and principles of two thousand years later? Socrates never set foot on a steamboat![141]

The "protectors of the faith" might well have countered that no one had attached as many new theories to Confucianism as Liang Ch'i-ch'ao himself. But, if Liang Ch'i-ch'ao realized that fact, he did not admit it. Proclaiming that "the spirit of Confucianism is freedom, not authoritarianism,"[142] he declared that "the greatest theme of this essay is that to protect the religion is to hinder free thought." But then again, he added ominously, "I love the ancients, but I love freedom even more."[143] Consciously or unconsciously, he was praising Confucius for his freedom, in order to free himself from Confucius.

The argument above all others that promised, or threatened, to cut Chinese free from Confucius, however, was Liang Ch'i-ch'ao's argument that morality itself was subject to the laws of evolution. If that was true, then truly the Old Confucius had no leg left to stand on (unless it was the leg so hopelessly bound by Yeh Te-hui and the Empress Dowager to the monarchy).[144] For it was as a moralist, not as a progress prophet, freedom advocate, or New World architect, that Confucius had been listened to for the last two thousand years. It was as a moralist that he was to have been "a model for all time."

But there *were* no models for all time, said Liang Ch'i-ch'ao. Only two weeks after he had published his essay attacking K'ang Yu-wei's call to "protect the religion," he published an essay, "On Public Morality," in which he finally let Darwin demolish Confucius's epithet. "Morality," he wrote, "is not something that, once established, never changes. It is not something that the ancients of thousands of years ago can set up in fixed form to encompass all the world for all time. . . . For morality comes half from nature, half from man, and it develops and progresses in accordance with the laws of evolution." For the good of our *ch'ün,* he went on, "we cannot hem ourselves in with a few words of the former kings and philosophers, so that we dare not advance. . . . The former philosophers are not *alive* today. How can they set a morality that is *fit* for today? If Confucius and Mencius were to arise again, they would have to revise their views."[145]

"You may threaten me with boiling or dismemberment," said Liang Ch'i-ch'ao, "but I will still dare state without fear that each and every one of the ethical principles of the Four Books and the Six Classics is not suitable for our use today."[146] "We must not be slaves of the ancients! . . . If Confucius had been a slave to Yao and Shun, there would have been no Confucius to survive through later generations. Do you fear my words? Just think of the destiny of the world. It advances ever upwards, and human knowledge gets ever clearer and more profound. Therefore the greatest of philosophers can speak only to the ills of one age; he can benefit only his own time. No one is great enough to set rules for the people of millions of years to come."[147] Not even Confucius.

So Confucius was put in his place. Liang Ch'i-ch'ao still praised him as the wisest man of his day—but *only* of his day. As far as any timeless wisdom was concerned, Liang Ch'i-ch'ao could only praise him for his alleged belief in progress and his commitment to freedom. Therefore, all Confucius seemed to have to say to the Chinese of Liang Ch'i-ch'ao's time was, "Farewell. Be free. Progress." Liang Ch'i-ch'ao had succeeded in doing to Confucius what Confucius had done to ghosts and spirits—he had "respectfully put him at a distance."[148]

But the revolution against Confucius did not stop there. "We must not be slaves of the ancients" might already seem a revolutionary cry, but it was not yet an accusation against the ancients, not a denunciation of them. Accusations and denunciations, however, were soon to come. Confucius was to be dragged from his sorry enough new position of very distant respect to be soundly thrashed, long before either "the new youth" or the Communists were yet on hand to get their hands on him.

Beating and Loss. It was probably Yen Fu who first set out on this darker side of the revolution against Confucius. He, at least, first gave Confucius a specific label that for the first time treatened to make him a scapegoat for all modern China's ills. Yen Fu wrote, in the preface of his 1904 translation of Edward Jenks's *History of Politics,* that "Confucius was the Sage of a patriarchal society. It is his moral pronouncements that have influenced our people the longest and penetrated their hearts and minds the most deeply."[149] This was no bland footnote. Nor was it praise. It was labeling with a vengeance, for *patriarchal* was becoming for Yen Fu as damning a word as *feudal* was to become to Chinese Marxists, and for the same reasons. *Patriarchal* was to make Confucius a "reactionary."

Yen Fu had just accepted Jenks's theory of social evolution. "The peoples of the world," he wrote, "if we examine the stages of their evolution, begin without exception in totemism, proceed to patriarchy, and come to completion in the nation-state. . . . The certainty of this sequence is like that of the four seasons, or like that of human life, as it passes from infancy, youth, and adulthood to old age. The stages may be long or short, but they can never be confused."[150]

Evolution's stages could *never* be confused. Therefore China's evolution was the same as Europe's. There was only one Way. But what, then, had gone wrong? Yen Fu looked at the militant, upstart nation-states of Europe and then at China, the oldest civilization in the world, and cried out in anguish: "We are still only a patriarchal people! That stage of evolution has dragged on and on and has never left us. It has been in our land for over four thousand years. Alas![151] . . . For over four thousand years we have been stuck in this cycle of order and disorder without ever progressing a single step."[152]

Yen Fu ultimately blamed this tragedy on "the Sages."[153] He admitted, of course, that the phenomenal early "success" of the Chinese people as a whole was partly to blame. China had been so brilliant in her adolescence, and had so overawed her pathetic neighbors, that she had culturally "enthralled" them, thus depriving herself of that challenge of competition so necessary for continued growth. But China's Sages, in their *seeming* brilliance, had enthralled the Chinese themselves. Too soon they had sold out to the ideal of stability and peace, thus killing in their countrymen that ultimate dynamic of successful evolution, the will to struggle. Yen Fu had so blamed "the Sages" in his very first essays. Now, he blamed Confucius, naming him as the "patriarchal Sage" most responsible for keeping China "patriarchal."

This argument deserves recognition as one of the most revolutionary in all modern Chinese intellectual history. Yen Fu had seemed revolutionary enough to Liang Ch'i-ch'ao back in 1896, when to Liang Ch'i-ch'ao's amazement he had "dared to speak" of his belief that Confucianism "[could] not be protected and need not be protected."[154] But *this* argument, of 1904, was *really* revolutionary. It was "daring to speak" with an intensity that might have warmed the heart of Mao Tse-tung himself. Indeed, it almost undoubtedly did, if not in 1904, then only a few years later when Mao Tse-tung was old enough to understand Yen Fu. For Yen Fu's argument was no longer just an argument that undermined the authority of Confucius. It attacked that authority. It condemned the official faith of China as unfit for modern times. It called it the cause of China's backwardness. It virtually called it China's sorrow. Yen Fu's argument that Confucianism had retarded China's natural evolution led a host of Chinese intellectuals for the first time into open revolt against Confucius. His argument, more than any other, led disgruntled patriots into the first namable, datable Anti-Confucian Movement, the first movement to *p'ai K'ung*, to "get rid of Confucius" forever.

In December 1904, late in the year in which Yen Fu's translation of *A History of Politics* was published, Liang Ch'i-ch'ao wrote a parenthetical "note" (of 1800 characters!) which gave testimony that this first *p'ai K'ung* Movement did indeed exist. The odd thing was, however, that he gave testimony to its existence by *decrying* its existence: "The new scholars of our modern age now flaunt at every chance their ability to condemn Confucius. . . . but I ask you, is this daily pommeling of Confucius really good for the future of our intellectual world?"[155] Liang Ch'i-ch'ao was beginning to hold back. Something very significant was beginning to happen. One can see in this one footnote that he had suddenly realized his own words might incite people to "go too far." He admitted that it would be "hard to avoid going too far in our criticisms of a great system that has controlled people's thought for two thousand years, when the doors are so suddenly thrown open to free thought," but the prospect frightened him. "Inflammatory language" in general began to frighten him:

> The American president, Roosevelt, once said in a speech, that journalists who make inflammatory statements win general acclaim, even though their words do their country no good, while those who make honest and loyal statements win general disdain, even though their words actually benefit the country in the end. I think that this Anti-Confucian talk, and all talk like it, is inflammatory. I admit that, in the past, I myself have liked to use such language, but, hereafter, I would rather suffer the majority's disdain than win acclaim that will do us no good.[156]

How prophetic were his fears that inflammatory language would win out and that he himself would come to know disdain! In sounding the trumpet of restraint, he lost a segment of China's youth—a segment that would grow. He also admitted that his own battle was now clearly on two fronts, one against the old forces of reaction, the other against new forces of "over-reaction." It was a hard battle to fight.

He sounded the trumpet of restraint and sounded, for the first time in his life, like an older man (he was, it is true, one whole year over thirty). He began to criticize the younger generation, decrying the existence in its ranks of Anti-Confucian "fanatical youths," most of whom had probably hitherto worshiped him. Very obviously such youths disturbed him, but he preferred to dismiss their "wild and irresponsible curses and insults" as words "of no harm to the sun or the moon," and therefore "not worth talking about."[157]

If he refused to address himself to "fanatical youths," however, he did feel obliged to answer the "one or two leading scholars who insist on challenging Confucius." Unfortunately, he did not name them, and we cannot, therefore, be certain that he meant to criticize Yen Fu himself, but certainly the first "challenge" that he mentioned was Yen Fu's:

> [These scholars], being dissatisfied with the old ethics, say that Confucianism made the family-clan its basic unit, and that therefore it was Confucius who so bound our country to a patriarchal form of society that it could not advance to a national-democratic one. They say that Confucius granted authority to the monarchs, and that therefore it was Confucius who served the people-ravagers of two thousand years as a protective charm.[158]

Reading only that far, one would conclude that Liang Ch'i-ch'ao disagreed with the above thesis, but his very next sentence read, "Granted, this is so!" Suddenly it became clear that his only objection was that these "one or two scholars" stressed only one side of Confucianism, ignoring most unfairly the other side, the lost (and alleged), progressive, *Ta t'ung* side. Liang Ch'i-ch'ao lamented that Yen Fu's argument was leading "fanatical youths" to lose all respect for Confucius, but he admitted Yen Fu's argument!

His admission proved much more powerful than his lament. This was equally true of his admission that he had "cast aside the superstitious belief in protecting the religion," which followed his lament that the young did not appreciate K'ang Yu-wei.[159] Liang Ch'i-ch'ao did appreciate K'ang yu-wei, and he still revered Confucius. He also felt guilty about hurting them both. But his late defense did little good to either.

"It is as clear as fire in a cave," he exclaimed in exasperation, "that hereafter the thought of our people can never again be tied down by two-thousand-year-old books. Therefore Confucius's doctrines, no matter what, can be no obstruction to our future progress. How, then, can those who attack him feel that they *have* to?"[160]

What a defense this was: "Confucius cannot hurt us anymore!" From the silence that followed, even a sympathetic reader might conclude, "But he cannot help us either," and there was also the unmistakable implication that Confucius *had* hurt China for the inevitably repeated "last two thousand years." There was, it is true, one hint of a stronger defense of Confucius in his pregnant footnote, but, strangely enough, there was *only* a hint. Liang Ch'i-ch'ao suggested that, if Confucius were totally repudiated, Chinese morality might totally collapse; "Alas, would this not leave our land open to 'floods and savage beasts'? [And to the law of the jungle?] Already we move in that direction. Can the good [*jen jen chün-tzu*] not be afraid?"[161] It is interesting that Liang Ch'i-ch'ao, who was about to lead a search for a "new morality," should nonetheless cling to something in Confucian morality, though the fact that he did not explain himself probably reflects his deep uncertainty about just how much could be clung to. It is also interesting, however, to see that the great "Social Darwinist" Liang Ch'i-ch'ao showed a moral concern that totally rejected the view that the "survival of the fittest" need be our only law.

As far as poor Confucius's struggle for survival was concerned, however, Liang Ch'i-ch'ao continued in other articles to make things difficult for him. He himself could not bear to curse Confucius, but he cursed Confucians ("fake" Confucians) with such inflammatory language that their master's "face" was almost impossible to preserve. In an essay, "On Progress," for example, he wrote, "I dare not blame Confucianism, but I must hate and despise those who have distorted Confucianism, who have used Confucianism, who have cheated Confucianism, to rob themselves and our people."[162] What they had robbed them of was not so much livelihood, happiness, or justice, as "progress." Liang Ch'i-ch'ao's basic argument was indeed Yen Fu's: "From Ch'in and Han times on, Confucianism unified [our thought]. Now Confucianism, of course, was good. But, even so, there is nothing more harmful to progressive evolution than to force the thinking of everyone in a country to follow one path."[163]

His attack on Confucian ethics or, as he insisted, on the tragic, but historical, "distortion" of Confucian ethics, was also related to China's alleged failure to evolve—because "we Chinese have no aggressive daring in our natures!" This was partially the fault of the Taoists (another of Yen Fu's old arguments), but it was also the fault of self-styled Confucianists who had ruinously indoctrinated generations of Chinese with a one-sided and wrong-sided version of Confucianism: "Even the so-called followers of Confucius have more often than not seized upon isolated phrases and ignored his greater meaning." Going wild with Confucian fragments and with the new word, *ism*, Liang Ch'i-ch'ao explained their tragic error: "They seized his 'restraint'-ism and ignored his 'wild'-ism. They seized his 'don't'-ism and ignored his 'do'-ism. They seized his 'female'-ism and ignored his 'male'-ism, and they seized his 'fatalism' and ignored his 'dynamism'." They quoted only such lines as the following: "'Avoid many words, for many words mean many troubles. Avoid many affairs, for many affairs mean many failures.' 'Enter not a dangerous state; stay not in a disordered one.' 'A filial son will not ascend the heights nor stand too near the depths.'" Confucians preached such maxims, but "when did Confucius himself ever take such maxims as sufficient rules for the world?"[164]

The awful truth, moaned Liang Ch'i-ch'ao, was this: "The 'Confucians' . . . paraded a Taoist nag under a Confucian hide, and turned a Confucian land into a Taoist camp. Because of this, our aggressive and daring spirit was utterly extinguished. Just look at the biographies in our Seventeen Histories. Hunt for men like Columbus or Livingstone. Are there any? No! Hunt for men like Cromwell or Washington. Are there any? No!"[165]

Thus did Liang Ch'i-ch'ao decry his Confucian heritage. He still defended Confucius. To the end he insisted on asking, "How is it Confucius's fault that those who have called themselves Confucians have opposed his spirit?"[166] But fewer and fewer of his readers seemed to appreciate that distinction. It was easier simply to say, as did one, Ma Hui, "Alas, the day Han Wu Ti exalted the study of Confucius was the day China turned from progress to regress."[167] Liang Ch'i-ch'ao might plead with China's youth to leave the Sage alone, but when he and Yen Fu had decided to abandon the effort to *pao chiao* (preserve Confucianism), he had openly agreed to stand back and abandon Confucianism *and* Confucius to trial by evolution. How then, after all his arguments, could he be surprised that "fanatical youths" should decide that China, to progress, must first *p'ai K'ung*?

"*P'ai K'ung,*" "*P'i K'ung,*" the cry was in the air. The revolutionary Liang Ch'i-ch'ao might have his second thoughts about revolution, but the revolution against Confucius had begun.

But what did it matter?

It mattered more than anyone can say, and more than Westerners, perhaps, can ever appreciate, for the revolution against Confucius really did lead most Chinese to lose the faith of their fathers. Westerners en masse have not been forced from their faith since the Christian conquest of Europe. They have known what it is like to lose the faith of their fathers as individuals, but never as a people.

The Darwinian revolution was supposed to have led Westerners to such a fate, of course, but, when all was said and done, it never did. The great challenge to Christianity proved a false alarm. The Darwinian deathblow to the Bible never struck home, for Jews *or* Christians. Darwinism revolutionized the way that most people would thereafter read the Bible, but it never caused Western civilization to throw the book away. After the initial Darwinian shock, most would-be believers happily found God as unprovable and as undisprovable as ever, and most concluded that they could have their faith and Darwin too. Darwin did, of course, shatter the faith of many. He shattered his own, almost (although he never shattered his wife's). But these "shatterings" were rarely traumatic. Westerners who lost their faith because of Darwin did not go on to lose their faith in themselves, in their people, in their civilization. On the contrary, proud Englishmen, Frenchmen, Germans, and Americans could, thanks to Darwin (or so they thought), have even greater faith in their "races," flags, cultures, and "systems." Thus, even if they did lose their fathers' faith, there was never the slightest danger of their having to accept some foreigner's faith in its stead. Darwinism itself was perfectly Western. So, for that matter, was "atheism." There was a host of ready-made, local faiths that one could turn to if one had to. And, in any case, there was the same old secular security and good cheer there had always been.

In China, however, things were very different. That Darwinism came as a far greater shock to Christian Europe than to Confucian China was perfectly true—in the beginning. Except for Yeh Te-hui's objection to the implication that human nature was not yet perfect, few Chinese saw any conflict between the biological theory of evolution and the Confucian Way, which did not seem perched on a Book of Genesis. But why, then, did Judaism and Christianity seem able to weather the Darwinian shock, while Confucianism did not (pace, Confucianists—Confucianism, to be sure, is not dead, but neither is it well)? The trouble lay not at the heart of Confucianism but with China's political weakness, the causes of which were so multifarious and coincidental that only with the grossest injustice could full blame be laid at Confucius's door.

Nonetheless, that is where the blame was laid. China was weak and Europe was strong. China was "backward" and Europe was "advanced." Europe had "evolved" and China had not. Once Chinese believed that, and once, rightly or wrongly, they believed that it was Confucianism that had kept them from evolving, Confucianism was doomed. Darwin doomed Confucius, not by revealing "the immense journey"[169] from amoeba to man, but by making it appear that, in the world of men, the Way of Confucius had not kept China fit.

The anguish that accompanied this conclusion, however, came from the fact that Darwin had destroyed the Confucian Way just when China most desperately *needed* a Way to survive in a world that Darwin, again, said was *naturally* an arena for ruthless imperialism. It was the combination of Darwinian theory and imperialist fact that was so devastating. For, when the Chinese lost their Way, they were compassless on the stormiest seas they had ever known.

That is why the revolution against Confucius was not carried out with rejoicing. Chinese were indeed slowly liberated from their "great tradition." They were liberated from the Three Bonds. But they were also cast adrift. Liang Ch'i-ch'ao had written so often of the glories of free thought, and yet, when he had it, when he broke free from the confines of Confucian instructions, he found himself so free it hurt. What to believe in? Which Way to follow? "Liberated" individuals had to stand on their own, or search for new support. They had to find their own Way or find someone else's.

They did have freedom of thought. Chinese in the first half of this century had greater freedom of thought than they had had at any time since the classical period itself. They did not necessarily have any parallel freedom of expression. Only those in power could grant that, and they never did, and still have not. But the horrible history of Manchu, warlord, Kuomintang, and Communist suppression of dissent should not fool us into forgetting the fact that for fifty years, in between the first attacks on Confucian orthodoxy and the imposition of the neo-orthodoxy of the Thought of Mao Tse-tung, Chinese were free to think for themselves. Indeed they almost *had* to think for themselves, for they had no orthodoxy. They *had* to come to new faiths, for men like Liang Ch'i-ch'ao had made it impossible, for all but the most strongly, stubbornly, or eccentrically independent, to maintain belief in the old ones. Chinese were forced into freedom—and forced freedom hurt.

It might not have, if China had been secure (though "forced freedom" would then not have come), but China was not secure, and Chinese intellectuals were not secure, and very few, therefore, found it possible to sit back over their tea and enjoy fresh, new vistas of the universe, thankful beyond words that traditional scales had been struck from their eyes. Some, of course, found moments of excitement in their new freedom. Some, perhaps even knew moments of great joy—but most seem to have known many more moments of worry, uncertainty, and anger.

It was hard to stand free of tradition, and very hard to stand alone. Only a handful, therefore, managed to create their own new

Ways. Most began frantically searching for a ready-made Way that was already marked. But the only marked Ways that seemed "viable" were Western! That is what really hurt—more than Westerners can ever know. Chinese felt forced to look West for a Way. This was the ultimate ignominy, the ultimate surrender, short of physical surrender, to the hated intruders who had shattered China's peace. This was the seeming end of that gloomy progression that had begun after the Opium War, when Lin Tse-hsu privately recognized a need for Western weapons. This went even further than the Reform Movement's cry for Western institutions. For the revolution against Confucius now repudiated the Reformers' cry that China needed Western institutions but a Chinese *chiao*. Now it seemed that China, to survive, must have a Western *chiao*, a Western doctrine, and Chinese, to live, must find a Western Way of life. If the *Tao* itself was in the West, then the price of free thought was one's civilization.

That was the first reason why the revolution against Confucius brought little rejoicing. But there was a second. The revolution against Confucius launched a true "Hundred Flowers" period. As individuals lost faith in Chinese Ways, they became converted to a plethora of different Western ones, and, in time, China came to have not only Republicans and Constitutional Monarchists, but Anarchists, Marxists, Leninists, Communists, Pragmatists, Romanticists, and Platonists, Kropotkinites, Ibsenites, Deweyites, and Russellites, and Christians, Kantians, Hegelians, Bergsonians, even Nietzscheans. Almost every conceivable kind of Western philosophy, personal and political, was seized upon by someone, and a "hundred schools" did indeed "contend."

But Mao Tse-tung was not the first Chinese to be unhappy with a Hundred Flowers. Almost no one in China rejoiced in the new diversity of Chinese thought. There seemed to be a ground swell of opinion, long before Mao Tse-tung was ever heard of, that freedom of thought was no end to be welcomed in itself. It was vital to be free from the shackles of the past, but, if "pure freedom" led only to diverging paths, then it was something China could ill afford. What China needed, all seemed to agree, was a new *unity* of thought, and contention was only good if it led to such unity, and if it made clear *the* fittest Way.

Thus, although individuals felt forced to seek for new personal faith, they at the same time longed to find *one* faith that was good for all China, not *a* new way, but *the* new Way. A common longing for a Grand Unity of thought reflected a Chinese "prejudice" that went all the way back to classical times, back, ironically, to the Confucian tradition itself, back at least to the thought of Hsun Tzu, who lived in the first "Hundred Flowers" period and hated it, who firmly believed that "the world has not two Ways."[170]

The pervasiveness of this conviction proved that there was at least on this one issue a unity of thought even among opponents in the classical age. Mo Tzu, the first great anti-Confucian, had been equally convinced that diversity of thought was no blessing:

> In ancient times, when men first lived, before there was yet any government or law, it is said that men had different views of what was right. One man had one view. Two men had two views. Ten men had ten views. There were as many views as there were people. For each thought his own view correct and the views of other men false. So each condemned the other. Even within the family, fathers and sons, elder brothers and younger brothers, all came to hate each other, and so broke apart, unable to live in concord. And the people at large took up fire and water and poison to do each other injury. . . . The world was as anarchic as the world of birds and beasts.[171]

So freedom of thought led to the law of the jungle. Perhaps Chinese Darwinists should have said, "Fine!" Perhaps they should have said, "But of course! Darwin has proved that the worlds of man, bird, and beast are all one." But they did not. "Liberated" by Darwin from all classical thought, they nevertheless, in their hearts, still hated intellectual anarchy (or freedom) as much as Mo Tzu did. "Let a hundred schools contend" should have been a perfect slogan for Chinese Social Darwinists. Liang Ch'i-ch'ao had decried Confucian unity, had praised the classical age for its freedom, had argued that contention was the only way to progress. But even he, when the chips were down, was happier with a hundred schools in theory than in practice. For even he was afraid that a hundred schools would pull the *ch'ün* to pieces.

Chinese Darwinists had come upon a true contradiction in their Darwinism. In their past, Darwin told them, unity had aborted their evolution. But in their present, Darwin said, unity was strength. What was the explanation? There was a need for a solution that almost looked forward to the theory of "the unity of opposites." But, as it was, Chinese Darwinists broke out of the unity of the past unhappy with disunity as a way of life. And so they soon came back to the belief that the Darwinian truth for *their* day was still "stop struggle within for the struggle without."

The search for new faith, therefore, became a grim search for a new unifying faith. Chinese looked West for a faith that could unite them against the West. They looked West for an "ism" to replace their *chiao.* That is why Hu Shih's famous plea, in 1919, "Raise more problems and talk less of 'isms'"[172] fell on deaf ears. For Chinese wanted an "ism." They were not comfortable each on his own. They wanted to *ch'ün.* They wanted a Way. They wanted a Sage. They wanted a book to replace the *Analects.* They wanted a compass, a polestar, a helmsman.

The revolution against Confucius in both political and personal philosophy had created a vacuum, a vacuum abhorent to "Chinese nature," at least to the nature of Chinese who thought themselves struggling to survive. It was a vacuum that had to be filled. But it took religious wars—the worst kind—to fill it. For, as men found new faith, the words of Mo Tzu proved hauntingly prophetic: "Each thought his own view correct and the views of other men false. So each condemned the other—and took up fire and water and poison to do each other injury." This was the irony that proves that there is something worse at work in the world than "nature red in tooth in claw." For the intensely shared longing for unity itself helped lead to such tragic and terrible disunity.

THE NEW MORALITY AND ANTI-IMPERIALISM

Liang Ch'i-ch'ao did more than any other individual to create the philosophical vacuum that invited such disunity. But he was also the first to rush to fill it, although he did not succeed. For years, in his journal, *A New People,* he led the search for a new faith, but he never really found one. He offered a thousand and one ways to make his people new, but he never found an "ismic" answer to China's problems, neither in the West nor anywhere else. Nor did he ever succeed in making his own eclectic thought into an "ism" that might light China's way. "The Thought of Liang Ch'i-ch'ao" was not destined, or fit enough, to be a rallying cry for a desperate nation.

And yet "the Thought of Liang Ch'i-ch'ao," more than the thought of any other man, did help prepare the way for "the Thought of Mao Tse-tung," for the "ism" that *would* fill China's philosophical vacuum. "The Thought of Liang Ch'i-ch'ao" prepared the way in many ways, but nowhere more clearly than in calling for a "new morality."

Liang Ch'i-ch'ao insisted that the search for a new *tao-te* (morality) must lie at the very heart of any search for a new *Tao.* This insistence, that China's Way must be a moral way, that morality was the only way to national survival and salvation, was quintessentially Confucian. But, even so, a campaign in China for a new morality could only seem anti-Confucian. Liang Ch'i-ch'ao knew perfectly well that his cry for a new morality would meet with virulent Confucian opposition. "I know," he said, "that talk of a 'morality revolution' will be decried by the entire nation." But he was willing to stand against the entire nation. As he put it, a bit dramatically, "I hate the fact that my talent is insufficient, but nonetheless, even though I must challenge to mortal combat the whole crowd of my generation, I will not fear such a task nor refuse it. Are there any in our day who with truly fervent hearts love our people, love our country, and love the truth? I will gladly hold the whip for them and begin a study of this problem."[173]

The problem was to find a *fit* morality quick. However reluctant Liang Ch'i-ch'ao might be to curse Confucius, he was fully convinced that he must launch a search for new morality and start a morality revolution, because the law of the survival of the fittest demanded it—Darwin demanded it.

DARWIN FOR MORALITY. It may seem odd that Darwin should be hailed in China as a champion of morality, when he had been so often reviled in the West as a champion of amorality. Droves of Western moralists had frantically warned that the "Darwinian revolution" would be the death of morality. "Evolution is force," cried one, "the struggle for existence is force, natural selection is force. . . . What will become of the brotherhood of man and of the very idea of humanity?"[174] What would happen, asked another, if "our reading youth entering upon life . . . are told in scientific lectures and journals that the old sanctions of morality are all undermined?"[175] And how could morality not be undermined, if survival was to be the only law? There was in the West, as Huxley claimed, "a general consensus that the ape and tiger methods of the struggle for existence are not reconcilable with sound ethical principles."[176] Most Western moralists took their stand with Mencius (although most had never heard of him): "Life I desire. Righteousness I also desire. But if I cannot have both, I will give up life and take righteousness."[177] What could be more absurd if "the first and great commandment" was to "survive"? If survival *was* Darwin's first commandment, how could he be anything but the arch-enemy of ethics?

He could very easily be something other than the arch-enemy of ethics. He could be dubbed the most High (Scientific) Defender of Ethics. All one had to do was declare ethics necessary for survival. That is what Kropotkin did. And that is what Liang Ch'i-ch'ao did, before he ever heard of Kropotkin.

If men were to be faced with Mencian-like dilemmas, it still made no sense to say that ethics were essential to the survival of the individual. For, even though an ethical world should, of course, be a "safer place to live in," until the natural world should rid itself of "floods and wild beasts"[178] there would still be Mencian dilemmas in which the ethical way would lead to self-sacrifice. But if one thought only of the "tribe," or the "species," or the "race," and if ethics meant individual willingness to sacrifice oneself for the good of one's tribe, species, or race, then, in a non-ethical world of warring *ch'ün,* ethics for the *ch'ün*'s members could indeed seem essential to the *ch'ün*'s survival. This was Kropotkin's argument, and Liang Ch'i-ch'ao's argument; and it was a Darwinian argument.

The morality that mattered to Liang Ch'i-ch'ao was *kung te,* "*public* morality," by which he did not mean a morality to be followed by the nation in its relations with other nations, but a morality to be followed by citizens for the good of their nation, a morality of the people *for the People. Kung te,* best seen, so he said, in "the new ethics" of the West, stressed "the relations between an individual and his group [*t'uan-t'i*]." *Ssu te* (private morality), "the old ethics," perfected in ancient China, stressed only "the relations between one individual and another."[179] *Kung te* was most necessary for the group's survival. But it was *kung te* that China did not have:

> One of the things our people most sorely lack is *kung te.* What is *kung te*? It is that which makes a *ch'ün* a *ch'ün,* and a nation a nation. It is the morality on which *ch'ün* and nation stand. Man is an animal good at grouping (this is a saying of the Western scholar Aristotle). If men do not group, how are they better than birds and beasts? But success in grouping does not befall those who simply raise empty shouts of "Group! Group!" There must be something that runs through a group and ties it together before the actuality of a group can appear. That thing I call *kung te.*[180]

Kung te was the cement that could make concrete out of China's "plate of loose sand."[181] *Kung te* was the *ch'ün*-ing agent that alone could make possible the solidarity Liang Ch'i-ch'ao had long since identified as the *sine qua non* for any people's evolutionary survival. *Kung te,* therefore, whatever its specific injunctions, had somehow, by definition, to be a unifying morality of patriotism and public-spiritedness or, to use Kropotkin's terms, just waiting to be discovered, a morality of "sociability" and "mutual aid."

Liang Ch'i-ch'ao's argument that *kung te* was a Darwinian necessity was virtually identical with Kropotkin's argument that "mutual aid" was a Darwinian necessity. "Sociability is as much a law of nature as mutual struggle," wrote Kropotkin. "If we . . . ask nature: 'Who are the fittest: those who are continually at war with each other, or those who support one another?' we at once see that those animals which acquire habits of mutual aid are undoubtedly the fittest. They have more chances to survive. . . . [182] 'Therefore combine . . . practice mutual aid!' . . . . That is what Nature teaches us."[183]

But Liang Ch'i-ch'ao did not have to learn that lesson from nature or from Kropotkin. Nor did he really have to learn it from Darwin either—and certainly not from Aristotle. For he knew perfectly well that Hsun Tzu had taught that very lesson without any Western help whatsoever. Here was a case where Chinese seemed indeed to have a right to say of a modern Western doctrine that China *ku i yu chih* (already had it long ago), for Hsun Tzu, quite long ago, had said not only that it was man's ability to *ch'ün* that made him win out over other creatures, but that it was his sense of *i* (moral duty) that made him "good at *ch'ün*-ing" (as "Aristotle" put it—in Hsun-tzian Chinese!).[184] Liang Ch'i-ch'ao had openly—and proudly—acknowledged as much in his earliest essays, at least he had proudly acknowledged the first part of Hsun Tzu's argument, ignoring at the time Hsun Tzu's stress on morality. Now, however, he did not acknowledge Hsun Tzu, even though his own decision that morality was "what made a *ch'ün* a *ch'ün*" had led him back to the crux of Hsun Tzu's argument, and even though his article not only borrowed Hsun Tzu's language but even echoed Hsun Tzu's contention that it was morality that made men different from birds and beasts.[185]

Liang Ch'i-ch'ao did not cite Hsun Tzu, because he was losing his interest in the *ku i yu chih* argument. Although he was still to fall back on it himself on occasion, he knew that, all too often, it was a reactionary argument, employed to resist the introduction of Western doctrines. He also refused to acknowledge Hsun Tzu, however, because China's weakness and Darwin's theory had destroyed his faith in ancient authority—especially ancient Chinese authority. For almost any contention, Western authority was now better than Chinese authority, and modern authority better than ancient authority. Therefore Aristotle was better than Hsun Tzu, and Darwin better than Aristotle. If the Western scientist, Darwin, said China needed morality, then China needed morality.

There was, of course, one more reason why Liang Ch'i-ch'ao failed to mention Hsun Tzu. Hsun Tzu insisted that a *ch'ün* must have morality, and the morality, *i,* that he insisted upon was without question a form of *kung te,* of "public morality." But it was not the form of *kung te* Liang Ch'i-ch'ao had in mind, for in his eyes it was an outmoded form of *kung te,* a form no longer fit.

This was Liang Ch'i-ch'ao's second point, and his second Darwinian reason for insisting on the necessity of a "morality revolution." China, to survive, needed not only morality but a "new morality":

> Our modern gentlemen who talk of reforms dare talk of all sorts of new things, but they do not dare talk of a new morality. This is because our scholars have not rid themselves of their slavish natures, nor has their love for *ch'ün,* country, and truth become sincere. People who think that morality, like the sun and the moon that course in the heavens and like the great rivers that flow on the earth, has had nothing added to it nor taken away since before the beginning of time, and people who think that our former saints and Sages have completely revealed the mysteries of morality to guide all who follow, such people simply do not realize that morality comes half from nature and half from man and that it develops and progresses according to the great law of evolution. . . . They have no idea that the law of the survival of the fittest cannot be escaped.[186]

Old moralities themselves could not survive . . . how much less could peoples who clung to old moralities? Darwin's demand for morality, therefore, was indeed a demand for a morality revolution (even for a continuous morality revolution) and not just a demand for a moral revival. The ways of the ancestors could not preserve China, not even their moral ways. "Moral rearmament" would require *new* weapons.

So Darwin "demanded" a "new morality." But did he provide

one? *Was* there any "ethic implicit in Social Darwinism," (or in Darwinism proper— there as yet being no distinction in China between the two)? Was there in "Darwinism" an answer, the "ismic answer" that Chinese were beginning to long for? Was there an "evolutionary ethic"[187]—already prescribed—that could be "just the thing" for China?

Clearly there was not. Darwinism was indeed the first Western "ism" widely to be accepted by Chinese, but it was not accepted in an "ismic" manner, as an ideology with clearly defined instructions for every endeavor—because it was not one. *The Origin of Species* offered very few instructions—least of all moral instructions. It offered instead, or seemed to offer, a description of newly discovered natural laws that would have to be taken into account by all who hoped to make up new ideologies fit for the future. But ideologies or, in this case, moralities, would still have to be made up. Liang Ch'i-ch'ao saw no work by Spencer, Huxley, Kropotkin, Kidd—not even by Darwin himself—that alone could serve China as a bible in her hour of need. If Darwin said China needed a new morality, *Chinese,* said Liang Ch'i-ch'ao, would have to "invent" one.

"If we wish to change our people with a new morality, no one little Western theory will be enough," he wrote.[188] For "we are born in our own *ch'ün*'s present. We should broadly consider the great trends of world events, calmly decide what is best for our *ch'ün*, and invent a new morality, as a way to strengthen our *ch'ün,* and advance our *ch'ün.* . . . If we do *not* hastily pick and choose from moralities past and present, foreign and Chinese, and invent and proclaim some new morality, then I fear that hereafter, the more our academic instruction advances, the more our moral instruction will decline. Western material civilization may be imported into China *in toto,* but our four hundred million people will treat each other like birds and beasts."[189]

This statement recognized that there *was* no available Darwinian manual for moral, *ch'ün*-strengthening behavior. But it also, in most significant proto-Maoist fashion, implied that, even if there were such a manual, it would have to be "Sinicized" before it could save China. A new Darwinian morality to make fit a *ch'ün* must first *fit* that *ch'ün.* Or, as Mao Tse-tung would later say of Marxism, a new Darwinian morality would have to "conform to the concrete peculiarities of our country, and take on a specific national form before it could be actualized."[190] Therefore, just as Mao Tse-tung would later urge Chinese to "study how to apply the theory of Marxism-Leninism to the concrete environment of China,"[191] so Liang Ch'i-ch'ao urged his countrymen to study how to apply the theory of Darwinism to the concrete environment of China. He urged them to seek for a Darwinian morality "fit" for our *ch'ün,*" to be made from fit pieces from moralities of "*ku chin Chung wai*" (past, present, China, and abroad). Thus, even in his choice of words he anticipated the slogans of Mao Tse-tung, "*Ku wei chin yung, Yang wei Chung yung*" (The past for the use of the present, and foreign for the use of Chinese).[192]

It would be more accurate to say that Mao Tse-tung, in *his* choice of words, revealed once again his indebtedness to Liang Ch'i-ch'ao. Mao Tse-tung shared Liang Ch'i-ch'ao's beliefs that foreign ideas would have to be "Sinicized" before they could prove fit for a Chinese environment. Liang Ch'i-ch'ao's Darwinism, therefore, lay behind Mao Tse-tung's doctrine of the "Sinification of Marxism."

One more significant sentence in the above statement of Liang Ch'i-ch'ao reveals, among other things, yet another affinity between the Thought of Liang Ch'i-ch'ao and the Thought of Mao Tse-tung—the sentence in which Liang Ch'i-ch'ao suggests that, even with "Western material civilization," Chinese might "treat each other like birds and beasts." Mao Tse-tung would agree that "Western material civilization," even with its promise of "wealth and power," was not enough, that such "wealth and power" would not even be worth anything without some sort of new morality. Liang Ch'i-ch'ao, when he wrote the above sentence, still longed for the importation of "Western material civilization," so that we are not yet justified in translating his phrase as "Western materialistic civilization," as it is best translated when it appears in his post World War I attacks on that civilization, but, even so, his statement gives a hint of his later "conservative" reasons for opposing the proponents of "complete Westernization" (*ch'üan p'an hsi-hua*), and shows, perhaps, that his alleged "reversion" to conservatism was not that much of a reversion. For his insistence that "Western material civilization" was not enough was proof of an abiding nationalism and of a moralism (Confucian?) which were always parts of the Thought of Liang Ch'i-ch'ao and which would become integral parts of the Thought of Mao Tse-tung.

At any rate, the new morality of Liang Ch'i-ch'ao was not to be simply "the new morality of the West," nor something that could be simply called "Darwinism" or "Social Darwinism." And yet it *was* Darwinian, as Darwinian, in its way, as any of the evolutionary moralities prescribed by Spencer, Huxley, or Kropotkin. And so it deserves to be considered seriously in its own right alongside them.

UTILITARIANISM AND MATERIALISM. Before considering the specific list of fitting and fittening virtues Liang Ch'i-ch'ao prescribed for his new evolutionary—and revolutionary—morality, we should consider the even more revolutionary new Darwinian *basis* for his new morality, which was utilitarian and materialistic, and which therefore most clearly helped in the paving of an intellectual way towards Marxism.

It should already be clear that there was pure Darwinian utilitarianism behind Liang Ch'i-ch'ao's desire for a new morality. He wanted a morality that would work, and he was against the old morality because it did not work, because it had not made China strong. In non-utilitarian terms, therefore, his quarrel with the old morality was not a moral quarrel at all.

This distinguishes Liang Ch'i-ch'ao's "morality revolution" from "the second morality revolution" launched by Lu Hsun with his "Mad Man's Diary" in 1919. Lu Hsun hated the old morality, because it "ate people." He revolted against it because he was revolted by it. He was sickened by Chinese inhumanity to Chinese, which he saw all around him, and all through history, being carried out in the very name of "humanity and virtue" (*jen-i tao-te*).[194] Liang Ch'i-ch'ao, however, rarely showed such "moral outrage" (or sensitivity). He was outraged at the old morality, not because it was cruel, but because it was unfit. Again he was blinded by the *ch'ün,* so concerned with his People's plight that he showed little concern for the plight of his people. His revolt against the old morality, therefore, seems far more utilitarian than moral.

Liang Ch'i-ch'ao's new morality, however, was utilitarian not only in practical inspiration but also in theory, which should somewhat absolve him from the above complaint. It was utilitarian both in its ultimate standard of right and wrong and in its basic explanation of what morality in general was and where it had come from. "Do you know," asked Liang Ch'i-ch'ao, "whence arose morality? Morality was established to benefit (*li*) the *ch'ün.* Therefore, although moralities that are suitable to different degrees of barbarism or civilization are often not the same, each, nevertheless, in the last analysis comes back to that which is able to strengthen its *ch'ün,* make virtuous its *ch'ün,* and advance its *ch'ün.*" Therefore "That which is of benefit to one's *ch'ün* is good, and that which is not of benefit to one's *ch'ün* is evil."[195]

China had probably not heard such an unabashedly utilitarian theory of morality since Mo Tzu tried to sell "universal love" to the warring Chinese of the fifth century, B.C. Mo Tzu, of course, believed that there was more to morality than human ideas of utility. He believed that morality was the will of Heaven. But, when critics said of his beloved theory of universal love, "Granted it is good, but how can we possibly use it?" he replied, in exasperation, "If it would not work if we used it, even I would be against it, but where is there any good that cannot be used?"[196]

Liang Ch'i-ch'ao, too, seemed to be saying that the good was the useful and that the useful was the good. Indeed, he seemed to be saying that the whole idea of "goodness" had been "set up" (and most significantly in his case by *man,* not by Heaven) because it seemed useful. The trouble was that he did not say what he meant by *use* or *benefit.* In his essay on *kung te* (public morality), he was woefully negligent in defining his terms. He did not even show that he recognized that to say that *the good* is *the good for* is meaningless unless one makes crystal clear "the good for what."

Did he mean when he said *good for the ch'ün,* good for its stomach or good for its "soul"? *Good for* could cover a multitude of virtues (or sins).

All that Liang Ch'i-ch'ao did to clarify what he meant by "that which is of benefit to one's *ch'ün*" was to imply that it meant "that which is of benefit to one's *ch'ün*'s strength, virtue, and progress." But even here he did not admit any possibility of conflict between these "goods." Nor again did he define them. *Strength* and *progress* were undefined enough. *Virtue* in his formula was a positive puzzlement. How could the statement that "that which is good for the *ch'ün* is that which makes good the *ch'ün*" help anyone understand either what *was* good for the *ch'ün* or what "good" was, or *why* "good" was good?

Liang Ch'i-ch'ao did offer a little help in a second essay, sought from Jeremy Bentham, whom he quite competently introduced in September 1902, along with, for the first time, Bentham's English word *utilitarianism* (which Liang Ch'i-ch'ao chose to translate as *le-li-chu-i,* "pleasure-and-benefit-ism"). Liang Ch'i-ch'ao seemed to accept Bentham's choice of pleasure, or happiness, as the ultimate standard of good: "Bentham, in asking by what one should determine a standard of good and evil for all human moral action, has said that 'that which increases people's happiness is "good," and that which hinders or diminishes people's happiness is "evil."' This is a doctrine that will hold true anywhere across the four seas."[197] And, to cement the importance of Bentham's theory, Liang Ch'i-ch'ao claimed that "in the last hundred years one of the most forceful dictums in society has been the phrase 'the greatest happiness of the greatest number.' Its influence on all intellectual theories has probably been as valuable as that of the dictum 'Things contend; Heaven chooses. The superior wins; the inferior loses.'"[198]

This time, however, Liang Ch'i-ch'ao did show an awareness, which he probably owed directly to Bentham, of the awful complexity of utilitarian moral standards, indeed of the unwieldiness of even such a pleasant-sounding standard as "happiness." He quite faithfully gave a summary of Bentham's elaborate system of measuring blessings, mixed blessings, and mixed unblessings, a summary which should have given his readers sufficient warning that the task of making up a new morality was going to be no picnic (or, as Chairman Mao would have said, "no dinner party"). He even added a measurement category of his own, "the relative time sequence of pain and pleasure," claiming that pleasure that comes after pain should be weighted much more heavily than that which comes before it.[199]

Yet, he came no closer to explaining what kind of pleasure or happiness he meant to hold up as a standard of good for his *ch'ün*. Did he look for political happiness, material happiness, or spiritual happiness—whatever those might be—or somehow all in one? Although he acknowledged a headache's worth of different happinesses, he made no effort, at least for the time being, to choose amongst them. This may have been a shrewd move, as any such choice is most hazardous for would-be utilitarians, forcing them almost inevitably to fall back on some non-utilitarian standard, but it did not help clarify the new morality.

The only benefit or happiness Liang Ch'i-ch'ao was *clearly* for was that equated with "survival." Even survival was left quite undefined, but apparently, under any conditions, "survival" was a happier prospect than its alternative (note the non-utilitarian assumption that life is good). At any rate, Liang Ch'i-ch'ao's revised standard of good seemed to come down in the end to little more than "that which is of benefit to the *ch'ün*'s survival." It looked as if the new morality might be based on the somewhat less-than-exalted principle "Anything goes for the life of the *ch'ün.*"

One must admit that Liang Ch'i-ch'ao never put it quite so bluntly. He never said that absolutely anything was all right—and good—if it was necessary for survival, but neither did he deny it. He never mentioned any unacceptable means of ensuring survival. Nor did he acknowledge the possibility of "moral dilemmas" when survival was at stake. All he offered was what seemed to him a "simple" and sufficient utilitarian standard of good.

But did not Liang Ch'i-ch'ao's new standard threaten to legitimize the very kind of "ruthless self-assertion" Huxley had railed against in *Evolution and Ethics*, which he feared the "evolutionary ethics" of Social Darwinists would condone?[200] "In Chinese fashion," Liang Ch'i-ch'ao was *not* in any danger of condoning "ruthless self-assertion" for the individual. His definition of good, after all, was a social definition. So far, he had only defined good in terms of what was good for the *ch'ün*, and he clearly agreed with the spirit of the old idiom, that a *hai ch'ün chih ma* (a horse that hurt the herd), however self-assertive, was unspeakably immoral. "Fanatical individualism" was still as anathema to Liang Ch'i-ch'ao as it was to Huxley.

But Liang Ch'i-ch'ao's next contention was that the seeming selflessness required for the good of the *ch'ün* was actually based on self-interest. This was the second and more startling part of his new Darwinian and utilitarian theory of morality.

In presenting this idea, Liang Ch'i-ch'ao for once acknowledged some of his sources of information and inspiration, Bentham, of course, and the Japanese scholars through whose studies and translations he had met him, and once again Yen Fu. But, most important, he acknowledged the separate help of one Japanese scholar, Katō Hiroyuki, whose thought he felt worth studying on its own, because he was "a leader of the Germanic school, who has concentrated on the theory of evolution and who believes self-love to be the standard for morality and law."[201]

Actually, in July 1901, in a little piece entitled "Li chi yü ai t'a" (Self-benefit and love for others), Liang Ch'i-ch'ao had offered a first version of what was in essence Katō's theory without acknowledging Katō, admitting only that he was presenting what "I have gained from my reading and thinking."[202] But a little over a year later, in his September 1902 article on Bentham, he admiringly introduced Katō, in Bentham's defense, and quoted a lengthy passage from one of Katō's works, significantly entitled *Tao-te fa-lü chin-hua chih li* (The principle of the evolution of morality and law).[203] Moreover, in November 1902, with the express purpose of introducing Katō's thought to China, he translated several more selections from another of Katō's works.[204]

If Liang Ch'i-ch'ao's new theory of morality was not, therefore, strictly his own, it was, nonetheless, tailor-made for him. Katō Hiroyuki's theory did not revolutionize Liang Ch'i-ch'ao's thinking, it supported it; and therefore Liang Ch'i-ch'ao seized Katō's theory and made it his own. For it was just what he had been groping for, the perfect "refinement" for his have-your-cake-and-eat-it argument in favor of Darwin and morality both. It allowed him to start where he felt one *had* to start, with the Darwinian truth of the necessity of self-assertion, and still end up where he *wanted* to end up, with a cry for the necessity of "public morality"—and self-restraint.

This "perfect" Katō-Liang argument was actually quite simple. The initial proposition was that "there has never been a system of morality or law in the world that has not been established for self-benefit." This was because "the teeming myriads of species are competing equally for survival in the world of evolution. Those who are able to benefit themselves must prove superior and triumph. Those who are not able to benefit themselves will prove inferior and suffer defeat." This indeed was the true meaning of the Western saying "Heaven helps those who help themselves."[205]

However, "Man cannot stand in the world all by himself. Therefore there is the *ch'ün.*"[206] This was the great human breakthrough. (Or so Hsun Tzu would have said. Kropotkin, of course, adamantly insisted that animals—other animals—had made the same "breakthrough" eons before.) Men, in the beginning unconsciously, had stumbled upon the great truth that "the common good and private good are one and not two."[207] One's own survival depended on one's *ch'ün*'s survival: "Therefore, he who is good at benefiting himself must first benefit his *ch'ün.*"[208] Loving others was a necessity: "If I do not love others, I can gain no benefit for myself and must needs join the ranks of the defeated unfit."[209] Love for others was really only "self-love in a different form."[210] For "he who is truly good at loving himself has no choice but to extend that love to love of family and love of country. He must extend it to love of his family members and love of his countrymen. From this is born the 'moral duty' of loving others. All loving of others is in the end for oneself."[211]

This theory was amazingly close to Kropotkin, who, totally unbeknownst to Liang Ch'i-ch'ao, had written in that very year, 1902: "It is not love and not even sympathy upon which Society is based in mankind. It is the conscience—be it only at the stage of an instinct—of human solidarity. It is the unconsious recognition of the force that is borrowed by each man from the practice of mutual aid; of the close dependency of every one's happiness upon the happiness of all.[212] . . . In the practice of mutual aid, which we can retrace to the earliest beginnings of evolution, we thus find the positive and undoubted origin of our ethical conceptions."[213]

At the earliest beginnings of *human* evolution, "primitive man" had instinctively recognized that "the common good and private good are one and not two." "The Savage," admitted Kropotkin, "is not an ideal of virtue . . . but the primitive man has one quality, elaborated and maintained by the very necessities of his hard struggle for life—he identifies his own existence with that of his tribe [a perfectly valid translation of *ch'ün*], and without that quality mankind never would have attained the level it has attained now."[214]

Liang Ch'i-ch'ao desperately wanted each *Chinese* to "identify his own existence with that of his tribe." And now he had found a proto-Kropotkinist way of proving that Darwin's law of self-assertion demanded such identification. Both Kropotkin and Liang Ch'i-ch'ao believed that Darwin must be right, but both believed that "mutual aid" must also be right. And both found "Darwinian proof" that it was.

Liang Ch'i-ch'ao was, by his own admission, somewhat wary of his proof, and therefore for a long time also wary of introducing either Bentham or Katō to China. He said of Katō: "I have long enjoyed reading his books, [but] I have been reluctant to introduce his scholarship to China because I have been worried that the advantages might not be up to the disadvantages."[215] And he said the same thing about Bentham: "I have long wanted to present Bentham's theories, but I have been afraid to."[216]

He was afraid that Bentham and Katō would be misunderstood and their theory of utilitarianism misused. "'Love for others,'" he wrote, "is the great source from which a *ch'ün* is formed. We cultivate it and nourish it every day, still fearful that it may not take root. Why should we now wipe it out and make it a mere appendage of egoism? The people who advocate this theory are apparently afraid that the human race does not know how to be selfish and must be taught to do evil!"[217] Liang Ch'i-ch'ao, of course, knew that that was *not* what Bentham and Katō meant, but he was afraid that, if people were simply told to pursue their own interest, without being taught how to reckon it, they and their country would come to grief. "There are too many people in the world who are no good at mathematical calculations," he wrote. "By nature, they are greedy for pleasure and avid for profit, but they do not know where real pleasure and real profit lie. If once they hear of utilitarianism, they will use that theory as an excuse and drown themselves in pursuit of what the shallow and addlepated call profit, and from that pursuit shall flow endless ills." Therefore, he concluded, "until education is universal, we must on no account advocate utilitarianism."[218]

But, in the end, if he did not actually "advocate" utilitarianism, he at least overcame his fears enough to introduce it with great praise, lending it the full power of his prose even as he claimed, "I am under no obligation to exhaust myself in its defense."[219] "How," he asked, "can our young intellectuals afford not to study the words of a great teacher whose influence has already extended over a hundred years and changed the entire world?"[220] For his part, Liang Ch'i-ch'ao was obviously convinced that, rightly interpreted, utilitarianism made very good sense indeed, and he accordingly made it a part of his morality revolution.

But was a switch to a utilitarian standard revolutionary? And, even if it was, did it matter? Utilitarians (cynically?) believe that all moral standards are, and always have been, utilitarian. Therefore they would argue that Liang Ch'i-ch'ao was morally revolutionary, if at all, only in his *acceptance* of utilitarianism (a contention which, if true, would still make him significantly revolutionary). Modern sociologists, moreover, mindful of "the Chinese family system," "the Chinese social system," and "Chinese religion," might reasonably argue that Liang Ch'i-ch'ao's dictum "Good is what is good for the *ch'ün*" was also anything but revolutionary, being instead "quintessentially Chinese," for the Chinese, they would say, have always defined good in social terms, in terms of the *ch'ün*'s good, be the *ch'ün* family, clan, or nation, and they have always subordinated the individual's good to it.

Surely one must admit that utilitarian moral arguments have abounded in Chinese tradition. "That old time religion," which (minus human sacrifice) in one form or another (certainly in later popular Taoism and popular Buddhism) proved good enough for most Chinese from the days of the *Shu ching* and *Shih ching* up to (not including) the days of the People's Republic, long ago distilled its most potent encouragement of morality into the utilitarian idiom *Shan yu han pao; o yu o pao* (Good is rewarded with good, evil is rewarded with evil). Or as the Sage-King Yü put it of old: "Auspicious it is to follow morality, inauspicious to go against it."[221] Elsewhere, the *Shu ching* put it even more clearly: "The way of Heaven is to bring down blessings on the good and disaster on the wicked."[222] A word to the wise was to be sufficient.

We have already seen the utilitarian side to Mo Tzu's thought. There was also a utilitarian side (one side) to Taoist thought. The *Tao te ching,* was not about *tao-te* (the modern word for morality) which it claimed to eschew. It did preach a "right" way to live, but, in many passages, that right way was a way to survive. Chuang Tzu, too, if we can speak of him as if he existed, talked often of survival, extolling above all (in the non-struggle to survive) "the usefulness of the useless."[223] The Taoist Sage was to "use this uselessness"[224] as did an ancient, useless tree that no one saw fit to cut down: "For, being of no use, how should it come to grief?"[225] In love with paradox, Taoists (like Marx) loved to "turn the tables" on their philosophical opponents, but their paradoxical theory of the survival of the unfittest was still "survival-oriented" and, however inverted, still utilitarian.

Buddhists likewise could sound utilitarian, even in their most extreme moments of "self-denial," as they sought to lose themselves to save themselves. Liang Ch'i-ch'ao, when he introduced

Bentham, actually gave the Buddha a higher grade in utilitarian expertise than he gave Bentham himself:

> The Buddha knew that earthly pleasure is transient and that being transient it would be followed by pain made thereby all the greater. Therefore he thought it better to break the roots of mental anguish and endure a lesser pain for an everlasting joy. . . . Hence the Buddha is the best at mathematics, the best at using Bentham's methods of calculation.[226]

The very morality of the Super Calculator's Fourth Noble Truth could seem utilitarian. The Eightfold Path was the way out of suffering. That was why so many Neo-Confucians thought Buddhists selfish.[227]

But of course there was much utilitarianism in the Confucian tradition as well. Hsun Tzu, Liang Ch'i-ch'ao's favorite Confucian, and the favorite, more recently, of the Gang of Four, who "rescued" him from the Confucian camp altogether to make him a "respectable" Legalist, was almost as unashamedly utilitarian as Mo Tzu.[228] The *t'ien* that demanded morality for Hsun Tzu was nature not Heaven, but it still demanded morality because morality worked. (The real Legalists, utilitarian to the core, rejected it because it *did not*.)[229] "If the moral way survives," Hsun Tzu wrote, "the country will survive."[230] Rites and righteousness would strengthen the state.[231] Morality, indeed, had been set up by the Sages to bring order out of chaos.[232] It was not just right, therefore; it was useful.

Actually, that argument was used not only by Hsun Tzu but by Mencius as well, and even by Confucius himself. Both Mencius and Confucius believed that morality worked, or at least they wanted to believe that it worked, even though it never seemed to work in their time. They believed that right should be might. For Confucius, goodness had all the force of wind over grass.[233] For Mencius, "Goodness conquers evil as water conquers fire."[234] Therefore '"The good man is invincible,'" and "could be king of the empire, though he start from a territory of only a hundred square miles."[235] In politics, at least, Mencius echoed "that old time religion," as he urged king after king to adopt the moral way. For "those who go with Heaven," he promised, "survive. Those who go against it perish."[236]

Even in the realm of private ethics Confucius and Mencius could sound utilitarian. The Confucian version of the Golden Rule, "Do not unto others what you would not have them do unto you," could as easily be made to sound utilitarian as the Christian version. Mo Tzu's version made that perfectly clear. Describing the road to success for a filial son, he announced, "I must first set myself to loving and profiting the parents of others, and then they will repay me by loving and profiting my parents."[237] The Golden Rule could appear a ploy.

But Mo Tzu missed the point (if he was not merely lowering himself to the level of his opposition). Neither Confucius nor Christ meant their Golden Rules to be ploys. They were not reasons for doing good but "rules of thumb," for people who wanted to do good. Utilitarianism was not the spirit of Confucianism, however much utilitarianism utilitarians might find there. The word *li* (benefit, advantage, profit), which Liang Ch'i-ch'ao used in his translation of utilitarianism, was a bad word for both Confucius and Mencius, used almost always in opposition, not apposition, to morality.

The book of *Mencius* actually opens with an attack on *li*:

> Mencius had an audience with King Hui of Liang, and the King said, "Old gentleman, you have not deemed a thousand miles too far to come. Have you something with which to profit my country?"
>
> And Mencius replied, "King, why must you say profit? Morality is enough."[238]

Mencius was only echoing the teachings of his master, for Confucius had said, "The noble man knows what is right, the petty man knows what is profitable."[239] Of course one could argue that Mencius meant to profit the King's country with morality, but he himself would surely have objected if the King had come back with an "Aha! So you think morality will prove profitable, do you?" Mencius did not think that way. He did think that morality would "work," but he never thought honesty just a policy. It was something more. It and all morality belonged to a realm above *li.*

Neither Confucius nor Mencius was very good at speaking of such a realm, but both clearly believed in one. Each had a sense of a "good" that should be followed even if it did not "make good sense." Mencius said that, if forced to choose, he would give up life and take morality. Confucius, too, said that "a good man of moral purpose will not seek life at the expense of goodness, but may give up his life to accomplish goodness."[240] Personal survival was not the ultimate Confucian standard. But neither was the *ch'ün*'s. Confucius and Mencius said over and over that the best thing that could possibly happen to a *ch'ün* was for it to be ruled by a truly good king. But Mencius also said that a would-be good king, "even if he could gain [and thus save] the world by committing only one unjust act or by killing only one innocent person, still would not do it."[241] For he said, "Before a man can *do* anything, there must be some things that he will not do."[242]

The Katō-Bentham-Liang-Kropotkin thesis, of course, could "explain" Mencius: Mencius's choice of morality over life was still a selfish choice. He himself had admitted that he "desired" morality and that he desired it more than life. Giving his life, if need be, for morality, then, was obviously the only way he could be "happy." Liang Ch'i-ch'ao had pointed out that Bentham was quite right in insisting that such an exalted sense of pleasure was still utilitarian.[243]

Likewise, Mencius's "baby-in-the-well" argument, his famous proof that all men have unselfish moral instincts, could "finally" be explained. Mencius had insisted that any man would feel an urge, even though he might suppress it, to leap to the rescue if he saw an infant tottering on the brink of a well, and he would feel this urge, Mencius insisted, "not because he wanted an in with the child's parents, not because he wanted to win fame in the community or among his friends, not because he detested the sound of the child's crying." He would feel an "innate sense of pain [that was] the root of compassion."[244]

The new thesis, however, while admitting "the Mencian feeling," would say that it was precisely what was meant by a "love for others" which was a "transformed love of self." It was the kind of "instinct" inserted into the individual by evolution for the good of the *ch'ün*, but it was also the kind of "social instinct" that worked to the good of each member of society, and that had only come about thanks to a primeval identification of self-interest with the interest of the group. Kropotkin, without realizing it, had already written a parody of the Mencian parable to establish this new thesis:

> It is not love to my neighbor—whom I often do not know at all—which induces me to seize a pail of water and to rush towards his house when I see it on fire; it is a far wider, even though more vague feeling or instinct of human solidarity and sociability which moves me. So it is also with animals. It is not love, and not even sympathy (understood in its proper sense) which induces a herd of ruminants or of horses to form a ring in order to resist an attack of wolves; not love which induces wolves to form a pack for hunting; . . . It is a feeling infinitely wider than love or personal sympathy—an instinct that has been slowly developed among animals and men in the course of an extremely long evolution, and which has taught animals and men alike the force they can borrow from the practice of mutual aid and support, and the joys they can find in social life.[245]

Here was the perfect, "scientific" explanation of the Mencian roots of morality. "Compassion," said Kropotkin, "is a necessary outcome of social life."[246] But Mencius did not see it that way—almost, perhaps, but still not quite. He held morality in more awe. It was, he would have agreed, related to the workings of the universe, but not that way. In one way, of course, it was perfectly natural. Both Mencius and Confucius thought that morality was of Heaven. But their Heaven was not as "natural" as Hsun Tzu's; hence their morality was not as natural as Hsun Tzu's or Bentham's or Kropotkin's. Like Everest, it was there, bigger than man, its own excuse for being, the polestar that should hold human beings in sway.

In the end, if Confucius and Mencius had been forced to answer the question "Why be good?" they would not have said "For your own good" or "For the empire's sake," but "For goodness's sake."

That would, of course, have been a non-answer, but, because it was a non-answer, it was also a non-utilitarian answer, and, because it was a non-utilitarian answer, Liang Ch'i-ch'ao's answer was revolutionary.

The Confucian mainstream, despite occasional influxes of naturalism from Hsun Tzu, Wang Ch'ung, and admittedly many others, had held true to Confucius's bedrock belief in a moral universe. Liang Ch'i-ch'ao's new theory of morality (which he later in part disowned) theoretically, at least, rejected that belief. It rejected Heaven (*t'ien*) and the ancient faith that "Heaven does not change and the Way does not change" (*t'ien pu pien, tao yi pu pien*). Almost forever in Chinese history, one of the worst things one could say about a person had been that he "had neither law nor Heaven" (*wu fa wu t'ien*). But, according to the new thesis, nobody had any Heaven. There was only man and the world and natural laws (which led to those man-made) of self-preservation. Morality no longer rang true even to the music of the spheres, but only to the demands of survival for a given group, at a given time, in a given place. Morality was of man and for man, and there was no mystery about it.

It was in this sense that Liang Ch'i-ch'ao's new theory of morality was not only utilitarian but also materialistic. He himself did not call it materialistic, although, interestingly enough, it was he, to the best of my knowledge, who first introduced to China the Western concepts "idealism" and "materialism" through their highly Buddhistic Japanese translations, *wei-hsin* and *wei-wu* (in the Chinese reading). He mentioned idealism and materialism at least as early as the October 16, 1902 issue of the *Hsin min ts'ung pao*. Probably he had mentioned them somewhere before, for he gave no explanation of their meaning, and yet he did imply that materialism was the better and that it was winning out over idealism, thanks to Darwin. "How great," he wrote, "is the world of the last twenty-four years, a world belonging to the theory of evolution. Materialism has arisen and idealism has cowered in a corner. Science has flourished and religion is nearly at its last gasp."[247] And in the next issue, of October 31, 1902, he mentioned them again, in a line that has become gospel for the Chinese Communists: "Philosophy . . . has [only] two great schools, the materialist school and the idealist school."[248] But as he went on, in the fascinatingly confused article in which he presented this "either/or" philosophical choice, it became clear that his purpose was to laud idealism, not materialism—even though he again implied that materialism was closer to the truth—for idealism, he said, led to action.

"Idealism" for Liang Ch'i-ch'ao (his concept of the term was very obviously strongly colored by the Buddhist "consciousness only" [*wei-shih*] connotations of the term in Chinese)[249] was "something close to religion."[250] Indeed, he seemed to praise it as if it meant "religion without superstition," which in turn seemed to mean "religious spirit," pure and simple. Whatever that might be, it was a "spirit" that got things done, and that sort of spirit he had long wanted China to have.

In this essay in defense of idealism, in his openly expressed worship of "religious spirit," Liang Ch'i-ch'ao seemed to revert to a position he had repudiated ten months before, to his K'angian—if utilitarian—*pao chiao* belief that there was something about religion that "worked." But this "reversion" too was not that much of a reversion. For he could admire "religious spirit" without admiring "all the rest" of religion. The "religious spirit" he worshiped was surely nothing more than "the human spirit" (whatever that is), and it was an idealistic constant in his thought just as it would be in the Thought of Mao Tse-tung.

Worship of "the human spirit" perhaps should simply prove that Liang Ch'i-ch'ao and Mao Tse-tung (Confucians both) were idealists, not materialists, but at least it proves that there are deep and similar contradictions in the thought of both men. Faith in "the human spirit," or in "human will power," clearly threatens to contradict the militantly self-proclaimed materialism of Mao Tse-tung, but it also contradicts the unproclaimed materialism of Liang Ch'i-ch'ao. For, except for Liang Ch'i-ch'ao's strange defense of idealistic spirit, the rest of his morality theory, not to mention the rest of his Darwinian world outlook, clearly pointed towards materialism.

Liang Ch'i-ch'ao was telling China that man and his morality were both products of purely natural forces. Human truth, moral as well as any other, was solidly down to earth, as natural as the truth of sticks and stones and animal bones. Nor did he let his praise of religious spirit challenge that contention, for his praise was archly backhand. He wrote: "In the past when discussing learning, I have most disliked religion, because it has held to superstition and obstructed truth, but nonetheless, although men of religion are not up to philosophers in truth-seeking, in doing things, philosophers are not up to men of religion."[251] Similarly, "materialists," he said, "can only produce learning. Idealists can sometimes produce men as well."[252] In both cases, the implication was that idealists and men of religion "had something," but that philosophers, and especially materialist philosophers, "made more sense." This was obviously the case in one of his examples of a man of spirit, William Gladstone: "Gladstone was a heroic figure in nineteenth-century England. The depths of his superstition were unparalleled. (On every holy day, Mr. Gladstone went to church without fail, throughout his entire life. Moreover, although he once conversed for an entire day with Darwin, and although Darwin tirelessly expounded his new biological theory, Mr. Gladstone apparently understood not a whit of its significance.) But it was religious thinking that made him able to persist with his program, move public opinion, and reform his country."[253] Surely the lesson was that to be truly powerful China should embrace Darwinian sense *and* Gladstonian spirit.

Therein lay the significance—or part of it—of Liang Ch'i-ch'ao's revolutionary theory of morality. Despite his abiding—and quite metaphysical—faith in "the human spirit," he pointed the way to the present day in which the word *metaphysics*, in China, means "nonsense." His utilitarian and materialistic new theory of morality did not necessarily, in practice, make Chinese any more "utilitarian" or "materialistic" than they had ever been, but made it easier for Chinese to accept other theories that were also—in theory—utilitarian and materialistic. His theory made it easier for later Chinese to accept Marxism. And the fact that his theory—logic be damned—held on also to idealism, probably made it especially easy for later Chinese to believe in the idealistic Marxism of Mao Tse-tung.

SELF-ASSERTION. But what, at long, long last, *was* the new morality, the Darwinian, utilitarian, materialist new morality Liang Ch'i-ch'ao thought so vital to China's survival? Liang Ch'i-ch'ao never presented any neat list of "Ten New Commandments" or "Sixteen New Maxims." The "fitting and fittening" virtues he extolled he presented in a host of articles and essays written and published over several years' time, mostly in his two famous series, "On Liberty" and "Towards a New People." A "complete" list of these virtues would be hard to assemble, but it is easy to draw up a list of "most important virtues," which falls quite neatly into two parts, corresponding to Liang Ch'i-ch'ao's own two categories of *kung-te* (public morality) and *ssu-te* (private morality). In seeming accord with the "natural laws" of Mao Tse-tung, "one divided into two."

This division entailed what seemed to be an enormous "contradiction," which, among other things, provides us with a perfect opportunity to study further the influence on Liang Ch'i-ch'ao of *T'ien-yen lun.* For *T'ien-yen lun* had in one way or another convinced Liang Ch'i-ch'ao that Darwin demanded from each member of any *ch'ün* that wanted to survive adherence to a new morality that comprised both *kung te* and *ssu te.* Supposedly this was not a self-conflicting demand. For "*kung te* and *ssu-te,*" said Liang Ch'i-ch'ao, "are not mutually exclusive terms, they are mutually inclusive terms."[254] And yet, if one actually draws up a divided list of the new morality's *kung te* and *ssu te* virtues, one finds that the *kung te* side is adamantly in favor of "self-assertion," while the *ssu-te* side is just as adamantly in favor of "self-restraint." If this was not a contradiction, it was at least a "double standard."

The fascinating thing is that, in *Evolution and Ethics,* "self-assertion" and "self-restraint" were Huxley's two great opposites, representing the opposite demands of "the cosmic process" and "the ethical process."[255] Huxley did admit that no group could survive if it absolutely adhered to either one without the other, and he thus did imply that some sort of balance was necessary, but the "balance" he had in mind was heavily weighted in favor of self-restraint. Ethics *meant* self-restraint. Ethics existed to *check* self-assertion. The "good fight," the ethical fight, if any fight was good, was the fight to restrain one's admittedly "natural" self-assertive instincts.

Liang Ch'i-ch'ao, however, wanted to make virtues of those instincts. He wanted to raise the level of self-assertive consciousness of every living Chinese. And yet he seemed to worship *T'ien-yen lun.* He used in essay after essay the terms *tzu ying* and *k'o chi,* the words Yen Fu had used to translate Huxley's *self-assertion* and *self-restraint.* Huxley's terms were enemies, yet Liang Ch'i-ch'ao wanted them both, and called them both virtues. Had he misread Huxley? Or did he so obviously disagree with him? Or had he been misled (or convinced) by Yen Fu, who perhaps *had* turned *T'ien-yen lun* into a Spencerian rebuttal of *Evolution and Ethics*? Who had listened to whom?

Kung te, or public morality, does not sound as if it could be the name of an ethic of self-assertion, and yet it was. For the self-assertion Liang Ch'i-ch'ao had in mind was of the *ch'ün* and of each member of the *ch'ün for* the *ch'ün.* This was admittedly a considerable qualification of the term, but the virtues extolled even for that kind of self-assertion still seemed to be the very "virtues" Huxley so deplored. The *kung te* virtues were very *tough* virtues. They were "gladiatorial," because they were indeed based on the very "gladiatorial theory of existence"[256] that Huxley wanted ethics to repudiate. They were based on *the* Darwinian article of Liang Ch'i-ch'ao's faith, on the belief that China was caught in a worldwide arena. The *kung te* virtues, therefore, were all variations of the theme "Struggle to survive."

Liang Ch'i-ch'ao wanted his people to quit lounging—or crouching—on the arena's floor. He wanted them to stand up and fight. He wanted every man, woman, and child in China to *want* to fight—before it was too late. And he wanted them to respect fighting. Therefore he wrote a militant essay, "On Respecting Militancy" (Lun shang wu), in which he made "respect for militancy" itself the first and foremost of his "gladiatorial" virtues.[257] It was foremost because a lack of respect for militancy, he protested, lay at the heart of China's weakness.

"People are always saying," he lamented, "that 'barbarians value brawn while civilized men value brains.' Alas! Such a statement sees two fives without recognizing ten. How stupidly blind it is to the real workings of the world. . . . A warlike spirit [*shang wu*] is the original creative force of any national people. It is that which the nation-state relies on to be formed; it is that which civilization relies on to be maintained. Bismarck has said, 'What you can relie on in this world is not international law but black iron and red blood, and nothing else.' So it is not only international law that is unreliable. If the founders of a state do not have a warlike people, and a blood-and-iron philosophy, then, even if they have civilization and learning, great numbers, and vast territory, they will still have no way to hold their own on this world stage of violent competition."[258] And China, said Liang Ch'i-ch'ao, despite its territory, its numbers, its learning, and its civilization, did not have a warlike people.

One cause of that sorry fact, he said, could be traced back to a longstanding social prejudice against the military profession: "Our Chinese habit of looking down on the military goes back to ancient times, as can be seen in the vulgar proverb 'Good iron is not used for nails. Good men do not become soldiers'"[259] (a proverb and prejudice that the Chinese Communists were still combating in songs of the 1930s).[260] But beneath that social prejudice, claimed Liang Ch'i-ch'ao, lay an even more insidious "moral prejudice" of "benighted Confucians" [*chien ju*] who "thought that the weak were the good, and that 'forbearance' was the highest of virtues."[261]

"Forbearance" (*jen*) was the virtue that would do China in. For, in the name of forbearance, "Chinese had borne insults that slaves would not bear and sufferings that beasts of burden would not put up with, and all without ever daring to resist with glaring eyes and flailing fists. . . . Alas," cried Liang Ch'i-ch'ao, "'To suffer aggression without response' was indeed the ultimate virtue for our ancient saints, but to follow such a code in the struggle for existence, in this world where the weak are the meat of the strong, to use such a code against fast and furious men who would attack you like hawks and tigers—that would be like discussing ethical subtleties with a robber who has broken into your house and has a knife at your throat. Not only would it not be fitting for survival, it would only add to your shame."[262] Chinese would have to learn the "virtue" of fighting back.

Liang Ch'i-ch'ao's implication that *not* fighting back might have been a "viable virtue" once, but was one no longer, seems at first glance a rather un-Darwinian echo of Han Fei Tzu's third century B.C. statement (equally strongly un-Legalist), "Morality worked in ancient times, but will not work today."[263] A "demoralizing" struggle for existence should not have cropped up only in modern times. Barring Eden, it should have been there from the beginning. But, as Darwin surely believed that the *intensity* of the struggle for existence could vary greatly in different periods, perhaps Liang Ch'i-ch'ao did not have to endorse an un-Darwinian fall-of-man theory to believe that his period—like that of Han Fei Tzu— was a warring-states period of such unusual intensity that human niceties that had once proved compatible with survival, and might one day prove so again, were definitely not compatible with survival in the present. His first moral argument, after all, had been that morality must fit the times. And the Darwinian lesson for his times he felt was clear: In the twentieth-century struggle for existence, one could not afford to turn the other cheek, or love one's enemies. The only "good" thing to do, was to fight with all one's might.

Now surely Chinese had known how to fight before Liang Ch'i-ch'ao started giving lessons, and surely Chinese before Liang Ch'i-ch'ao had called "fighting back" a noble virtue. Throughout their long history, Chinese were no more famous than anyone else for turning their cheeks or loving their enemies. When Mao Tse-tung in 1942 said, "We cannot love our enemies," he was able to add immediately, "That is everyday common sense."[264] For he knew that it was a "common sense" that went back throughout all human history to the beginning of time. Ad hoc appeals to "the right of self-defense," with its corollary, "the right to shoot before one's shot" ("it was either them or us"), must be *older* than many of the hills. In that regard, the new morality was not that new.

And yet the Darwinian justification for the virtue of self-defense *was* new, and raised it (or lowered it) to a new level. Self-defense was no longer a last-resort excuse for violent action; it was now the scientific ultimate in modern moral theory, cause for pride, not apology. And therefore it gave "common sense" an uncommon power.

Liang Ch'i-ch'ao's new Darwinian virtue of fighting back would become an all-important virtue in the new morality of Mao Tse-tung. The Communist Party's attitude, Mao Tse-tung wrote in 1939, is that, "if any trespass against us, we will certainly trespass against them."[265] Mao Tse-tung's basic argument was the same as Liang Ch'i-ch'ao's: The old-time morality was suicidal. "We are not like Duke Hsiang of Sung [an ancient general who refused to sound the charge until his enemy's troops were all out of the river and ready for battle, and was then wiped out]. We do not want any such stupid-pig type of virtue and morality."[266] Mao Tse-tung was tough, and his reasons for being tough were Liang Ch'i-ch'ao's Darwinian reasons: it was a tough world. "We must study Wu Sung of Chin-yang Ridge," he wrote, ". . . Either beat the tiger to death or be eaten by him. It must be one or the other."[267]

There was nothing odd or even important about these statements, written as they were during war time. But disciples still believe that the world is a Darwinian jungle, "peopled" by tigers, in the guise of countries or classes or both, Chairman Mao's paper-tiger argument notwithstanding.

But making a virtue out of fighting back is not the same thing as making a virtue of fighting. Before Mao Tse-tung said, "If any trespass against us, we will certainly trespass against them," he said, "If people do not trespass against us, we will not trespass against them," and he added, "We occupy a strict position of self-defense."[268] How strictly he occupied that position in practice is open to question, but in theory his position was clear, and so too *in general* was the theoretical position of Liang Ch'i-ch'ao.

When Liang Ch'i-ch'ao urged his people to *shang wu*, to "revere militancy" or to be warlike, he usually sounded as if he meant them to be warlike only in defense of their country. Therefore his militancy should not have seemed as "offensive" a form of self-assertion as the unrestrictedly "ruthless self-assertion" of which Huxley spoke. And yet sometimes Liang Ch'i-ch'ao's distinctions between offensive and defensive martial virtues were not clear. He was terribly impressed, for example, with the Spartans, lauding their "harsh military education," even though it seemed to make them so warlike that they fought for the fun of it: "Their entire nation, men, women, young, and old, made light of death and loved to win. Habit becoming nature, they would go off against their foes as if to practice calisthenics or to attend a banquet."[269]

In the contemporary world, he praised the Germans. "Bismarck's blood-and-iron policy," he wrote, "produced nationalism. . . . The present Emperor has carried on with heroic, martial bearing forcefully to extend his idea of national imperialism . . . so that Germany is the world's one and only military state."[270] And, even closer to home, he praised the Japanese for their successful revival, now on a national scale, of their ancient *bushidō* (Way of the warrior), of their samurai spirit. "In the 1900 expedition [against the Boxers]," he wrote, "their army's fierce bravery and fighting strength was by far the best of all the allies. They made the white men bow their heads in shame" (a particularly sad example of Chinese longing for proof of "yellow power," even if from the Japanese: Yellows were better than whites—at beating up Chinese!). "It has been only thirty years since their Restoration," he went on, "and yet those three little Japanese islands have been able to win a war against us in one battle. They have shown their might and stand unrivaled as the reigning power in the East . . . all because of their reverence for militancy."[271]

What shocked Liang Ch'i-ch'ao more than anything else into respect for the Japanese fighting spirit was a banner he saw in Japan at a send-off ceremony for new army recruits. The banner read, "We pray that we may die in battle."[272] He had loved the words of the Spartan mother, "May you return either bearing your shield or borne upon it,"[273] but here were words more glorious still. But they only made him bewail, of course, the "spirit" with which Chinese went off to war: "Chinese songs and poems throughout the ages have all spoken of the agony of joining the army. Japanese songs and poems all speak of the joy." And he quoted Tu Fu's "War Chariot Ballad":

> *The chariots rattle, the horses neigh.*
> *The men on foot strap bows and arrows to their waists.*
> *Fathers and mothers and wives and children*
> *All crowd along to see them off,*
> *Raising such dust that Hsien-yang Bridge is lost.*
> *They cling to their clothes.*
> *They stamp their feet.*
> *They block the road.*
> *And weep.*
> *And the sound of their weeping rends the skies.*

"Now compare this with the banner's words, 'We pray that we may die in battle.' Could anything be more opposite?"[274] And it was Tu Fu, alas, who was supposed to suffer by comparison.

Liang Ch'i-ch'ao's scorn for Tu Fu's poem, however, reflected a respect for militarism that threatened this time to go considerably further than a respect for self-defense. For the poem he rejected rejected imperialism. He did not choose to quote Tu Fu's later lines: "Blood enough has flowed at our borders to form an ocean, yet still would our martial Emperor extend them."[275] Liang Ch'i-ch'ao rejected the poem and praised the martial Emperor of Germany.

He was no longer, therefore, simply praising those who "would not refuse competition"[276] if it were thrust upon them, or those who would "resist with all their strength and struggle with might and main to defend themselves," because they knew that "the slightest show of fear or the slightest step backward would lead to defeat with no hope of survival."[277] When he damningly praised "all the peoples who embrace Christianity" for their "fierce love of warfare,"[278] he not only made doubly clear, with unintended irony, his rejection of the cheek-turning notion that the meek would inherit the earth; he suggested that an aggressive as well as a defensive love of warfare could be a virtue. He suggested that Chinese should not only not shrink from warfare, they should welcome it. For war could be good for one, good for one's moral Darwinian health: "If a country engages in foreign wars, and conquers, its people's virtue [*min te*] will go up a level with every year. The patriotism of the Germans increased with their campaign against Austria; it increased with their campaign against France. And the patriotism of the Japanese did likewise in their campaigns against Korea and China."[279] If self-defense was still *the* virtue, it seemed already to have acquired the corollary: "The best defense is a good offense."

Liang Ch'i-ch'ao's essay, "Lun chin-chü mao-hsien" (On aggressiveness and daring) did little to dispel that notion. Chinese were urged to be "go-getters" ("go-get" being a pidginly literal translation of *chin-ch'ü*) and risk-takers. They were urged to advance, braving all dangers, and seize for themselves a position of power and safety. For they were reminded that aggressive advance was the only way to safety: "There is no standing still on this earth. You either savagely advance or go backwards."[280] They were also reminded of a "truth" expressed a year earlier in a poem published in the *Ch'ing-i pao,* perhaps written by Liang Ch'i-ch'ao himself:

Towering pressures bear upon us, yet our people's will is weak.
Hesitate no longer! Band together to protect the race!
Barbarism has to be supplanted by civilization.
Evolutionary progress comes about through braving
danger . . .[281]

The "proof" of all this lay, as usual, in the West: "There is more than one reason why European peoples are stronger than the Chinese, but the most important is that they are richly endowed with a daring and aggressive spirit."[282] Liang Ch'i-ch'ao had made that point before: "Why are whites better than other races? Other races love tranquility. The whites love action. Other races are addicted to peace. Whites do not avoid conflict. Other races guard what they have. Whites go out for more [*chin-ch'ü*]. Therefore other races can only produce their civilizations. The whites can propagate theirs."[283]

"White virtues," therefore, were *not* defensive virtues. They were aggressively self-assertive. Liang Ch'i-ch'ao, in embracing them, did seem to be embracing the very self-assertiveness, that Huxley so deplored. And yet, when he introduced his next virtue, *i-li* (perseverance), which he felt to be an indispensable concomitant to the "gladiatorial virtues" described above, he relied heavily on Huxley, borrowing an argument from *T'ien-yen lun* almost in Yen Fu's own words:

> Human endeavor and the workings of nature are often in conflict. Indeed they wage an endless competition. For nature often moves counter to human hopes. As its reactionary force is thus immense, and as the human race, with its beautiful instinct towards upward progress, can never be content with its present position, human life becomes like a voyage of several decades on a boat going against the stream. There can be no rest even for a single day.[284]

Huxley *was* for perseverance, in man's "struggle." He himself used "gladiatorial" language. He spoke of man "combating" the cosmic process. He spoke of "[setting] man to subdue nature to his higher ends."[285] He said that "that which lies before the human race is a constant struggle to maintain and improve, in opposition to the state of nature, the state of art of an organized policy; in which, and by which, man may develop a worthy civilization."[286] He did want perseverance in an upstream fight.

But Huxley was really, of course, talking about a fight against *human* nature. He wanted men to pit their "better selves" against their selfish selves, to struggle to restrain themselves. Liang Ch'i-ch'ao, however, for the moment at least, chose to ignore that fact. He used Huxley only to paint a picture of man struggling against physical nature, a picture in which "those who have perseverance succeed and those who do not are defeated."[287] Perhaps Liang Ch'i-ch'ao was honestly confused, by Huxley's gardener-against-the-weeds metaphor, as to which nature Huxley was combating. Perhaps he was not. But, in either case, he used Huxley to praise that which he, like Yen Fu, most admired in Huxley—his willingness to persevere in an uphill—or upstream—fight.

For Liang Ch'i-ch'ao was at the moment not talking about where the battle was, or "where the game was being played." He was talking about what virtues would lead to victory, and he agreed with Huxley that no virtues could lead there without perseverance. It was as a paean to perseverance, therefore, as well as to aggressiveness and daring, that he offered the following appalling poem (in English, without translation, and, humanely, anonymous):

Never look behind, boys,
When you're on the way;
Time enough for that, boys,
On some future day.
Though the way be long, boys,
Face it with a will;
Never stop to look behind
When climbing up a hill.
First be sure you're right, boys,
Then with courage strong
Strap your pack upon your back;
And tramp, tramp along.
When you're near the top, boys,
Of the rugged way,
Do not think your work is done,
But climb, climb away.
Success is at the top, boys,
Waiting there until
Patient, plodding, plucky boys,
Have mounted up the hill.[288]

The rest of the assertive virtues virtually all came down to "pluck," or, as we could now say, freed from the demands of alliteration, "spirit." For Liang Ch'i-ch'ao had independently discovered that truth known so well to high-school coaches: "It is spirit that wins ball games"—and struggles for existence.

Liang Ch'i-ch'ao used many different words for "spirit": *ching-*

shen, the closest Chinese equivalent to the English word, *ch'i,* that favorite, traditional, last resort "stuff," that not even Mencius could define ("Might I ask what is 'all-embracing *ch'i*?'" 'That is hard to say,' he replied."[289]), *hun* (soul), and finally *yen-shih-p'i-li-ch'un,* which was as close as Liang Ch'i-ch'ao could get, in characters, to "inspiration" (his Cantonese probably got him closer than does Mandarin). All these words were different, and yet they obviously all referred to the same mysterious thing—"spirit," mysterious in any language.

"Spirit" was clearly the magic ingredient in any formula for survival—"When spirit arrives, what task cannot be done?"[290]—but what was it, and how did one get it? It was clear that the spirit he wanted was "fighting spirit," for he wrote in his essay, "On Reverence for Militancy":

> I have heard that our country has pursued a militaristic policy for several decades. [Our officials] have bought ships and trained soldiers, established factories and manufactured weapons, reorganized armies and planned for war, all most diligently and for a long time. And yet, in the end, all went up in the smoke of one defeat. Why? Because what they meant by militarism was form. What I mean by militarism is spirit. If you have not the spirit but only the form, then you send sheep in tiger skins out to fight with wild beasts. You only serve them up to be pounced on and devoured.[291]

Apparently, what the Chinese needed was the kind of fighting spirit the Japanese thought was their "soul": "It is commonplace," said Liang Ch'i-ch'ao, "for Japanese to say, 'What is Japan's soul? It is *bushidō* [the Way of the warrior, the samurai spirit].' Thanks to that, the Japanese have been able to establish their country and carry out their reforms. And that is why I have searched for what I call 'China's soul.' Frantically I have searched throughout our whole four hundred districts, and yet I have not been able to find one! Alas, the pain of it! How on this earth can there be a country without a soul? It makes me afraid."[292]

In the end he decided there was still hope, for he discerned at least "the seeds of a Chinese *bushidō*" in the fighting spirit of feuding Chinese villages. But it would take work to make these makings of a Chinese soul into a real one: "Today our most pressing task is to fashion a Chinese soul. What do I mean by a Chinese soul? I mean a soldier's soul. For only if we have soldiers with souls will we have a country with a soul."[293] And a soldier's soul, he said, could only come from a heart that loved country as well as self.

If Chinese were to have a national soul, therefore, and become a people with spirit, they needed militant patriotism—something that foreigners, scornful enemies and exasperated friends alike, had repeatedly told them they did not have. They also needed two more ingredients which foreigners also said they did not have, self-respect and self-confidence. These too were "new virtues" in praise of which Liang Ch'i-ch'ao wrote impassioned essays. Although his readers might well have wondered whether or not he was confusing cause and effect, he marshaled a host of evidence from the current scene to prove that self-confidence led to success and that self-respect led to the conquest of positions of power that could demand respect. His examples read like a *Who's Who* of world imperialism: England, Russia, France, Germany, America, Japan—"Self-respect," he wrote, "is the source of these six nations' great power."[294] And he added his usual lugubrious warning, this time in a pathetic line that would have seemed ludicrous a century or even half a century before: "How sad that we Chinese have no natural self-respect. . . . For a country without self-respect has never been able to stand."[295]

But he had no way to tell Chinese how to gain or regain self-respect and self-confidence. All he could offer, and here again he sounded like a high-school coach, were the moral imperatives "Be confident!" "Show self-respect!" This was true of his whole new virtue of "spirit." He could not say how to get it. He could only say, "Have it!" or at most, "Will it!"

For, in the end, "spirit" came down to "will," the "will to win," as he had made perfectly clear at the very beginning of his essay, "On Self-Respect" ("Lun tzu-tsun"): "A Western philosopher has said, 'Each man stands where he wants to stand.' And Yoshida Shōin has said, 'If a man born today wants to be a willow, then he can be a willow. If he wants to be a pine, then he can be a pine.' "[296] Yoshida Shōin, confident though he was, himself being already shaded by pines, was not arguing that modern man, by will power alone, could turn himself into the tree of his choice (the better, perhaps, to lose himself in sylvan retreat). He was arguing in classical Chinese allusions that any man could show moral fiber if he wanted to. The Chinese allusions he used were even better for Liang Ch'i-ch'ao's uses, for there was something Darwinian about them. When all the other trees had lost their leaves, then one would see that the pine was still green, that it was the strongest, the fittest, that it survived. Hence the moral, Yoshida Shōin's moral, as Liang Ch'i-ch'ao read it, was that one could be the fittest if one wanted to (and one had a moral responsibility to one's *ch'ün* to want to).

This made no biological sense, but that had never kept Social Darwinists down. Liang Ch'i-ch'ao could bring virtue, will, and nature all together in a typically un-Darwinian tour de Darwinian force:

> Today, anyone with even a little modern knowledge knows the Darwinian slogans, "The superior win, the inferior lose," and "The fit survive," and so on. These seem to make survival a matter of fate. And yet urging oneself on to be superior and fit so that one can avoid being eliminated with the losing *un*fit is purely a matter of effort [*li*]. It has absolutely nothing to do with fate. The animals of the desert were not necessarily all yellow in the beginning, but the yellow ones have survived while the non-yellow ones have become extinct . . . The difference between survival and extinction might seem a matter of fate, but, if one really gets right down to it, why should others be able to become yellow while I alone should not? . . . Because I did not expend effort enough![297]

There it was. Those who survived were not necessarily the biggest or even the most strong. They were those with the most competitive spirit, who put in the most effort, who showed the greatest *desire* to survive. In that sense, Liang Ch'i-ch'ao could conclude: "Desire is truly the spring of social evolution."[298]

In other contexts, we have already seen this belief of Liang

Ch'i-ch'ao, that it was will power that makes the world go round—and man grow up. What was new here was simply his open espousal of will, indeed of *desire,* as a virtue. He ran all over both the religious and philosophical lots to prove that, contrary to popular belief, desire was not the enemy of morality. For all great moralists—Confucius, Mo Tzu, Moses, Luther, even Sakyamuni—had been men, to their greater glory, full of desire: "Did not Confucius say, 'If I desire goodness, then goodness will arrive?'[299] Then I ask you, did Confucius have desire? The answer is Confucius had more desire and greater desire than anyone in the world."[300]

In what was obviously a good day for Confucius (which was only fitting, as Liang Ch'i-ch'ao did indeed owe his own voluntarism to him) Liang Ch'i-ch'ao even patched up the Mencian-Hsuntzian contradiction over desire:

> Mencius says, "In cultivating one's mind, there is nothing better than making few one's desires." Hsun Tzu says, "Any man with many desires will have many uses." These two sayings each make clear a truth, and they go together, they do not go against each other. For with material desires one fears only that they be many. But with spiritual desires one fears only that they be few.[301]

To cement his argument, however, he left the Confucians and shockingly sought "moral support" from the hero and heroine of the popular, but not yet altogether respectable, novel, *Hung lou meng* (*The Dream of the Red Chamber*): "Alas, where can we find one who will fill our people with the infatuated love and obsessive desire of Pao-yü and Tai-yü?"[302]

Liang Ch'i-ch'ao himself was trying to do just that, to arouse his people's desire, to fire them up with fighting spirit, to fill them with the spirit and desire to struggle to survive. He therefore sounded like a coach (or like a field general, the difference in rhetoric being slight). His entire many-essayed introduction of his new morality's *kung te* virtues was one long "pep talk." In our silly modern idiom, he was merely trying to get his people "psyched-up" for the coming contest. All he was really saying was, "You've got to fight. Get in there. Give it all you've got. You can do it. Want to win. You can win it if you want to!" (Later others would carry on with, "Push 'em back! Push 'em back! Way Back! Hit 'em again! Hit 'em again, harder, harder!")

Were the new virtues not, then, the "gladiatorial" virtues? Was Liang Ch'i-ch'ao's new morality not the ethic popularly ascribed to be that "implicit in Social Darwinism"? Did it not offer justification for the "ruthless self-assertion" Huxley feared? And did that not mean that Huxley's warning, despite (or because of) *T'ien-yen lun* had gone unheeded?

The answer to all those questions is "yes"—on the national level, but on the individual level, decidedly "no." For there was one all-important qualification of Liang Ch'i-ch'ao's espousal of the self-assertion virtues that brought him charging back to Huxley. Liang Ch'i-ch'ao was for self-assertion, but for country not self. What Liang Ch'i-ch'ao really wanted was for China to assert *itself,* and therefore, and only therefore, he wanted individual Chinese to assert themselves—not for themselves, but for China. He would have said, of course, that "for China" *was* "for themselves," but that was the only kind of self-assertion "for themselves" that he would have called good self-assertion. His new morality—to an almost frightening degree—was for "selfless self-assertion." "Selfish self-assertion" was the very opposite of virtue.

Self-Restraint. When Liang Ch'i-ch'ao praised the "struggle virtues," he was almost always thinking of struggling countries or struggling races. When he cited, as he often did, the Western adage "Heaven helps those who help themselves," he was almost always talking about peoples, not people. When he extolled the virtue *tzu li* (self-effort), what he usually had in mind was collective self-effort. When he congratulated the self-yellowing animals of the desert for their *tzu li,* for example, he was congratulating species, not individual lions or camels. So too when he quoted Yoshida Shōin's protestation that any man could be a pine, he was not thinking, as Confucius had been thinking, of the individual standing firm in an evil world (the pine standing green while the crowd —the *ch'ün*—turned yellow); he was thinking of *his crowd* standing, while others fell, standing on its own ten hundred million feet, thanks to its own *tzu li.* Liang Ch'i-ch'ao could have invoked the virtue of *tzu li* in the name of "rugged individualism," but he did not, no more so than did the followers of Mao Tse-tung when they invoked their very obviously related slogan *tzu li keng sheng* ("Self-reliance," or "Life through self-effort.")

Similarly, when Liang Ch'i-ch'ao urged his countrymen to fight for their rights, he almost invariably meant China's rights. He was interested in civil rights, but sovereign rights came first—and civil duties came before civil rights as well. He wrote a special essay, "On the Concept of Duty," to make clear the latter point, arguing that "rights" were nothing more than rewards for duties performed. If one considered, as one should, the "natural history" of rights, it would be clear that, in the struggle for existence, rights were the fruits of victories won by those who fulfilled their duties, by those, that is, who recognized the vital nature of duty. "Duty," he wrote, "is the cause of victorious superiority. . . . If one gives up one's duties, one can have no more rights. That is the law of evolution."[303]

Here he was already moving back towards Huxley. Huxley had angrily decried the fact that, as the "fanatical individualism of our time attempts to apply the analogy of cosmic nature to society . . . the duties of the individual to the state are forgotten, and his tendencies to self-assertion are dignified by the name of rights?"[304] Huxley's notion of "the duties of the individual to the state" may have been very different from Liang Ch'i-ch'ao's, but the two men most emphatically agreed that any exaggerated notion of the individual's "right to self-assertion" *within* his state was bad for it or, as Huxley would have put it, just plain bad.

The secret was that, whenever Liang Ch'i-ch'ao looked within China, instead of at the world outside, he could sound very much like Huxley. Liang Ch'i-ch'ao bewailed the fact, for example, that "our modern, so-called 'heroes cognizant of the times' . . . look at Darwin's theory of the struggle for existence and use it not to consolidate us as one body but to give birth to internal strife."[305] Huxley did not talk of "using" Darwinism for consolidation—that was Kropotkinism—but he was certainly against "misusing" Darwinism to justify contention. (Liang Ch'i-ch'ao's admission that

Darwinism could be misused was a very early one for China [October 31, 1902], undoubtedly inspired by the "revolutionary misuse"—thanks to many of Liang Ch'i-ch'ao's own arguments—that Darwinism was beginning to suffer at the hands of Sun Yat-sen and Company.) Huxley and Liang Ch'i-ch'ao were agreed, therefore, that within a society selfish self-assertion was public enemy number one. They were also agreed that morality was the answer to that enemy, the kind of morality that Liang Ch'i-ch'ao this time called *ssu te* (private morality).

But their reasons were different. Their concerns were different. Their goals were different. In his essay "On Private Morality" ("Lun ssu te"), in a section on "The Need for Private Morality," Liang Ch'i-ch'ao made it absolutely clear that the ultimate need for private morality came from the need for the solidarity necessary for success in the outside struggle. "What do we rely on to vanquish our enemies?" asked Liang Ch'i-ch'ao, "On our ability to band together [and here he used, I think for the first time, what was to become the classic Communist term for "solidarity"—*t'uan-chie*] into a firm and powerful organism, and on nothing else. . . . If we want to become [such an organism], what way is there except through morality?"[306]

Huxley might have agreed with Liang Ch'i-ch'ao that "internal disorder is the most inauspicious thing there is,"[307] but, for Liang Ch'i-ch'ao, "inauspicious" seemed to mean only one thing, "suicidal in the struggle for existence," for it was "a law of sociology," he pronounced, "that only those with internal solidarity can compete externally."[308] Surely, Huxley had said as much. He had said of his "ethical process" that "it tends to make any human society more efficient in the struggle for existence with the state of nature, or with other societies,"[309] and he had said of Western colonists (without advocating colonialism), "If they waste their energies in contests with one another, the chances are that the old state of nature will have the best of it. The native savage will destroy the immigrant civilized man."[310] But the solidarity survival value of morality, or of the ethical process, was not Huxley's main point.

This was Huxley's main point:

> The practice of that which is "ethically best"—what we call goodness or virtue—involves a course of conduct which, in all respects, is opposed to that which leads to success in the cosmic struggle for existence. In place of ruthless self-assertion it demands self-restraint; in place of thrusting aside, or treading down all competitors, it requires that the individual shall not merely respect, but shall help his fellows; its influence is directed, not so much to the survival of the fittest, as to the fitting of as many as possible to survive. It repudiates the gladiatorial theory of existence. . . . Laws and moral precepts are directed to the end of curbing the cosmic process and reminding the individual of his duty to the community, to the protection and influence of which he owes, if not existence itself, at least the life of something better than a brutal savage.[311]

Huxley, perhaps, a cynic could certainly argue, only because he was a nineteenth-century Britisher, unafraid of international competition, did not put international competition in the forefront of his mind. He was not like Liang Ch'i-ch'ao almost always looking outward towards threatening national enemies. He was almost always looking inward, either from an Englishman's point of view at England, or from a "civilized man's" point of view at "civilization," or—and surely his argument by extension went that far—from a human being's point of view at humanity. Always he was concerned most of all with the creation—the human creation—not of a stronger England but of a less savage form of human life, and he did not mean by that the establishment of an unsavage life over the dead bodies of all "savages." His main concern was with "curbing the instincts of savagery in civilized men."[312] He wanted people to respect and help each other, but not just in the cause of fitting their *ch'ün* for battle, nor even just in the cause of "happiness," for "the practice of self-restraint and renunciation," he said, "is not happiness, though it may be something much better."[313] It was that "something much better" that was his main concern. He never named it, but it proves that his "ethical process" was not utilitarian. Morality was not for Huxley, as it was for Liang Ch'i-ch'ao, a matter of national life or death. It was a matter of brutality or "decency." And yet, Huxley's most important virtue, "self-restraint," the virtue that alone, for Huxley, offered hope of "curbing the instinct of savagery in civilized men," was precisely the virtue Liang Ch'i-ch'ao wanted—to gird his people for gladiatorial combat. That is why Liang Ch'i-ch'ao liked Huxley. Huxley was for self-restraint, and so was Liang Ch'i-ch'ao, just as much, although in a different way, as he was for self-assertion.

Liang Ch'i-ch'ao made "self-restraint" the most important virtue of his "private morality," and the Chinese word he used for it was *k'o chi,* the word Yen Fu had used to translate Huxley's word. Yen Fu had borrowed *k'o chi* from Confucius's famous (or, to the Gang of Four, infamous) line: *K'o chi fu li wei jen* (Goodness is self-conquest and a return to decorum).[314] And Liang Ch'i-ch'ao acknowledged that borrowing. He quoted Confucius's line in full and prefaced it with the classical version of "Confucius said," even though he did so at the end of the revolutionary essay, "Lun tzu-yu" (On freedom), in which he urged Chinese "not to be slaves of the ancients," and in which he virtually urged them not to be slaves of Confucius.[315]

Liang Ch'i-ch'ao really only wanted half of Confucius's formula for goodness. He ignored the injunction to "return to decorum," and concentrated on "self-conquest" (or self-control, or self-restraint), on *k'o chi.* And it was obvious that, even there, what Liang Ch'i-ch'ao had in mind was not at all what Confucius had had in mind. Confucius had been talking about the ideal ruler, who by curbing his selfish instincts and respecting others could show such virtue that the people of the world would flock to his allegiance of their own accord.[316] Liang Ch'i-ch'ao was talking about individual citizens, who could so curb their selfish instincts that they could actually "take the country's business as their own"[317] and so make the country strong. He said of *k'o chi:* "What is conquered is the self, but what conquers it is another self. For the self to conquer the self, that is to 'overcome oneself' [*tzu sheng*], and to overcome oneself is to be strong."[318] To make his point, Liang Ch'i-ch'ao had to switch schools in mid-sentence and throw in a paraphrased passage from the *Tao te ching,* "He who overcomes himself is strong,"[319] but the strength he thereby praised was no more the private strength of a Taoist Sage than the magnetic strength of a Confucian king, it was the strength of a militantly self-sacrificing patriot.

In one respect, the self to be conquered was the same for Confucius, Lao Tzu, and Liang Ch'i-ch'ao, the selfish self, the self that harbored selfish desires and had clay feet, the traditional self that moralists (pace Lao Tzu) had always designated as the enemy in each man's own civil war. But, for Liang Ch'i-ch'ao, "self-conquest" of such a self was not for the sake of a "personal victory." No man's private war was *the* war.

His readers may have been confused, at least those who had taken to heart his praise for "the infatuated love and obsessive desire of Pao-yü and Tai-yü" may have been confused, when he sighed, "To what depths have love and desire poisoned men!"[320] So, too, those who remembered that "desire is truly the spring of social evolution" may have been confused when he praised people who could "conquer and keep down their loves and desires, so that their minds are not imprisoned by their hard shells of impure flesh."[321] They may have been confused when he quoted Wang Yang-ming: "It is easy to get rid of the brigands in the mountains, but hard to get rid of the brigands in the heart."[322] He sounded as if he was about to go back and praise Buddhist self-conquest, as if the battle was all in one's head, and all one had to do was sit like the monk or poet in Wang Wei's poem, "peacefully meditating, conquering poison dragons."[323] But that was not what Liang Ch'i-ch'ao intended.

He showed what he really meant in the lines that followed his quotation from Wang Yang-ming: "We call ourselves men of serious intent [*chih shih*], and yet we all have demons lurking in our breasts and cannot overcome ourselves. No wonder, then, that a country should have demons ever lurking in it that cannot be overcome."[324] He was still thinking first of the country. Personal victory over demons within oneself was not the purpose of existence, but a preparatory contest before the fiercer fight against demons within the country, itself only a preparatory struggle for the real struggle against demons without the country—"foreign devils." (Herein, perhaps, lies a hint as to why the Thought of Liang

Ch'i-ch'ao proved less fit, as a package, than the Thought of Sun Yat-sen or the Thought of Mao Tse-tung. Liang Ch'i-ch'ao never clearly identified the demons "lurking within the country," at least not clearly as one recognizable breed. Sun Yat-sen had the Manchus, and Mao Tse-tung had capitalists and feudal landlords, all of whom could be fought to forge unity while it was still impossible to fight foreigners, but Liang Ch'i-ch'ao's demons were more like the "five demons" of Hu Shih: "poverty, disease, stupidity, corruption and disorder,"[325] real but disembodied. It proved easier for Mao Tse-tung to rally support with his battle cry, "All demons will be annihilated,"[326] because his demons were in human form.)

Private victory over demons, or original sin, was meaningless, therefore, if it did not lead to self-dedication to one's *ch'ün.* For private victory was selfish self-conquest, and what Liang Ch'i-ch'ao wanted was selfless self-conquest, self-conquest not for oneself, but for one's country or, as Chairman Mao would put it, "for the people."[327] Desire was not just "one man's poison." It was the people's poison, unless in each and every person it was sublimated to the people's needs. At the present stage of history, desire *had* to be so sublimated before it could flow on as a salubrious "spring of social evolution." And "self-conquest" had to lead to such sublimation. Chinese had to desire China's survival more than their own. They had to so conquer themselves that they could willingly die for their country.

SELF-SACRIFICE. "Self-restraint" was turning inexorably into "self-sacrifice," and to a degree that left Huxley behind. Huxley was not against self-sacrifice, but neither was he for it in the manner of Liang Ch'i-ch'ao, for he was against the kind of world Liang Ch'i-ch'ao thought made it necessary. He did not preach self-restraint to inspire his people to die for their people. He prayed that self-restraint might make it possible for people to live with people. But Huxley, a cynic could argue, lived in England, and England seemed safe, while China did not, and England was not yet restraining itself. Indeed it was largely because England had not and was not restraining itself that Liang Ch'i-ch'ao felt forced to echo the ancient adage "*Gloria est pro patriā mori*" in the first place. There was nothing strange or extremist, therefore, in his doing so, as that adage has been echoed in almost every dark hour of the world's warring history. Liang Ch'i-ch'ao's nationalistic espousal of self-sacrifice would hardly be worth mentioning except for one fact: He went so far beyond Huxley that he tried, with peculiar, new, Darwinian help, to make "self-sacrifice" not only a cry for a dark hour, but an entire "philosophy of life."

The peculiar help came from Benjamin Kidd, "an obscure English government clerk"[328] whom Liang Ch'i-ch'ao hailed, in 1902, as a great "revolutionist."[329] Kidd was a self-styled "evolutionist," who had leapt from obscurity to sudden, if not undisputed, fame in Britain and the United States with the publication of a book called *Social Evolution,* but he was not the kind of evolutionist one would immediately imagine Liang Ch'i-ch'ao to have welcomed with open arms. For Kidd's book tried in one-stone-many-birded fashion to employ "the principles of evolutionary science" to refute Spencer, Huxley, Mill, Bentham, and almost everyone else except Darwin and a Professor Weismann (who had just reaffirmed the necessity of competition and selection)[330] to defend religion, democracy, and imperialism, and to disparage German socialism, all in the ultimate cause of predicting that "the path of future progress"[331] was leading to "the final filling up by [the Western peoples] of all those tracts in the temperate regions of the earth suitable for permanent occupation,"[332] and to the ultimate control by those peoples, following their realization "that the tropics must be administered from the temperate regions,"[333] of all those regions not suitable for permanent occupation (except by "colored races" of "low social efficiency").[334]

Surely, Liang Ch'i-ch'ao could not have accepted such a thesis, unless he was willing to watch palefaces move en masse into all but the hottest spots of China, and clearly he did not. He rejected it most effectively, and shrewdly, by ignoring it. But there was nothing peculiar about that. Liang Ch'i-ch'ao had demonstrated before that he could abstract "principles of winning" from Western wise men without accepting their contentions that the West would win. What was peculiar was his enthusiasm for Kidd's principles of winning, for they were peculiar even by late-nineteenth-century standards, and they were not in actuality just what Liang Ch'i-ch'ao was looking for, however much he seems to have thought so. Liang Ch'i-ch'ao's enthusiasm for Kidd could only have come from hasty reading, misreading, or very selective reading.

He would, it is true, have been excited by Kidd's basic premise, with which he was in complete agreement: "The law of life has been always the same from the beginning. . . . ceaseless and inevitable selection and rejection, ceaseless and inevitable progress."[335] He would also have agreed with the next proposition, perhaps, though reluctantly, even with its final contention: "All this, the conflict of races before referred to, the wasting of the weaker, nonetheless effective even when it is silent and painless, the subordination or else the slow extinction of the inferior, is not a page from the past or the distant; it is all taking place today beneath our eyes in different parts of the world, and more particularly and characteristically within the pale of that vigorous Anglo-Saxon civilization of which we are so proud, and which to many of us is associated with all the most worthy ideals of liberty, religion, and government that the race has evolved."[336] And he would have agreed with great excitement with Kidd's conclusion: "We have in the altruistic development that has been slowly taking place amongst the European peoples the clue to the efficiency of our civilization."[337]

Here Kidd seemed to be saying just what Liang Ch'i-ch'ao wanted to say, that it was morality that made a people fit, that it was altruism, *li-t'a-hsin,* that restrained each person's devisive egoism, *li-chi-hsin,* and kept one's *ch'ün* together. But, if Liang Ch'i-ch'ao thought he had found in Kidd a perfectly kindred spirit, he must have missed or misunderstood Kidd's peculiar use of the word "altruism."

Kidd liked altruism. It was one of those "most worthy ideals" that distinguished the Anglo-Saxon race. Therefore he did not like the argument of certain Spencerians that altruism killed competition, and hence progress. But neither did he like Huxley's contention that altruism *should* kill competition, or at least a good part of it, for he did believe that without competition there could be no progress. He had, therefore, to dream up a way to have altruism and competition both—and he did. He defined altruism as the spirit which led men to protect competition even at the risk of their own self-interest.

Altruism did not express itself in stopping contests, but in permitting them. Altruism was the spirit that drove the King of the Mountain to reopen the contest, wishing the while, "May the best man win," instead of pronouncing as his first decree "The game is over." One can see Kidd's position if one imagines his response, for example, to the following famous exchange between an American Social Darwinist, William Graham Sumner, and one of his skeptical students:

> "Professor, don't you believe in any government aid to industries?"
> "No! It's root, hog, or die."
> "Yes, but hasn't the hog got a right to root?"
> "There are no rights. The world owes nobody a living."
> "You believe then, Professor, in only one system, the contract competitive system?"
> "That's the only sound economic system—all others are fallacies."
> "Well, suppose some professor of political economy came along and took your job away from you. Wouldn't you be sore?"
> "Any other professor is welcome to try. If he gets my job, it is my own fault. My business is to teach the subject so well that no one can take the job away from me."[338]

Kidd would have applauded Sumner for welcoming competition and for not trying to secure his position against competition by rights of tenure. He would also, most likely, have agreed with Sumner, that men had no natural rights. But he would have said that Sumner, by refusing to claim any right to his position, was in effect granting others the right to compete for it. By not hogging his post he was giving lesser hogs "the right to root." And such a granted right was right and good, and altruistic, because it threw open one's own door to progress.

Kidd's insistence on the altruistically granted right to compete kept him from becoming an "individualist," because "individualism," as he saw it, had led, and could only lead, to the obstruction of competition. For, if everyone was out for himself, it would be in the interest of both the strong and the weak to stop competition, the strong thereby securing their positions and the weak at least surviving. Individualism, he argued, in a section in which he sought to give Karl Marx "his due," had led to feudalism, a system by which the fit of one age sought to insure power and privilege for their children, fit or unfit, for all ages. Such a system, which was still, he said, too much with us, could only shackle the progressive talent of the masses, as it had for centuries. Fortunately, however, the "altruistic development" in the Western countries, especially in England, which had therefore been able to avoid such unseemly excesses as the French Revolution, had led feudal powers gradually to give up their privileges. "It is this development," Kidd wrote, "which . . . by its influence in breaking down an earlier organization of society, and by its tendency to bring, for the first time in the history of the race, all the people into the rivalry of life on a footing of equality of opportunity . . . has raised our Western civilzation to its present position of ascendency in the world."[339]

But here was a new danger. Just as democracy was at last welcoming everyone "into the rivalry of life," Karl Marx and his comrades urged the masses to seize control and stop that rivalry, to make the best of their world by sharing instead of competing. But that, moaned Kidd, would be as fatal to progress as had been feudalism. For the masses, by refusing to compete, would for their own security be selling out the future to the present.[340]

Again the answer was "altruism," this time needed on an even greater scale. The liberated masses, to keep their people going, would have to resist the temptation to enjoy the lowly present. They would have to keep competing.

The peculiarly interesting thing about Kidd's argument was that he admitted that such continuous competition did not "make sense" for most people then alive. Anyone cognizant of his own interest, and in his right mind, would choose either the road to socialism or the road back to feudalism, for either one led far more assuredly to happiness than did the road to progress. The human intellect had evolved at last to such a height that men could know that progress was not good for them. It was good for those who were to follow. For looking forward as no other "creatures" had ever been able to do before, men could see that "progress everywhere is evident, but the way is strewn with the unsuccessfuls which have fallen in the advance."[341] Having seen that, one could only conclude, intellectually, that "the facts present themselves in this wise. Throughout the whole period of development hitherto the conditions of progress have necessarily been incompatible with the welfare of a large proportion of the individuals comprising any species."[342]

Therefore, Kidd went on, seemingly ruling out any religious promise of self-sacrifice for progress being rewarded anywhere else, "If our conscious relationship to the universe is measured by the brief span of individual existence, then the intellect can know of only one duty in the individual, namely his duty to himself to make the most of the few precious years of consciousness he can ever know. Every other consideration must appear dwarfed and ridiculous in comparison. Every pain avoided, every pleasure gained in these few years, is a consideration, beside which the intellect must count any aspiration to further a process of cosmic evolution in which the individual has no interest as mere dust in the balance."[343]

But men *had* furthered the process of cosmic evolution! They had not been swayed by their calculating intellects. They had not listened to reason. "For no good reason" they had given up feudal privileges to allow progressive rivalry, and for no good reason they would keep up that rivalry and not sell out to socialism. Progress would progress, for progress, Kidd exulted, would continue "to secure, in an increasing degree, the subordination of the present interests of the self-assertive individual to the future interests of society, *his expanding intellect not withstanding* [my italics]."[344]

The "expanding intellect," to be sure, was a formidable enemy: "The most profoundly individualistic, anti-social, and anti-evolutionary of all human qualities is one which, other things being equal, tends to be progressively developed in the race, namely, reason." However, "the central feature of our evolution has always been the supreme struggle in which the control of this disintegrating influence is being continually effected in the interest of society first, and of the race in the next place."[345] Reason was "effected" by altruism. Reason was defeated by altruism.

Kidd immediately admitted that, if reason could be defeated by altruism, altruism must be "ultra-rational." So too was religion, which was institutionalized altruism, but in neither case was the term *ultra-rational* an insult. For it was only because they were ultra-rational that altruism and religion had been able to protect progress. The religions of the world, Kidd insisted, and none more so, of course, than that of the Anglo-Saxons,[346] had protected progress by providing "an ultra-rational sanction for the sacrifice of the interests of the individual to those of the social organism."[347]

Religion protected progress, and it protected the progressive. "The religious character" became the distinguishing mark of the fit. "The winning races," wrote Kidd, "have been those in which, other things being equal, this character has been most fully developed. Amongst these again the races that have acquired an ever-increasing ascendency have been those which have possessed the best ethical systems."[348]

So Kidd came out where Liang Ch'i-ch'ao wanted him to—arguing for ethics, which was what Huxley had been doing. But Huxley, Kidd, and Liang Ch'i-ch'ao—three evolutionists for ethics—were not saying the same thing. Kidd, for one thing, had come to ethics by way of an evolutionary contradiction that was at least as troublesome as Huxley's contradiction between the ethical process and the cosmic process, but which was only deceptively similar to it. Kidd admitted that reason and religion, pitted against each other though they were, in the "supreme struggle," had both been spawned by evolution. Therefore, like Huxley, Kidd seemed to have the cosmic process at its own throat. And yet ethics were for Huxley to check the cosmic process and for Kidd to keep it going. Huxley wanted reason and ethics to stop the cosmic process; Kidd wanted ethics to stop reason from stopping the cosmic process.

Obviously their cosmic processes, both called evolution, were different beasts, and consequently, so were their ethics.

For Huxley, evolution had made man, but it was for man to make something of himself—and by his reason, not despite it: "In virtue of his intelligence the dwarf bends the Titan to his will. . . . Intelligence and will. . . . may modify the conditions of existence. . . . And much may be done to change the nature of man himself. . . . The intelligence which has converted the brother of the wolf into the faithful guardian of the flock ought to be able to do something towards curbing the instincts of savagery in civilized men."[349]

For Kidd, evolution had made man on the way to something better, or at least it had made present society on the way to a better society, if not a perfect one at least one close to perfect (Kidd expressed dismay that one of the "dignitaries of the Anglican Church" should still argue "that the *Regnum Hominis* can never be the *Civitas Dei*").[350] Evolution had brought man where he was and it would make man move on, whether he wanted to or not. Evolution would not let man drag his feet.

But Huxley *wanted* man to drag his feet, to hold his own, to carve a garden out of the jungle. For Huxley, "the supreme struggle" was a human struggle. It was man's struggle to break free from the bloody process that had produced him. For Kidd, the supreme struggle was a supra-human struggle. It was evolution's struggle, or "progress's" struggle, to get man, by playing, in almost dialectical fashion, one of his evolutionary traits against another, where evolution wanted him. Kidd's "progress" was almost like Hegel's "Spirit"; it worked "in the grand manner" and had its own purposes.

Huxley, as we have seen, simply did not believe in any such "progress." Consequently, he did not believe that the struggle for existence must be continued at all costs for the sake of such progress (which would make it, *nota bene,* no longer just a struggle for existence). Consequently, his ethics were ethics of self-restraint designed to reduce such struggle. But Kidd's ethics were just the opposite. They were ethics of self-restraint designed to get men to refrain from *avoiding* such struggle. Ethics and religion were for

Kidd things marvelously produced by evolution to keep the home fires of progressive competition burning—and burning brightly. "A marked feature," he wrote, ". . . of all the most advanced and progressive societies is the high pitch at which the rivalry of life is maintained within the community, the freedom of the conditions of this rivalry, and the display of energy and the constant stress and strain which accompany it."[351] Ethics, religion, altruism, all were evolution's mechanisms for keeping up the pitch.

But surely that was not what the ethics of the new morality were supposed to do! Liang Ch'i-ch'ao most heartily agreed that only a good ethical system could make China a "winning race," but he did not agree that a good ethical system was one which kept at high pitch "rivalry . . . within the community." His ethical system was supposed to stop such rivalry, so that Chinese would all fight as one team. He did not relish any "display of energy" in internal rivalry. He agreed instead with Huxley that races would lose if their members were to "waste their energies in contests with one another."[352]

Liang Ch'i-ch'ao's view, which reflected Yen Fu's, that democracy was good because it loosed the energies of the people, did seem close to Kidd's. So too did his hope that the Emperor, if only he could get rid of the Empress Dowager, might grant democracy and loose those energies in a Kidd-like altruistic manner. But the people's energies were to be loosed only so they could be reharnessed. The game was not won once "all the people [had been brought] into the rivalry of life on a footing of equality of opportunity." Liang Ch'i-ch'ao wanted his people to harness themselves, with unbreakable self-restraints, to China. He wanted them to pull together so they could pull together.

Liang Ch'i-ch'ao had missed Kidd's main point, as he had missed Huxley's, and therefore did not see that in actuality he was ethically still closer to Huxley than to Kidd. But Kidd did excite him, and Kidd did help him "beyond Huxley" to a new philosophy of life—for better or for worse.

Kidd gave him a "perfect" evolutionary explanation of self-sacrifice and death. It was perfect, of course, only if Liang Ch'i-ch'ao ignored or remained ignorant of the self-sacrifice Kidd had in mind. Kidd meant the sacrifice of private security for the sake of free rivalry. Liang Ch'i-ch'ao meant sacrifice of private freedom for the sake of national security. But Kidd *was* for "self-sacrifice." He spoke over and over again of the evolutionary necessity of "the sacrifice of the interests of the individual to those of the social organism."[353] Quoted out of context, Kidd was eminently quotable.

It is impossible to say whether or not Liang Ch'i-ch'ao knowingly quoted Kidd out of context. Perhaps he did—he knowingly ignored Kidd's prophecies about the Anglo-Saxons. Or perhaps he was misled by the translation (almost assuredly, as usual, Japanese). Most likely, however, he was simply so struck by hearing what he wanted to hear that he did not hear anything else and did not take time to make sure he had really heard what he thought he had heard.

What he did hear was the following evolutionary law: "This process of evolution cannot but sacrifice the individual for the good of society; it cannot but sacrifice the present for the good of the future."[354] This was necessary, meet, and right because the creation of the society of the future was the raison d'être of evolution. "Kidd believes," said Liang Ch'i-ch'ao (without realizing that such a belief re-admitted, without admitting it, a designer into Darwinism), "that the goal of natural selection is to bring the greatest number within a given race to the very fittest kind of existence. But the greatest number is not in the present but in the future. Therefore, the interests of everybody as an individual and the interests of the entire present body as a whole must all be sacrificed in order to reach this future goal."[355] For as Kidd put it, quoted Liang Ch'i-ch'ao, "The meaning of evolution lies in the creation of the future. The past and the present are only convenient avenues for transition.[356] . . . To speak honestly, the present is a sacrifice to the future. If one simply says there is only the present, then the present has not an iota of meaning or an iota of value. Only as it meets the needs of the future does the present have meaning and value."[357]

It was the pronouncement of this philosophy, which Kidd

called "futurism," (or so said Liang Ch'i-ch'ao—I have been unable to find such a word in Kidd's works) that made and makes Liang Ch'i-ch'ao's brief and misleadingly incomplete introduction of Kidd's thought so deeply significant. For futurism became Liang Ch'i-ch'ao's faith, and the faith of Lu Hsun, and Hu Shih, and Mao Tse-tung—and in time, *mutatō nomine* (as Marx would say)[358] the official faith of the People's Republic of China. That was not, of course, all Kidd's doing. Futurism was a nineteenth-century Western faith shared by Kidd with a sorry host of others, including Karl Marx. But in China, Kidd pronounced it first.

Kidd's inspiration was not Marxian, although it was Kidd who introduced Marx to China;[359] it was purely, however impurely, Darwinian. His futurism, however, was actually only the simplest, or most simple-minded, logical extension of the impurely Darwinian evolutionary-progressionism we have been talking about all along. That is why it seemed so perfect to Liang Ch'i-Ch'ao. But Kidd's statements were still important, because they made explicit what had only been implicit (or unrecognized) before. It was Kidd, therefore, thanks to Liang Ch'i-ch'ao, who gave China its first clear statement of the revolutionary new concepts of "the meaning of life," "the meaning of death," and "the value of the individual," towards which China's own Social Darwinists had been "progressing."

This was the new dogma, as Liang Ch'i-ch'ao extracted it:

> The good for which things have life cannot be in their own persons. Things have life only to be bridges to that great goal (the complete future). Death too exists as an important device for the reaching of that great goal. Therefore death is the great source of evolution.[360] . . . Death is the mother of evolution and a great event in human life. Everyone benefits the race with his death, and the present race benefits the future race with its death. Is not death's function noble? Clearly it is certain that, as it is only for the future that we have death, so it is only for the future that we have life.[361]

Death was necessary for progressive evolution. Progressive evolution was good (a "given"). Therefore death was good. Necessity once again became a virtue. But, just as suddenly, virtue became a moral obligation (by what authority it was not clear). Death became "a duty that everyone must fulfill."[362] And for some, apparently, the sooner it was fulfilled the better.

"A race's ability to develop," said Liang Ch'i-ch'ao, "depends sometimes, it is true, on its members' having long lives. But it also depends sometimes on their having short lives. For, when the changes met in the outside environment are especially violent, then only the short-lived are able to adapt to them. Why? Because only the short-lived can keep taking over from one another. Consequently, in habit, form, and nature they can change very quickly so that they can fit the times and survive." Dying early, therefore, could be a patriotic act and, if it was, death was nothing to be feared: "Such a death! Such a duty! What ground is there for hesitation, what ground for fear, what ground for dissatisfaction?" ("O death, where is thy sting?")

Seventeen years later, Lu Hsun made the identical argument, only even more forcefully, insisting that all men should perform their ultimate evolutionary duty not only fearlessly but joyfully:

> I think the continuation of the race . . . that is, the continuation of life . . . is indeed a great part of the work of the biological world. Why does it want to continue it? Obviously because it wants to evolve and progress. But on the road to progressive evolution the new must always replace the old. Therefore the new should joyfully go forward, to grow up, and the old should also joyfully go forward, to die. If everyone thus moves on, that is the road to evolution.
>
> Old people, get out of the way. Urging, encouraging, let the young go. If there are deep abysses in the road, then use death to fill them up. Let the young go.
>
> The young will thank them for filling the deep abysses, so that they themselves can go on; and the old will thank the young for going on across the abysses that they have filled. . . . On, forever on.
>
> If one understands this, then from youth to maturity to old age to death, all should pass joyfully, step by step; for most likely we will be new people surpassing our ancestors.
>
> This is the wide, correct road of the biological world! The ancestors of the human race have all trod it before us.[364]

Lu Hsun gave no explanation as to why one should be happy or how one could be happy throwing oneself into a rut for the sake of future generations, shouting joyfully, "over my dead body!" For Lu Hsun, the assumed goodness of the cause—progressive evolution, which like Liang Ch'i-ch'ao he accepted as a given—was to be cause enough for happiness. And that is still the official view in China today, where the problem of death is seemingly solved once and for all with two sentences: "If we die for the people, we die as we should. . . . If one dies for the benefit of the people, then one's death is weightier than Mount T'ai."[365] (Mao Tse-tung relied on Ssu-ma Ch'ien, just as Liang Ch'i-ch'ao—forty-two years earlier! —did in his Kidd essay,[366] to help inspire acceptance of man's newly recognized "duty.")

But this later official version was simpler than Liang Ch'i-ch'ao's. Liang Ch'i-ch'ao seemed more worried than either Lu Hsun or Mao Tse-tung that one's duty to die might not seem cause for joy. He did find some comfort in Kidd's contention that this ultimate evolutionary duty was in full accord with modern democratic principles—properly understood. Liang Ch'i-ch'ao appreciated the following ingenious (if diabolical) argument:

> The forces which are at work in the evolution of society are certainly, on the whole, working out the greatest good of the greatest number in a progressive community. But the earlier utilitarian conception of the greatest number has always related merely to the majority of the existing members of society at any time. The greatest good which the evolutionary forces, operating in society, are working out, is the good of the social organism as a whole. *The greatest number in this sense is comprised of the members of generations yet unborn or unthought of, to whose interests the existing individuals are absolutely indifferent* [original italics]. And, in the process of social evolution which the race is undergoing, it is these latter interests which are always in the ascendant.[367]

But Liang Ch'i-ch'ao clearly felt that such an argument still left something to be desired. He seemed to sense that it would be little fun for each generation, always a minority group, to have to sacrifice itself just because it was inevitably outvoted by the future.

Liang Ch'i-ch'ao could not have liked Kidd's statement that "existing individuals are absolutely indifferent" to the interests "of generations yet unborn." He did not want Chinese indifferent to China's future. He wanted existing Chinese to sacrifice themselves for future Chinese because "they cared." And he did not like Kidd's next contention either, that such caring, when it developed, was "ultra-rational." Liang Ch'i-ch'ao, after all, was trying to rationalize self-sacrifice. He seemed to want to combine Kidd's self-sacrificing altruism with the Benthamite notion that self-sacrifice could be in one's own self-interest. Yet again, he wanted both the fish and the bear's paws.

Liang Ch'i-ch'ao praised Kidd, therefore, for his theory of death. He said it was "thanks to this [theory] that Kidd can be [called] a revolutionary giant among evolutionists."[368] But he also said that Kidd's theory could "not yet be considered the last word,"[369] and admitting that "death is the thing most able to disturb men,"[370] he left the impression, at the end of his essay, that there was still something about death, and Kidd's theory, that disturbed him.

IMMORTALITY. It is not surprising, therefore, that in December 1904 Liang Ch'i-ch'ao came back to the problem of death and offered his own solution, in which he bolstered Kidd's—with Buddhism! He sought help from "religion," which would have pleased Kidd, but he did so to come up with an evolutionary theory of death that made self-sacrifice necessary, good, *and* rationally comforting—at least to his satisfaction.

The remarkable article in which he did this, "Yü chih ssu sheng kuan" (My view of life and death) is even more remarkable because it anticipated by fifteen years Hu Shih's famous theory of "social immortality,"[371] announced in 1919 in an article now much better known than Liang Ch'i-ch'ao's, "Pu hsiu—wo ti tsung-chiao" (Immortality—my religion). Again Kidd would have been pleased. Liang Ch'i-ch'ao enunciated "Hu Shih's theory," so perfectly, indeed, that one is tempted to accuse Hu Shih of plagiarism. For Hu Shih, the greatest second-generation Chinese Social Darwinist, did not acknowledge that the greatest first-generation Chinese Social Darwinist (aside from Yen Fu) had given him any help with his theory whatsoever. He simply claimed his theory had been "thought up at the time of my mother's funeral."[372] Perhaps he "thought up" Liang Ch'i-ch'ao's ideas without remembering how they had gotten into his head. But if he somehow never had read Liang Ch'i-ch'ao's article, if his ideas were the same by "pure coincidence," or because, as the Chinese also say, "great minds think alike," one must still admit that Liang Ch'i-ch'ao thought up Hu Shih's theory first. Liang Ch'i-ch'ao first pronounced in China the "good news" of a Buddhist-Darwinian doctrine of "social immortality."

We have seen "Darwinian Confucianism" and "Darwinian Taoism" in the thought of Yen Fu. Now we have an example of "Darwinian Buddhism" in the thought of Liang Ch'i-ch'ao. For the key terms in his doctrine were Buddhist, the *hsiao wo* (Little I) and the *ta wo* (Great I), while the "intimations of immortality" that led him to see those terms in such a bewildering new light came straight from one of Yen Fu's commentaries in *T'ien-yen lun*: "Mr. Yen of Hou-kuan, pulling together the words of past and present biologists, has said that 'in the body of an organism there is that which dies and that which does not die.'"[373]

Liang Ch'i-ch'ao faithfully quoted Yen Fu's hasty classification that "that which does not die is not what is called 'the soul',"[374] for he himself had already taken pains to point out that the concept of the soul had recently been largely discredited. Barbarian religions, he said, had all believed in the soul, and even "so-called civilized religions like Christianity" still did, but Confucianism had never put much stress on the word, and Buddhism had steadfastly opposed it, and now so did "modern European and American philosophers of the evolutionary school." And yet, soul or no soul, all those schools and all those religions, however different, past or present, scientific or superstitious, as Yen Fu now demonstrated so clearly, had "all in the end agreed on one point: When a man dies, something that does not die survives."[375]

All Yen Fu had meant by "that which does not die," as Liang Ch'i-ch'ao again explained faithfully, was heredity (*i-ch'uan*). Yen Fu, therefore, for all his Darwinian science, offered no more exciting an intimation of immortality than the traditional, commonsense observation that one could "live on" in one's descendants (if one could manage to beget some)—or at least one's eyes, ears, nose, and throat might. But Liang Ch'i-ch'ao saw in heredity something far more satisfying. It took some doing, but he managed it, by mixing Darwinian heredity with his favorite "scientific fact" of the "social organism" and with his newly remembered Buddhist truth of the "Little I" and the "Great I."

A cell, said Liang Ch'i-ch'ao, was to one's body as one's self was to one's race. Cells died, but the body lived on. Selves died but the race lived on. But cell and self—here came the leap of faith, or logic—also lived on in body and race even after their "Little I's" had died, for each lived on in its "Great I":

> I have a Great I and a Little I. It [one's "material" self] has a Great It and a Little It. What is the Great I? It is my corporate body. What is the Little I? It is my individual body. What is the Great It? It is the material entirety of my individual body. What is the Little It? It is each individual physical element within my individual body. Each Little It must die to keep the Little I alive. And the Great It must die to keep the Great I alive.[376]

The death of the Little I, being the same as the Great It, was not dwelt upon, but obviously that was because the point was that, as the Great I was mine, "I" would live on if it did. The great Darwinian-Buddhist news, therefore, originally printed in more than double-sized type was this: "We all die, and we all do not die. What dies are our individual bodies. What does not die is our corporate body."[377]

That this was indeed intended as good news, especially to those still facing with reluctance their evolutionary and patriotic duty to die, was proved by the exultant passage that followed: "I have a corporate body; therefore, my family does not die, so I do not die. My country does not die, so I do not die. My race does not die, so I do not die. My world does not die, so I do not die. My very universe does not die, so I do not die!"[378]

But how could that be so? How did the Little I live on in the Great I? How did a dead cell live on in a living organism. How did my being part of the Great I make the Great I part of me? How, in short, could Liang Ch'i-ch'ao's good news bring any comfort less cool than that which the theory of the conservation of energy could bring to a burning stick?

Liang Ch'i-ch'ao was trying to cheer up those who might have felt that he had taken Huxley's "self-restraint" too far—through "self-sacrifice" to almost literal "self-denial" ("It is only for the future that we have life"). But the Great I good cheer that so obviously excited him personally was not the original good cheer Buddhism offered in its Great I, although that had indeed been joyful "self-denial." That Great I was supposedly nirvana, which one had to lose oneself to find.[379] And yet the Great I as Buddha nature was in each Little I. Therefore, if one could deny oneself and see through oneself, one could see one's Self. If one could see through one's Little I, one could be one with one's Great I. There was immortality in extinction.

This was not the answer Liang Ch'i-ch'ao offered, although he toyed with it in the most unlikely form, for him, of the Christian doctrine of the Holy Trinity! Seemingly envisioning, not unreasonably, the "Holy Spirit" as something like the Buddha nature, he said that, according to the doctrine of the Holy Trinity, God, as the Holy Spirit, existed in each person as the "first life" of that person, and as such was called the "soul." What each person thought of as his life was really his "second life." Therefore it was only one's "second life" that came to an end. One's "first life," the oversoul, would naturally live on.[380]

This was a theory, said Liang Ch'i-ch'ao, that "not even those evolutionists most scornful of Christianity can disprove,"[381] but it was still not quite enough of a theory for him. His Buddhist self liked it, but his patriotic self wanted an immortality more down to earth. He seemed to realize that such a vision, Christian or Buddhist, of an absolute Humpty Dumpty breaking itself into pieces only to take them back again in the end, offered precious little immortality to each piece as a piece.

And yet there *was* a doctrine, exclaimed Liang Ch'i-ch'ao, that could do that—the Buddhist doctrine of karma—as long as it was ruthlessly kept "down to earth," down to the red earth from which earlier Buddhists had sought to escape. If one understood karma, said Liang Ch'i-ch'ao, one would see that one's immortality did not depend just on one's oneness with the universe. Nor did it depend only on the physical transmission of one's seed. One could be immortal through one's energy, for one's energy was indeed conserved, not only physically but socially, through the influence of one's deeds (the original meaning of karma). Every "unit" (and here he used Spencer's words *unit* and *aggregate*—borrowed from Yen Fu—for "individual" and *ch'ün*), every Little I, had (somehow) the power to "push," in a world in which force once initiated went on forever. Every deed was hence a cause, which had an effect, which became a cause, which had an effect, and so on to the end of time. Therefore, one truly lived on through one's deeds. One's "soul" could not go marching on—it did not exist—but one's "spirit" could, for spirit, as the Little I's energy, did exist. "I should like," said Liang Ch'i-ch'ao, "to call both Buddhist karma and evolutionary heredity *spirit*."[382]

Liang Ch'i-ch'ao first noted the immortality of the spirit of famous men, quoting from Yang Tu's preface to his, Liang Ch'i-ch'ao's, work, *China's Bushidō*, probably because he greatly appreciated Yang Tu's selection of famous men. But he did so only to note the most obvious of examples: "Confucius died, but the spirit of all the world's Confucians is his spirit. Sakyamuni died, but the spirit of all the world's Buddhists is his spirit. . . . Hence we can say that everyone in today's world who speaks of democracy is Washington. Everyone who speaks of militarism is Napoleon. Everyone who speaks of natural rights is Rousseau. And everyone who speaks of evolution is Darwin. Ever since Confucius, Sakyamuni, Washington, Napoleon, Rousseau, and Darwin were on this earth, from the past to the present, their spirits have been transmitted and spread abroad; there is no telling how many trillions of Confuciuses, Sakyamunis, Washingtons, Napoleons, Rousseaus, and Darwins have existed."[383]

"Darwin lives," Yang Tu seems to be exclaiming, which was good news, but not *the* good news. *The* good news, said Liang Ch'i-ch'ao, was that everybody lives—forever and ever, amen, through karma: "It is not only Sages, saints, and heroes who do not die.

The most stupid and most immoral also do not die."[384] But therein lay not only the promise but also the rub. For evil influences were thus as immortal as good. Saints and sinners were "alike in their immortality, but the former, bequeathing deathless good deeds, cause the Great I to enjoy good fortune, while the latter, bequeathing deathless bad deeds, cause the Great I to endure great suffering."[385] The Little I, therefore, endowed with the (absolutely unexplained) power of free will to create either good or bad "causes," had a fearsome responsibility, once he realized that, "although ancestors die, their deathless good deeds and bad deeds are both passed down to their descendants."[386]

Something had happened to the original Buddhist concept of karma. The Little I had always been "fearsomely responsible" for his actions, but only in the sense that he was "responsible to himself." For, in the old days, one's karma was on one's own head. One suffered the consequences of one's actions, for better or for worse, in this life and the next. One could create good "causes" or bad, but the "fruits" of those causes, which were always one's just deserts, were to be eaten only by oneself. "Society" did *not* suffer the consequences of one's actions, despite appearances to the contrary. If a barbarian chieftain put a city to the sword, he was planting a seed that would bear him bitter fruits in lives to come. But those put to the sword were not slaughtered innocents, for they were all partaking of the fruits that they themselves had sown in lives before. They and not their killers were "responsible" for their suffering. Strictly speaking, the worst person in the world was not "to blame" for society's suffering any more than society was to blame for the suffering of the best person in the world. For all suffering was self-inflicted. Therefore, according to the original concept of karma, there could only be "individual responsibility"; there was no "social responsibility."

Liang Ch'i-ch'ao could speak of "social responsibility" only because he had "socialized" karma—in two ways. For one thing, he insisted, whether on his own authority or with the backing of some unspecified Buddhist school I am not sure, that societies, families, clans, races, and countries, all had collective karmas.[387]

The karmic deeds of one generation would affect the "destiny" of the next. This, of course, made good common sense, but questionable Buddhist sense, as did his second contention. Denying the doctrine of reincarnation, at the same time that he tried to cheer people up by promising them immortality, he said that it was *only* society that could suffer or enjoy the fruits of one's deeds. One lived on *only* in one's deeds. There was no self reincarnate to taste them from outside. Therefore, quite to the contrary of the original concept, it was to society, and to future society, and not to oneself that one was responsible.

But what could such "responsibility" mean? If one would not necessarily suffer, personally, the consequences of one's actions, could one not afford to act "irresponsibly"? If the only immortality possible was the immortality of the influence of one's deeds, and if one were assured of that much immortality whether one's influence was for good or ill, why should one bother to be "socially responsible?" That, of course, was Kidd's question, and, not being able to answer it, he felt forced to declare that social responsibility, a progressive characteristic selected by evolution, was "ultra-rational."

But Liang Ch'i-ch'ao wanted a rational answer, and by rationalization he found one. He realized that, without reincarnation, he needed a new basis for responsibility, and he found one in the Little I's dependence on the *continuous* life of the Great I, which he defined for practical purposes—and surely this was one—not as the universe, but as his people. Granted, he said, the Little I lived on in one's deeds whether they were good or bad, but those deeds lived on only if the Great I lived on, and the Great I lived on only if there were enough good deeds to let it live on. For good deeds, socially responsible deeds, were, by definition, only those deeds that let the Great I live. Bad deeds, by definition, could make the Great I die. But, if the Great I died, then every influence of every one of the Great I's Little I's could come to nought, and then the Little I's would be dead indeed.

So Liang Ch'i-ch'ao had moved from promise to threat. Social immortality depended on social responsibility, and social responsibility was still that originally fearsome responsibility demanded by the new morality, the responsibility to sacrifice one's Little I for one's Great I and its future. The threat was clear, even though he half denied it:

> The reason I so anxiously expounded this doctrine is not that I wish to urge people to pray that they may quickly die to fulfill their responsibility. It is just that people do not understand the principle that we die and yet do not die. Therefore they think that their business and happiness are limited to the minuscule size of their bodies and their several dozen winters and summers, and that outside of that there is nothing. And, because they think that, the two concepts of society and the future do not develop. But the concept of society and the concept of the future are precisely what differentiate men from birds and beasts! [A return to the words of Mencius with an "updated" version of his belief that it was mortality that made man man.] [388] Without them there is no difference. Even within the human race these two concepts may be deep or shallow, broad or narrow. That is what distinguishes the gradations of the barbarous and the civilized. That is what distinguishes the numbers of the fit and the unfit and the victorious and the vanquished. [389]

Karmic immortality must be taken seriously, said Liang Ch'i-ch'ao, or else. For China's success in the struggle for existence depended on good karma.

That was neither a Buddhist message, nor a Darwinian one. Liang Ch'i-ch'ao's Darwinian-Buddhism was an unholy mixture either way you looked at it. But Liang Ch'i-ch'ao did not see that unholiness. He did see that there were discrepancies in the Buddhist and evolutionist views of things. He admitted that Buddhists and evolutionists were quite opposed in their major concerns. Evolutionists struggled to survive; Buddhists struggled not to survive, they struggled, that is, to escape the painful cycle of transmigration. But, having noted that "minor" discrepancy, Liang Ch'i-ch'ao glibly went one: "Their conclusions are indeed quite opposite, but, in saying that all living things die and yet leave something deathless behind, they are as one, without the shadow of a doubt." [390] He did not see that there was more than a shadow of a doubt, not only about the oneness of the two deathless somethings left behind, but also about the accuracy of his description in each case of just what was left behind. "Socialized karma" was not what the Buddhists had originally had in mind, and neither was "karmic inheritance" what Darwin had originally had in mind—not at least in the *Origin.*

For "karmic inheritance" was not only an inheritance of acquired characteristics, but of characteristics that could be acquired on purpose. Consequently, it could not easily be supported even by the vestigial Lamarckian passages in *The Origin of Species.* Liang Ch'i-ch'ao misstated Darwin's original theory when he claimed, in answer to his own question, What is the deathless, inheritable part of any organism?: "It is named 'character,' which is translated *hsing-ko*, what evolutionists call heredity. It means that every living thing transmits to its sons and grandsons every influence it receives from its environment as well as the behavior, habits, and nature that it creates."[391]

Darwin did in *The Origin of Species* suggest that certain characters caused either by influences of the environment or by forces of habit could prove inheritable.[392] His whole theory of evolution, after all, rested, on the inheritability of "variations." Sub-species and "new species" were to be selected from variant strains of old species, which strains could only exist to be selected if their variants could be inherited. "Unless profitable variations do occur," said Darwin, "natural selection can do nothing."[393] But, when he spoke of the *causes* of those variations without which there could be no "descent with modification,"[394] although he did say, "Something may be attributed to the direct action of the conditions of life. Something must be attributed to use and disuse,"[395] he stopped far short of Liang Ch'i-ch'ao's contention that much could definitely be attributed to the *creation* of behavior, habits, and nature.

Darwin admitted that "variability is governed by many unknown laws,"[396] and admitted as well that "the laws governing inheritance are quite unknown."[397] But as often as he spoke of variations arising from "use and disuse," he spoke also of nature *giving* variations[398] and of modifications *chancing* to arise[399] (even though he once admitted that "due to chance" was "a wholly incorrect expression").[400] He stated, moreover, that, even though animal breeders, through artificial selection, seem to direct evolution, "man does not actually produce variability."[401] Of man's conscious attempts to effect the evolution of birds and beasts he exclaimed, "How fleeting are the wishes and efforts of man!"[402] And, with regard to man's own evolution, he stated flatly, in *The Descent of Man,* that man has "risen [*sic*]—not through his own exertions."[403]

If mutations did not seem quite as "random" to Darwin as to later biologists, neither did they seem like things that could be consciously produced. "Use" and "disuse" seemed to hold out the possibility that will power might be involved in the altering of heredity, but Darwin—in the *Origin*—never stressed such will power. He gave no promise that one could "will" an altered character to one's descendants.

But to Liang Ch'i-ch'ao the chance to change one's bequest to one's descendants, by wishful efforts and exertions, was *the* promise of "heredity." That was the key to descent with *progressive* modification. Liang Ch'i-ch'ao seized upon the all-important question of just what was responsible for the variability in inheritance that made natural selection possible, but his answer was not anything like random mutation, nor anything like Darwin's ill-defined concept of unconscious use and disuse. He agreed that the environment could be a factor, but the exciting factor was karma:

> This, then, is heredity. But how is it that heredity keeps changing? I receive the force of the deeds of my grandfather and my father, but all of the multifarious influences that I receive from present society during the several decades of my life, from conception to birth, to childhood, to maturity, to old age—these my grandfather and father did *not* receive. I have two endowments. Therefore I have my own special nature. In the course of several decades, I daily wield this special nature to create deeds, whether good, bad, conscious, or unconscious, with which I influence modern society until I die. Then I take (1) what I have received from my grandfather and father, (2) what I have received from my present society, and (3) what I have in my own special nature, and transmit them all together to my son. What my son transmits to his son and what my grandson transmits to his grandson will also be like this. Indeed,

> what the people of ages past, of the present age, and of ages to come transmit to their sons and grandsons is all like this. That is why, though there is no extinction, there is change.[404]

Karma *created* change. There was karmic creativity at work even in the partially determining environment, for the environment was itself partially determined by the creative karma of one's forefathers. One received the fruits of one's forefathers' karma from two directions—directly, through their seed, and indirectly, through the society their deeds had helped effect. It was that double endowment plus a new double endowment from one's own karma that one bequeathed to one's descendants. And that bequeathal made them either more fit or less fit to survive.

Liang Ch'i-ch'ao was a proto-Bergsonian—of sorts. His ingenious Darwinian-Buddhism led him to propound at least a kind of "creative evolution" four years before Bergson wrote his *Evolution Créatrice*—which was ten years before Li Ta-chao and Ch'en Tu-hsiu discovered Bergson's book, and fourteen years before Liang Ch'i-Ch'ao himself acclaimed it.[405]

But *Creative Evolution* was not the title of Charles Darwin's first book. One would have had to stretch the vestigial Lamarckian passages in *The Origin of Species* even further than the Lamarckian giraffes had stretched their necks to find support in the *Origin* for Liang Ch'i-ch'ao's creative karma. For Darwin, even when he had admitted the possible effects on heredity of use and disuse, had been talking about inheritable, *physical* "characters." Liang Ch'i-ch'ao was talking about "character." What he was actually talking about was China's moral character, which was what he wanted to change through inheritable karma.

Part of the problem was a mistranslation, or at least a misleading translation. The word *hsing-ko,* which Liang Ch'i-ch'ao cited as the translation of Darwin's word *character,* when used of people, as it almost always was, referred to one's "personality," not to one's physical traits. Talk, therefore, of inheritable *hsing-ko* may have made it easier for Liang Ch'i-ch'ao to leap from facts in the "biosphere" to conclusions in the "noosphere," (if we may borrow the spheres of Pierre Teilhard de Chardin anachronistically).

But Liang Ch'i-ch'ao had long before shown a typical, Social Darwinian ability to do that anyway, even without mistranslations.

In his discussion of heredity, whether because of his own misinterpretation or someone else's mistranslation, or both, he leapt so quickly out of the biosphere that he actually had to stop and point backwards to remind his readers that "'heredity' is to be seen not only in immaterial character, but also in material bodies." Having acknowledged that fact, however, he *dismissed* physical inheritance, as the cruder part of the inheritance package, and went back to talk of *hsing-ko* as a psychological inheritance: "Today the science of 'national psychology' and 'social psychology' is steadily developing. What is national psychology? What is social psychology? It is that deathless something that the dead of an entire nation or society bequeath to their descendants."[406]

Liang Ch'i-ch'ao was talking about "national character," about "Chinese characteristics," of the sort Arthur H. Smith had first delineated so unflatteringly in 1890 in his book of that name (at least one chapter of which, "The Absence of Public Spirit," had been translated into Chinese in the *Hsin min ts'ung pao* by March 1903).[407] And karmic inheritance offered the only hope of changing those characteristics, for, if the effect of deeds could become hereditary, then one could not only effect change in the national character in the present, but also preserve that change for the future. Generation after generation could—if they would—improve upon it.

To be able to believe that such a promise was all part of the Darwinian theory of evolution, Liang Ch'i-ch'ao had to lump together in his mind, without distinction, both "biological inheritance" and "cultural inheritance." He had to assume that "socialization," learning, education, and sexual reproduction were all part of the same process of reproduction—with modification. He had to assume, in effect, that one word, *heredity,* could explain the phenomenon of a redneck passing on both his red neck and his blue eyes.

But who would not admit that red necks are often "inheritable"? The whole world hopes the prejudices of fathers need not be passed on to their sons. Indeed, "enlightened" sons hope, and believe, that, by willfully rejecting the prejudices of their fathers, they can *change* the "inheritance" of *their* sons. That is all Liang Ch'i-ch'ao really hoped for, for he wanted to change attitudes and behavior more than he wanted to change skin and bones.[408] In the realm of biological science his theory of karmic inheritance perhaps looks silly, but in the realm of common sense and even in the fuzzy realm of behavioral science his theory does not look silly.

Liang Ch'i-ch'ao's greatest "error" was to claim to have Darwinian support for his theory. But the great irony is that, if there was no such support, except for the vaguest of hints, in *The Origin of Species,* the only Darwin Liang Ch'i-ch'ao knew at all, there *was* such support in *The Descent of Man,* and even more so in *The Expression of the Emotions in Man and Animals.*

In *The Descent of Man,* Darwin reaffirmed his belief that "habits . . . followed during many generations probably tend to be inherited."[409] But the habits he this time had in mind had nothing to do with the use or disuse of bodily parts. His claim was that "it is not improbable that after long practice *virtuous* tendencies may be inherited."[410] Indeed, he even promised that "looking to future generations . . . we may expect that virtuous habits will grow stronger, becoming perhaps fixed by inheritance. In this case the struggle between our higher and lower impulses will be less severe and virtue will be triumphant."[411]

Moreover, in *The Expression of the Emotions in Man and Animals,* Darwin insisted that "the mysterious power of the will" could have something to do with the acquisition of inheritable characteristics.[412] He said that "actions, which were at first voluntary, soon became habitual, and at last hereditary, and may then be performed even in opposition to the will."[413] Actions–karma–could lead to "instinct."

When Darwin wrote the above sentence he was not talking about "good instincts." He was talking only of certain expressive movements, originally voluntary but finally "instinctive":

> A man often wishes to make certain gestures conspicuous or demonstrative, and will raise his extended arms with widely opened fingers

> above his head, to show astonishment, or lift his shoulders to his ears [*sic*], to show that he cannot or will not do something. The tendency to such movements will be strengthened or increased by their being thus voluntarily and repeatedly performed; and the effects may be inherited.[414]

But, if one considers that statement together with another statement, from *The Descent of Man,* "Sympathy, though gained as an instinct, is also much strengthened by exercise or habit,"[415] Darwin might seem to be saying that one could by willful deed improve one's descendants' moral instincts.

And that, of course, was just what Liang Ch'i-ch'ao wanted to do, to so "instill" his new morality in his generation that it would become "ingrained" in generations to come. He wanted his contemporaries to so strengthen voluntarily their every tendency to *k'o chi* (control themselves) that their descendants would *k'o chi* "naturally." He wanted his people to so "practice" nationalism that their children would have *ai-kuo hsin* (patriotic hearts) that were as instinctual as the *ts'e-yin chih hsin* (sympathetic hearts) Mencius had thought all men to have *ab origine.* The ancient Mencian dream that "all men could be Yaos and Shuns" could be effected, in time, by changing Chinese heredity through creative karma.

This view blended traditional Mencian optimism with an unorthodox Buddhist activism and a seemingly Darwinian progressionism, and it was a view that convinced many. Mao Tse-tung certainly shared Liang ch'i-ch'ao's faith in improvable heredity, at least in his vintage years. In old age, he seemed to fear "bourgeois thought" as an almost inirradicable original sin, but he once believed, and certainly wanted to believe, that "proletarian consciousness" could be instilled and inherited. There was a hint of such a faith in his own poetic Mencian echo: "Six hundred million Chinese shall all be Shuns and Yaos."[416] He meant that the people would rule, but also that they would be good—with proletarian class nature.

Darwin, of course, would have dismissed both Liang Ch'i-ch'ao and Mao Tse-tung as most unscientific disciples, but there was no way he could have washed his hands completely of them or of any other Social Darwinists, for Darwin himself sounded more like a Social Darwinist the more he wrote. Certainly his final open step into the behavioral sciences was at least a little step towards a doctrine of karmic inheritance.

Liang Ch'i-ch'ao had almost assuredly not read either the *Descent* or the *Expression.* If he had actually read any Darwin at all, it was probably only the first five chapters of *The Origin of Species,* translated and published by Ma Chün-wu in spring 1904, some six months before Liang Ch'i-ch'ao published "My View of Life and Death." (Ma Chün-wu had published Chapters Three and Four, "Struggle for Existence" and "Natural Selection," in 1903. His complete translation—China's first—was not finished until 1919.)[417] Liang Ch'i-ch'ao might, of course, have read more Darwin in Japanese, but he had not yet quoted any Darwin from either the *Descent* or the *Expression.* He seems to have leapt to Darwin's most ill-founded conclusions largely unassisted.

It is not difficult to understand how or why he leapt to those conclusions, however, for there was something undeniably neat in Darwinian Buddhism. The confusion of karma and heredity was the perfect confusion to bring both comfort (to those now told that the moral Way led through self-sacrifice), and hope (to those who desired above all else that China be saved). Chinese who believed in the struggle for existence and in natural selection, and who believed that "Chinese characteristics" were for the most part now unfit, *had* to believe there was a creative element in evolution, for, if there was not, then they were lost. That is why Yen Fu and Liang Ch'i-ch'ao, as we have seen, "instinctively" assumed from the very beginning that it was those who *made* themselves fit who survived. They had to believe they could do something, and they wanted to believe the effects of what they did would last. Karma and Darwinian heredity "perfectly" explained how both could be. That is why Liang Ch'i-ch'ao could sigh in wonder that Buddhism and Darwinism, despite their differences, could at least in this regard "fit together like two halves of a tally."[418]

The neatest thing, however, in Liang Ch'i-ch'ao's essay "My View of Life and Death" was his final coup, in which he brought karma, heredity, and one *good* Chinese characteristic all together into one neat new package of saving grace. In arguing that karma affected the future life of a whole society, not just of an individual, he had argued that, at the family level, Chinese had accepted that principle all along. Chinese had always believed, he said, that the good or bad deeds of family members would bring down blessings or disorders on the family some time, even though it be several generations post facto. And they had been right to so believe. "Not only can our race's two-thousand-year-old belief in family and clan retribution not be disproven by modern science," he proudly stated; "modern science has made its bulwarks all the stronger!"[419] Because of that belief, Chinese had long since become accustomed to sacrificing private interests for family interests. Where the family was concerned, Chinese did know how to *k'o chi.* And therein lay China's hope: "Even our stupid fools know how to give up some of their present happiness to have happiness when they are old, how to sacrifice some of their private interests for the interests of their family, clan, and descendants. This kind of habitual nature [*hsi hsing*] is even stronger in our people than in people of other countries, and it is the element that will enable our country in the future to compete in the world."[420]

Chinese had one "hereditary" advantage! If only they could fit it to the modern world by conscious karma, so that Chinese would be willing to sacrifice themselves for their country as they had always been willing to sacrifice themselves for their families, then China "could compete." That was the ultimate promise of karmic inheritance. Liang Ch'i-ch'ao had set out to promise immortality to each individual. He ended up promising something even dearer (inexplicably) to his heart: the immortality of his people.

Such was the promise of the new morality, a morality that, if not yet as internalized and inheritable as Liang Ch'i-ch'ao would have hoped, is still nonetheless very much with us—in the People's Republic of China. Liang Ch'i-ch'ao, of course, is no longer acclaimed as even one of its authors. But if Liang Ch'i-ch'ao could somehow see the new new morality of the new China, he would recognize in it many of "his" virtues, and much of his moral philosophy, and at least a de-Buddhacized version of his doctrine of karmic immortality.

He could see all that in the "life" of Lei Feng, that surely half-mythical model hero who so personifies the very breed of new Chinese Liang Ch'i-ch'ao hoped would soon "evolve." Lei Feng, the perfect P.L.A. man, was a young soldier, happily an orphan, who was able to "serve the people" as his family ("What, have you no family?"—"No, I have a family. The Party and Chairman Mao are my new father and mother, the People's Commune is my family").[421] His hereditary sense of filial piety has been transformed to "class friendship."[422] He wanted to serve the people as "a rustless revolutionary screw in a machine,"[423] as selflessly, that is, as a cell in a social organism. And to do that he was ready both to assert himself and to restrain himself. He seemed instinctively to realize that Liang Ch'i-ch'ao's cardinal pair of virtues, self-assertion and self-restraint, did not reflect a double standard. The people's good was the standard that made both good. All men and women should *li cheng shang yu* (struggle upstream with all their might) and *tzu li keng sheng* (live by their own efforts)—they should assert themselves—but only *wei jen-min* (for the people). Therefore they must also *k'o chi* (restrain themselves) from every selfish impulse that could lead them from that duty.

Lei Feng was willing to "restrain himself" to death. "A person's life has a limit," he wrote, "but serving the people is limitless. I want to cast my limited life into the limitless service of the people."[424] The people would live on, and the Little I would live on through service that helped the people live on. Lei Feng did not talk about the Little I or about karma. But he did talk about "good deeds." He called them "bricks for socialism," but they were as lasting as karma, and their good was as far in the future, and it was only through them that one's life could have meaning. Lei Feng gave his life laying bricks for socialism, but, "although Lei Feng's life is over," we are told, "his spirit continues to shine forth undiminished."[425] Liang Ch'i-ch'ao would have sighed, "Amen."

The fact that the late Minister of Defense, Lin Piao, may have had much to do with the making of Lei Feng into a model hero whose greatest virtue was his selflessness should do nothing to destroy the irony of Lin Piao's being denounced in the campaign against him and Confucius for hanging up in his bedroom Confucius's words *K'o chi fu li* (Conquer yourself and return to decorum). If Lin Piao did hang up those words, he undoubtedly did so because he was for *k'o chi,* not for *fu li,* but his critics insist he was for *fu li* not *k'o chi,* claiming that by *fu li* he meant *fu pi* (restoration), restoration, of course, of capitalism. Poor *k'o chi,* therefore, has suffered disgrace through guilt by association, but most unfairly, for *k'o chi*—self-conquest, self-restraint, self-denial—has been, is, and always must be one of *the* most important virtues of the people's new morality—until, perhaps, humanity, now "still in its youth,"[426] according to Mao Tse-tung, so comes of age that all men, like Confucius at seventy, will be able "to follow their heart's desires without stepping out of bounds."[427]

The virtue *k'o chi* did come originally from Confucius. How strange that it should be brought back by Yen Fu and Liang Ch'i-ch'ao as an evolutionary ethic proclaimed by Huxley, and how silly that it should be thrown out, as babe with bath, by those very anti-Confucian Social Darwinian Marxist-Leninists who in social practice most insist on it.

MORAL IMPERATIVES AND IMPERIALISTS. There was one more imperative of Liang Ch'i-ch'ao's new morality that the People's Republic has not thrown out, either in theory or practice—the injunction "Oppose imperialism." Anti-imperialism was the very raison d'être of the new morality. The new morality was to preserve China from imperialism, and yet the Darwinian, utilitarian basis of the new morality raised a terrible problem. It seemed to make it impossible to mount a moral argument against imperialism. The anti-imperalist new morality could not call imperialism immoral.

Back in 1839, when Chinese were just beginning their long acquaintance with Western imperialism, Lin Tse-hsu could demand of the young Queen Victoria, then in the third year of her reign, "Where is your conscience?"[428] By 1901, however, when Queen

Victoria died, and when, to the best of my knowledge, Liang Ch'i-ch'ao first used the word *imperialism* (a new word in Chinese, made in Japan), Liang Ch'i-ch'ao had long since decided against appealing to the consciences of imperialists, not because he believed they had none, but because he felt they need feel no qualms of conscience about their imperialism, imperialism being "perfectly natural."

It is a sorry fact that Liang Ch'i-ch'ao first used the word *imperialism* to announce that "America is turning from republicanism to imperialism," America having just seized Cuba, Hawaii, and the Philippines. This dramatic turn of events he saw as a terrible threat to China, and yet he refused to blame America, for he insisted that America "in recent years has been forced by the struggle for existence; it *had* to extend itself far beyond the Western Hemisphere." For America, despite its seemingly splendid isolation, was inextricably embroiled in the international struggles of modern Europe, struggles far more serious than those waged of yore by *private* imperialists like Ch'in Shih Huang or Napoleon: "Their motive force arises from the struggle for existence between national peoples. To judge from the evolutionists' laws of the struggle for existence, natural selection, and the survival of the fittest, there are probably some who cannot stop even though they wish to."[429] If the blackguards could not help it, how could they be blackguards?

In the very first issue of his *Hsin min ts'ung pao,* Liang Ch'i-ch'ao was already propagating the almost Leninist doctrine that "nationalism" had naturally to evolve into "national imperialism."[430] That was the result of Darwinian law. But it was also, rather oddly, the even more direct result of Darwinian theory. The theories of Malthus and Darwin, he said, were the theories most responsible for "giving birth to national imperialism."[431] Darwin had demonstrated that "there are no equal rights in the world, there is only the right of force. Rights are sought and won, they are not bestowed by Heaven." Once Darwin had made *that* clear, "there were none who dared not strive to be strong and fit, so that they could win a place in this world of competition and natural selection.

. . . It is from this that has come the national imperialism of the modern age."[432] Therefore, said Liang Ch'i-ch'ao, the imperialist nations then converging on China were "all driven by the tide of this new ism; they cannot do otherwise."[433]

Idea had subtly replaced nature as the cause of imperialist aggression, but Liang Ch'i-ch'ao had not noticed. Nor did he question the idea. As we have seen before, he accepted the very worst message Darwin could possibly have brought to China:

> If a country can strengthen itself and make itself one of the fittest, then, even if it annihilates the unfit and the weak, it can still not be said to be immoral. Why? Because it is a law of evolution. Even if we do not extinguish a country that is weak and unfit, it will be unable to survive in the end anyway. That is why violent aggression, which used to be viewed as an act of barbarism, is now viewed as a normal rule of civilization.[434]

Liang Ch'i-ch'ao knew perfectly well that many Westerners were using precisely that argument to justify their very "worst" designs against China herself, and yet he accepted it. "Would you call it a crime?" he asked. "But of all living things in the universe who does not struggle for his existence? Can extending one's power as far as possible in order to survive be called a crime? Who forces you to be content with unfitness and defeat?"[435] The worst criminals, he said, are not those who take away others' freedom, but those who give up their freedom.[436] If one suffers imperialists to enter, one deserves to suffer.

So he returned to one of his earliest arguments, borrowed from Mencius, "A country must attack itself, before another can attack it."[437] And he returned to Confucius's teaching, "The noble man seeks for the fault within himself, the villein blames others."[438] Liang Ch'i-ch'ao told Chinese to blame themselves, not the imperialists.

Not only were imperialists not to be called "bad"; there were some suggestions in Liang Ch'i-ch'ao's journals that they should perhaps be called "good," although the amorality of the "need" to struggle should have vitiated "good" as well as "bad." A week after Liang Ch'i-ch'ao first used the word *imperialism,* he published the first installment of a Japanese article entitled "Imperialism," which presented a paean to imperialism that could not have been bested by the best (or the worst) of white imperialists anywhere:

> As for those primitive peoples who are swallowed up and assimilated by civilized nations, their plight, of course, is pitiable, and yet, as they do belong to the unfit races, had they *not* the help of the fit races, they would become extinct in the end nonetheless. As it is, the civilized nations, without any reward or thanks, train them and teach them and have no other designs but to mobilize and use them. So it is with the Japanese people's treatment of the Ainu race and the natives on Taiwan. If people call this the tyranny of the strong, if they say that the Japanese have no sense of morality, then they really do not understand what real morality is.[439]

The Japanese were learning fast. They had already joined "the Western Powers" in their siege of Peking against the Boxers in 1900. But Liang Ch'i-ch'ao, far from denouncing the "dwarf pirates" for presuming to assume a share of the "white man's burden," himself actually echoed their argument. "Ruling the territory of barbarian countries," he pronounced, "is an evolutionary right that civilized nations ought to enjoy; and enlightening the people of barbarian countries is an ethical responsibility that civilized nations ought to perform. China, though famed throughout the world as the first ancestor of civilization, has in several thousand years had only one or two men who have embraced this idea. That is China's shame."[440]

Liang Ch'i-ch'ao did not fully endorse the "white man's burden," but he did endorse the "imperialists' burden," and that endorsement led him, and many others, to dream of *Chinese* imperialism. If imperialism was the Way of the fit, then to be fit China too must become imperialist. And China *could* become imperialist. "If we wish to save China," cried Liang Ch'i-ch'ao, "there is no other way but to build a nationalistic nation," for then China could evolve not only into an imperialist nation but into the greatest imperialist nation: "If as the largest race on this planet we are able to build a country fit for evolution, then who will be able to usurp from us the title of the Number One Imperialist Nation on Earth?"[441]

Actually, he had dreamt the same dream even before he had a word for imperialism: "Someday, in the twentieth century we Chinese will be the most powerful race in the world.[442] . . . Who will have the actual power in days to come to open up the world? Our Chinese race! The whites are arrogant and not up to hardship. The blacks and browns are lazy and unintelligent. Therefore, except for us yellows, there is no race that could undertake such a task. North America and Australia today are areas for the white race's colonies. South America and Africa will someday be areas for the yellow race's colonies, . . . of that there is no doubt."[443]

One contributor to the *Hsin min ts'ung pao* even saw yellow hope for Australia: "When the strong meet the weak, force becomes reason. If we yellows have no force, then not only Australia may go white, Asia itself may go white, and what "reason" will we have to rely on. But, if we yellows do have force, then since when can a white Australia not become a yellow Australia?"[444] Another contributor had even wilder dreams: "If our race can really develop culture in places of colonization and establish new countries, then outside China we can gain innumerable new Chinas, and the whole world will be occupied by our race. These are not empty words! . . . If we advance, the whole earth will belong to us Chinese. If we retreat, the whole earth will no longer have any of us Chinese. There are two choices, . . . and we must choose one or the other."[445] So the only real choice was to run full steam ahead towards imperialism.

Liang Ch'i-ch'ao printed such dreams of global conquest, and he himself was not above them:

> China's land is not poor, and yet her country is poor. China's people are not weak, and yet her soldiers are weak. This is a strange phenomenon in our world. There must, therefore, be some devil at work behind her poverty and weakness. Our people should seek this devil and cast it out, for then, who knows, the twentieth-century stage could be ours alone.[446]

These were all dreams of glory inspired by near despair. If not absolutely "empty words," they were brave words of the "just-you-wait" variety that must only briefly have dispelled the fears even of their authors. Nonetheless, Liang Ch'i-ch'ao and his confederates defied the imperialists with the cry "We will out-imperialist you!" Anger and hurt pride combined with Darwinian theory to lead them to reject the cry "You are evil" as the cry of a weakling (the cry, Lu Hsun would tell them, of Ah Q). So, instead, they said to the imperialists, "Do what you must, and we will outdo you."

Gone for the moment was all talk of the *Ta t'ung.* No one yet suggested that "peaceful co-existence" could be the solution to the struggle for existence, and, as they did not, no one yet protested that "China would be no superpower" either. Instead Liang Ch'i-ch'ao's younger brother, Liang Ch'i-hsun, reminded their readers that "evolution has no heart,"[447] and that it therefore viewed the superpowers as the superfit. "Human races," wrote another confederate, "present natural barriers that cannot be leveled. . . . Consequently it is inevitable that the weak be the meat of the strong, and it is only reasonable that the superior should win and the inferior lose. Although men speak of equality and universal love, they are not easily practiced in the violent competition of evolution's realm."[448] That was why Liang Ch'i-ch'ao himself could survey the dog-eat-dog record of world history and exclaim: "Where do we see 'civilization'? Where do we see 'international law'? Where do we see those who loved their enemies or their neighbors as themselves?"[449]

But, if there was no morality to be seen, neither was there any morality to appeal to. The oddest thing was that no appeal was made even to the virtue of *k'o chi.* People were supposed to restrain themselves, but peoples did not have to. Indeed, they were told that it would be suicidal if they did. It was finally clear that *k'o chi* belonged indeed to "private morality." If there was no double standard between self-assertion and self-restraint for the individual, there was certainly a double standard between individual morality and national morality. Nations did not have to make a virtue of self-restraint. Morality had to rule within because amorality ruled without. Huxley's real message had not been heard. That is why Liang Ch'i-ch'ao did not shout back at Huxley, "Then tell your countrymen to restrain themselves!"

Instead he concluded in this wise:

> When European countries meet with other European countries, they all take reason to be force, but, when European countries meet non-European countries, they all use force for reason. And so they must, because of evolution. The struggle for existence makes it natural. So what cause can there be for blame? What cause can there be for hate?[450]

What cause could there be for moral indignation? That was the theory, but that was also its weakness. For patriotic Chinese were *seething* with moral indignation. Using the same quotation from the *Tso chuan* that Yen Fu had used in his anger in 1895, someone wrote in 1903 that after the Opium War "anger and hatred entered the bones of the Chinese, and they longed to 'devour' [the foreigners'] flesh and sleep on their hides." And that anger and hatred were still in their bones. The author lamented only that the Chinese were still not yet able "to do unto the Europeans precisely as they have done unto us, so that the words *the yellow peril* may again strike terror in their hearts!"[451] It is sad, perhaps, to have to find proof of an "innate moral sense" in indignation instead of in some sort of Mencian-like compassion, but expressions of indignation are enough to prove that most Chinese patriots felt that imperialism was *wrong.*

In Liang Ch'i-ch'ao's journals one can find a host of articles like that translated from a Japanese paper on "The Barbaric Behavior of a Civilized People," which described alleged atrocities committed by Russians in Manchuria.[452] Another entitled "Humanitarians or Enemies of Humanity?" demanded whether the Boxers or the Allied Powers were the real enemies of civilization. The author called the foreigners a bunch of brutal savages and finally growled, "The unspeakably barbarous Powers call themselves civilized and preach humaneness. Whom do they think they're kidding?"[453]

For a long time, there was in every issue of the *Ch'ing-i pao* a special section entitled "Meng hsing lu" (A record to make us fiercely examine ourselves) devoted to the listing of current imperialist insults and injuries, along with commentary. "The editor," who could have been Liang Ch'i-ch'ao himself, did, to be sure, often echo the new morality's (and Mencius's) theory that, "when other people oppress you, it is not their fault but your own," and in keeping with the section's title he asked, "What good does it do if one does not examine oneself but simply blames others?" But when he described the actions of those "others" one should not waste time blaming, those others who were "chopping up our Chinese territory, killing our Chinese people, raping our Chinese women, and robbing our Chinese property," his words, despite his theory, still burned with moral outrage.[454] The moral outrage that Chinese evolutionists theoretically had no right to have, they did have nonetheless.

The "amorality of imperialism" was hard to stomach and, because it was, a few modern thinkers began to find it hard to stomach Darwin—that is, all of the Social Darwinism of Liang Ch'i-ch'ao. It was not that they repudiated the theory of evolution or even the struggle for existence, but, if imperialism was "right," they felt that something must be wrong. Whereas many without anger acknowledged that Darwin's theory had given rise to national imperialism, in 1902 one brave soul finally *blamed* Darwin's theory for giving rise to national imperialism. "Europeans," he said, "pretty-up imperialism with excuses based either on Nietzsche's extremist individualism or on Darwin's theory of evolution—but if you say it straight, imperialism is brigandism!"[455]

Imperialism, therefore, the phenomenon that first made Chinese rush to Darwin, was ironically also the phenomenon that first made some at least back away. If Darwin was for imperialism, how could he be good?—And yet was he not right? That was the terrible problem. No one could yet mount a counter-argument against Liang Ch'i-ch'ao's conclusions. Darwin seemed to make such sense. And the world outside seemed so Darwinian. But something was wrong, something that one pen-named poet tried to catch in the last issue of the *Ch'ing i pao,* in a poem written in challenging response to a poem by Liang Ch'i-ch'ao, purposely echoing his rhyme, but not his reason:

Oh, how I dread
To talk of evolution!
For if fit flourish and unfit fail
Then whom do we dare hate?[456]

SEVEN

The Real Revolutionaries

The "real Revolutionaries" had very few ideas that were really revolutionary, at least about Darwin and revolution. And yet their revolutionary ideology was definitely Darwinian, and their leader, Sun Yat-sen, was an "independent" Chinese Social Darwinist, one who did not owe his first knowledge of Darwin to Liang Ch'i-ch'ao or to Yen Fu.

DARWIN AND THE REPUBLICANS

Sun Yat-sen did not put Darwin's name into print until two years after Yen Fu had done so (1897 vs. 1895). Moreover, he almost invariably used Yen Fu's or Liang Ch'i-ch'ao's Darwinian vocabulary whenever he mentioned Darwin, in Chinese.[1] Nevertheless, he must first have learned of Darwin in English and from Westerners in Hawaii or in Hong Kong, in the course of his formal "foreign" education, which he completed in Hong Kong in 1892, with his graduation from the British-run "College of Medicine for Chinese." For when he did first mention Darwin, it was in a description of that education, in a brief autobiography written in England in 1897, after his rescue from the Chinese Legation. "From an early age," he wrote, "I had lofty ambitions, and by nature I longed for the new and the strange. Therefore, most of what I studied was a broad, if impure, mixture of things. In Chinese studies I loved only the literature of the Three Dynasties and the two Hans. In Western studies I was most fascinated by the Way of Darwin. As for religion, I worshiped Jesus, and, as for men, I revered China's Kings T'ang and Wu and America's George Washington."[2]

This remarkable trinity of Western influences—Darwin, Jesus, and Washington—surely should have produced some ideas that were revolutionary, but, by the time Sun Yat-sen got his Darwinian arguments for revolution into print, they were no longer really revolutionary, because they were already old Liang Ch'i-ch'aoian hat.

This was certainly true of the real Revolutionaries' first Darwinian argument for revolution, which they presented, however, not as an argument but as a "law" that would brook no argument. Tsou Jung stated it most axiomatically, on the second page of his *Ko-ming chün* (A revolutionary army), published in 1903: "Revolution is a law of evolution."[3] But Liang Ch'i-ch'ao, as we have seen, had said the same thing a year before.

Somewhat later, the *Min pao* admitted, with some grace, that, "according to what biologists say, there are a great many paths to evolutionary progress," but it immediately went on to insist that, "for some, evolutionary progress can only come after a revolution,"[4] and so it would have to be for China. Sun Yat-sen said that, "in these times, if you do not struggle, there is no way to exist."[5] And Tsou Jung said, "If our China wants to exist in the new world of the twentieth century, we must have a revolution! If our China wants to be a famous nation on this earth, if it wants to be lord of this earth, we must have a revolution!"[6] "Otherwise," put in Chang Ping-lin, "we will be going against the tide of evolutionary struggle, we will be taking the position of the unfit, of slaves or beasts of burden, and in the end we will only go the way of the red savages and the brown barbarians!"[7]

The real Revolutionaries, therefore, simply equated revolution with struggle for survival, thus making it both necessary and natural—which Liang Ch'i-ch'ao had done before them. But Liang Ch'i-ch'ao had not had in mind what the real Revolutionaries had in mind—the violent overthrow of the Ch'ing dynasty—and of the Manchus.

RACIAL REVOLUTION. It was as "anti-Manchus" not Republicans that the Republican Revolutionaries gained their greatest notoriety, for their rivals, the Reformers, were also for a republic, at least as their ultimate goal. The main difference between the Reformers and the Revolutionaries was simply that the former believed China could and should get to a republic in an orderly fashion *by way of* a constitutional monarchy (even with a Manchu monarch), while the latter believed China could get nowhere until the Manchus were thrust aside.

But, even in their militant anti-Manchuism, the Republican Revolutionaries were not intellectually original, for the "best" reasons they managed to come up with for their first People's Principle of anti-Manchu "racism" (*min-tsu chu-i*) were the Darwinian reasons so generously provided them by their chief opponent, Liang Ch'i-ch'ao, in his angry first essays from exile.

The Republicans, of course, did not need either Darwin or Liang Ch'i-ch'ao to tell them to be anti-Manchu. Even Sun Yat-sen was anti-Manchu before he was pro-Darwin, for he had been inspired to hate the Manchus as a boy by an uncle who had fought for Hung Hsiu-ch'üan, the "Heavenly King" of the Taiping Rebellion.[8] The Manchus were, after all, foreign conquerors, and, even if foreign conquest was not by definition an atrocity, they had definitely committed atrocities in their conquest. Worse yet, they had been in power throughout all of China's nineteenth-century humiliations. They had not been able to prevent them, and they did not seem to be doing anything that would prevent the ultimate humiliation of outright partition under white rule. Surely they had no claim to the Mandate of Heaven. (And surely they were the perfect scapegoats for China's weakness.)

But the Republican Revolutionaries found to their dismay that not all Chinese shared their "natural" hatred of the Manchus. Many had to be reminded that the Manchus were foreigners, and that the (conquering) Manchus were guilty of atrocities.

But, as the Revolutionaries leapt to the task of reminding their countrymen of such things, and reawakening racial hatred, they found that such vengeful anti-Manchuism was easily subject to reformist attack, and they became sensitive to such attack, especially to the charge that there was nothing more to their anti-Manchuism than a bloodthirst for revenge two and a half centuries too late.

Vengeance, of course, *was* one of the Revolutionaries' first cries. It was the favorite cry of the ill-fated Tsou Jung, who shouted in his *A Revolutionary Army:*

> It says in *The Records,* "Share not the sky with your father's or brother's murderer." Even a three-foot-tall child knows what that means. Therefore, if a son cannot take vengeance on his father's or brother's murderer, he entrusts the task to his sons, and they to his grandsons, and they on to his great-grandsons, and great-great-grandsons. So our great-great-grandfathers' enemies are our father's and brothers' enemies. If we do not take revenge on our fathers' and brothers' enemies, but shamelessly continue to serve those enemies, talking the while of filial piety and fraternal respect, then I know not where filial piety and fraternal respect have gone. If our ancestors indeed have spirits, they will not be able to close their eyes in the Nine Heavens.[9]

The variety of vengeance to which true filial piety and fraternal respect was to lead was made clear at the end of the tract, in a "Drive out the Manchus Song":

Brothers, today I bitterly entreat you,
Keep this blood debt fixed in your heads!
In the beginning, when the Tartars crushed Nanking,
Your fathers were killed and your mothers raped.
Everyone says we should hate the Europeans.
But our grudge against the Manchus should be deeper still.
Brothers, you are sons of the race of Han.
If you kill not your enemies, how are you brave?
Don't listen to K'ang-Liang rant and rave.
Our number one enemy, Tsai-t'ien by name,
Is the Kuang-hsu Emperor, one and the same.[10]

"Ten thousand years to the revolution and independence of the glorious race of Han," shouted Tsou Jung: "Ten thousand years to the Chinese Republic!" And then, lest anyone forget that the Manchu blood debt must be paid in blood, he ended with the "Wheat Cake Song of Liu Po-wen," the early (not late!) Ming patriot, Liu Chi:

> *I grasp a great sword, chop, chop, chop,*
> *I'll kill every Tartar before I stop.*[11]

Poor Tsou Jung's writing stirred many, but just as many found his revolutionary rhetoric revolting, and to counter that revulsion other Revolutionaries began to change their tune. That was when they discovered in Liang Ch'i-ch'ao's "Darwinian" reasons for anti-Manchuism a godsend: "Today," wrote Hu Han-min, "as soon as one speaks of driving out [the Manchus], those listening suspect one of speaking in narrow terms of vengeance. Little do they know that the Manchus' standing in the ranks of the unfit, as a result of their own deeds, is just like our Han race's having to free itself from the control of an alien race if it is to compete with outsiders for its survival. It is all a matter of evolution and nature; there is no malice involved."[12]

If a thirst for vengeance was a bad reason to wish to be rid of the Manchus, Darwin gave a "good" reason. The Manchus were "unfit," and, being unfit, they had no right to rule. Darwinists, of course, had no right to talk about rights, but the Revolutionaries, just like Liang Ch'i-ch'ao, immediately leapt to the conclusion that *Yu sheng lieh pai* (The superior win, the inferior lose), was a description not just of what did happen but of what should happen. Being quite unable, moreover, even to conceive of the Manchus as having ever been anything but inferior to the Han, they could only conclude that the Manchu conquest and Manchu rule had to be an infraction of natural law, and they demanded, therefore, "evolutionary justice."

The nature of the infraction was clear, said Hu Han-min: "Our splendid, superior, majority race is under the control of an evil, inferior, minority race"[13]—And that was an evolutionary aberration. A *Min pao* contributor from Shansi province, who signed himself "A Gentleman from T'ai-yuan," established that fact, in a passage that showed that the Revolutionaries were no closer to "third-world solidarity" than were the Reformers:

> For the black barbarians and red savages to be subordinate to the whites, given the great differences in their degrees of civilization and intelligence, is in accord with the laws of the struggle for existence and natural selection and with the law of the weak being the meat of the strong. There is nothing therein that goes outside the standard way or the common rule. But for that beastly breed of eastern Huns to flog the descendants of the glorious Han, to slit their throats and sit on their heads, . . . where in that is there any "conquering superiority"?

The "Gentleman from T'ai-yuan" then listed all the ways in which the Manchus were inferior to the Han and drew the only "logical conclusion": "Hence we see that our glorious Han nation's falling to the Manchus was indeed a deviation from the fixed laws of evolution."[14]

The logical ease with which the revolutionaries assumed that Manchu dominance was unnatural (it never entered their minds that Manchu dominance proved Manchu fitness) was uncannily similar to that with which the Western Han statesman Chia I had two thousand years before assumed that Hsiung-nu dominance was unnatural. The Hsiung-nu had not conquered China, but they were intimidating China, and Chia I clearly viewed them as the "Manchus" of his day:

> The Empire is strung up by its heels. The Son of Heaven is the Empire's head. Why? Because he is the most high. The barbarians are the Empire's feet. Why? Because they are the most low. But today the Hsiung-nu with complete disrespect insult us and raid us and cause our Empire no end of trouble, while China every year gives them gold and floss silk and bright colored cloth. The barbarians give orders, which is the right of the Son of Heaven. The Son of Heaven gives tribute, which is the duty of his vassals. The feet are placed above the head! How absurd to be hung up upside down like this. And still we claim there are *men* in this country?[15]

If either the Hsiung-nu or the Manchus were on top of the situation, then natural law had been stood on *its* head—and that situation would, naturally, have to be righted. The unnatural thing, however, was that the Republican Revolutionaries claimed that they, the Chinese, had a natural *right* to right natural law, indeed more than a natural right, a natural duty. "Revolution," said Tsou Jung, "is a law of evolution." But he also said, "Revolution is our countrymen's heavenly appointed task (*t'ien chih*)."[16]

It was up to the Chinese to enforce natural law, and, once they woke up to that fact, claimed the Republicans, they would. And when they did—then woe to the Manchus:

> Today [wrote Wang Ching-wei], our race's thinking has advanced another step, and we no longer as before exalt ourselves and belittle others, but we know enough to take the preservation of the race and the struggle for existence as our highest duties. From now on, therefore, we know that there is no way, even with luck, for their [Manchu] race to survive. They may plot to assimilate us with all their might, but how can they? How can they?. . . . That being the case, if they are able, they had better get assimilated *by* us, for if they do not, like America's red savages, they will eventually disappear.[17]

Another contributor to the *Min pao* put it even more forcefully: "On that day when the greatest race on earth, our Han race, of unsurpassed intelligence, powerful as a sleeping lion, suddenly wakes from its deep sleep, looks to the skies and roars, not one of their five million [Manchu] savages will be left alive. How can they stand against us?"[18]

But the woe that would befall the Manchus would be on their own heads. It was "as a result of their own deeds," said Hu Han-min, that they stood "in the ranks of the unfit." No one could blame the Chinese for harboring a bloodthirst.—Here again we find an uncanny echoing of the ancients. Ch'i, son of the Sage King, Yü, had said (so said *The Book of History*) of a feudal lord he sought to exterminate, "He who holds Hu has brazenly insulted the five elements and has heedlessly abandoned the three right ways [of Heaven, Earth, and Man]. Heaven will exterminate him to revoke his mandate. So today I am simply executing, with reverence, Heaven's punishment."[19] And likewise Chou Wu Wang had said, as he set out to overthrow the last Shang king, "Shang's sins have come full measure. Heaven's order is to destroy it. Were I not to obey Heaven, then my sin would equal Shang's."[20]

Of old, those who sought to "change the mandate" (*ko ming*) said they would enforce the will of Heaven and execute Heaven's Mandate (*t'ien ming*); now, those who sought to "revolt" (*ko ming*) said they would enforce the laws of nature and execute *Evolution's* Mandate ("*t'ien-yen chih ming*"). "It is all a matter of evolution and nature," said Hu Han-min. "There is no malice involved."

So "Darwin" gave "good reason" for revolting against the Manchus—although nothing, of course, could be further from scientific Darwinism than the notion that natural law could be enforced, or, indeed, that it could be broken.

Nevertheless, once Hu Han-min and others had made it clear that they were obeying evolution's orders and not simply indulging their own desires for private vengeance, they could return with clear consciences to venting their (unscientific) anger and frustration in racial insults; for now their racial "insults" were not insults, but scientific "proof" of China's evolutionary aberation.

Ming sources therefore "proved" (surely Ming sources were in the know) that the Manchus were barbarians:

> The Nüchens' [Manchus'] land is extremely cold. They live in dens and clothe themselves in pelts. In summer they go naked, with only a strip of cloth to cover them in front. They are so filthy and stinking that they are quite unapproachable. They do their excretions in their homes and live surrounded by the stuff. They like to shoot and love to steal. They eat raw meat. When drunk they kill people, and they know not their own parents.[21]

That was what the Manchus had been like in the old days, and, despite their modern Chinese trappings, they were the same race in the present. The Emperor himself, after all, was "not a Chinese but a barbarous, sheep-stinking Tartar."[22]

Once that was "established," the Republican Revolutionaries were in a perfect position to dispose of a second troubling contention from the side of Liang Ch'i-ch'ao and the Reformers, the contention that revolution was now unnecessary—because the Manchus were at last both ready and willing to grant China a constitution, to open the way towards majority—Han—rule. But that, Chu Chih-hsin could now say, was impossible: "The Manchus could not grant a constitution if they wanted to. . . . The struggle for existence necessarily makes this so."[23] For the Manchus were unfit, and, knowing it, they knew it would be suicide for them to enter into a fair fight with the Chinese. "If the Manchus were to give up their privileges at the urging and demanding of the Chinese," wrote Wang Ching-wei, "if they were to step down and assume a position of equality with the Chinese, and engage in free competition, it could only mean that they wanted to exterminate their own race. The Manchus may be stupid, but would they be willing to exterminate their own race just because the Chinese asked them to?"[24] They would not and could not—unless they were lemmings—for to do so would be to rush against the tide of evolution.

This was an explicitly Darwinian version of a later maxim of Mao Tse-tung: "The enemy will not carry out their own annihilation. . . . [They] will not retreat from the stage of history under their own power," and of its corollary: "Nothing reactionary will fall unless you hit it."[25] This was therefore yet another way in which Republican and eventually Communist Revolutionaries alike convinced themselves that revolution was a natural necessity. For both, revolution appeared the *only way,* because the inexorable law of the struggle for existence seemed to preclude any "policy of giving way" (*jang-pu cheng-ts'e*).[26] Bound by the "laws" of Marx *or* Darwin, ("Do or die," "It's them or us,") the "enemy," racial or class, could not surrender.

As the Manchus' granting a constitution seemed synonymous with Manchu surrender, the Republicans simply ruled it out as a possibility, and with it they ruled out the possibility of bloodless change. In so doing, alas, they finally did come to a Darwinian conclusion of their own, albeit the saddest Darwinian conclusion of them all. The evolutionary necessity of violence was not a tenet of the Social Darwinism of Liang Ch'i-ch'ao, however violent his language had sometimes been. Despite the "evolutionary" violence he pointed to in the world—to frighten his people, and even more his government, into reform—he had consistently held to the belief that Chinese could make China fit through "self-cultivation," first changing their minds and then changing their ways. "Revolution" itself for Liang Ch'i-ch'ao was something that started in the head and became reform before it reached the streets. "Struggle" was an unavoidable law; violence was not.

But the opposite conclusion was the easier to come to even from the Social Darwinism he taught. The Republican Revolutionaries were learning to twist Liang Ch'i-ch'ao against himself, to twist his Darwinism against his Darwinism. They were learning, in short, to "wave the green flag against the green flag," if we may invent a green flag for Darwinism and then plant it in a Maoist phrase.[27] And yet they did not do this cynically. For most of them had learned from Liang Ch'i-ch'ao to take their Darwinism in deadly earnest.

This can be seen from their reaction to yet another argument against their anti-Manchuism, a Darwinian argument from Yen Fu. In the very first issue of *Min pao,* and in many thereafter, the Republican Revolutionaries voiced their dismay at Yen Fu's recently published translation of Edward Jenks's *A History of Politics,* for in it he seemed to wave the green flag against *them,* and their cherished first principle of *min-tsu chu-i* (racism).

Yen Fu, in his commentary, seemed to suggest that racism was an evolutionary throwback. It represented the very kind of "tribalism" symptomatic of the "patriarchal" stage of social evolution which China so desperately needed to escape if it was to survive. China had to evolve into a modern "military state"—and quickly. It had to pull its people—all its people—together into one social organism, bound together by "nationalism," not pulled apart by "tribalism." Tribalism looked backward, and would *keep* China backward—at best. Those who agitated for *min-tsu chu-i,* therefore, whether they were chauvinistic Manchus or Republican Revolutionaries, were all (in modern lingo) "backward elements."[28]

This was an appalling suggestion in any language to the self-recognized bearers of evolution's mandate, and they leapt to the task of refuting it. Wang Ching-wei, Hu Han-min, and Chang Ping-lin all wrote lengthy articles to combat the feared influence of Yen Fu's book. They tried to establish that racism was not at odds with nationalism, by arguing that all of the modern military states that Yen Fu so admired were also racial states. America, it was true, seemed an unholy, racial mishmash, and yet clearly it was politically only a white mishmash, despite its "reds" and blacks, and therefore it too was a racial state.[29] At least it was certain that, of all the fittest, modern military nations, not one was ruled by a minority race. Therefore it had to be an evolutionary rule (for fit states) that the majority race should rule, and Chinese would have to enforce that rule if China was to be fit. "Racism" was necessary for nationalism and China needed both.[30] And *min-tsu chu-i* really meant both.

Such counterarguments were simple enough, but the Republican Revolutionaries felt compelled, nonetheless, to argue them at great length, because they took Yen Fu—and Darwin—seriously, because, indeed, they accepted Yen Fu's premise that there existed a progression of evolutionary stages for states. For they had staked their own legitimacy on being progressive.

If the length of their arguments showed that they took Yen Fu seriously, however, so too did the remarkable deference with which Wang Ching-wei and Hu Han-min, at least, treated Yen Fu in those arguments. Wang Ching-wei, in an article in which he dismissed K'ang Yu-wei as "mentally ill" and Liang Ch'i-ch'ao as "even less worth talking about,"[31] expressed only deep "regret" at the most offending of Yen Fu's statements: "Will *min-tsu chu-i* be enough to strengthen our race? I have reason to believe it most definitely will not."[32] Hu Han-min went even further. He actually defended Yen Fu from those who saw in that statement proof that Yen Fu was against *min-tsu chu-i.* The only danger, he said, came from those "shallow people" who distorted Yen Fu's meaning (though he admitted that, if not set right, their distortions would have ten times the power of what they distorted).[33] Yen Fu himself was not against *min-tsu chu-i,* he simply believed—quite correctly—that it was not enough. Yen Fu was thinking ahead. He was worried that people would neglect the next necessary step, the actual construction of a "military state." But of course Yen Fu realized that China needed both *min-tsu chu-i* and militarism. He knew that neither could succeed without the other. Yen Fu's views differed from the Revolutionaries' only in degree, not in substance.[34] Those who viewed him as an enemy simply misunderstood his books.[35] Otherwise they would realize that he was actually trying to "drum up a *min-tsu* [racial *or* national] spirit."[36]

If Hu Han-min's description of Yen Fu's position was itself somewhat of a distortion, it was not altogether one,[37] nor in all probability was it intentionally cynical. For Hu Han-min himself probably *had* read *revolution* between the lines of Yen Fu's works. He was probably as much hurt as angry at Yen Fu for not joining the Revolutionaries, for he still respected Yen Fu and felt indebted to him, as he acknowledged with more sincerity than sarcasm:

> Ever since Mr. Yen's books came out, the principle of the struggle for existence and national selection has little by little fixed itself in people's minds, and the spirit of the Chinese people, because of it, has changed. As for those who [Mr. Yen disparagingly] says "speak of uniting the *ch'ün,* driving out the foreigners, and driving out the Manchus," granted many are only echoing "the fashion," but Mr. Yen's contribution to that fashion has certainly not been slight.[38]

Not all the Revolutionaries, of course, acknowledged their debt to Yen Fu (*none* acknowledged their debt to Liang Ch'i-ch'ao), and not all treated him with deference. In 1907, two years after the above efforts, the *Min pao* published a nasty little *ad hominem* attack on Yen Fu which certainly thanked him for nothing (it was written, moreover, in the vernacular, still rare at the time, perhaps to rub in the fact that its author had no use for Yen Fu *or* his eloquence):

> Yen Fu, as a person, knows only how to be selfish. He enjoys his privileges without fulfilling any of his duties. At the Anch'ing Upper School, he smokes opium every day. He meets with no one and does

no work. He finagles them into giving him five hundred foreign dollars a month for doing nothing. . . . He's indeed a slick operator. . . . He preaches [that the Manchus and Han are one race], which is not much different from honoring beasts as your brothers. Think it over. Are there any people on earth who treat beasts as their brothers? People who preach such things can't be human.[39]

But of the intellectual responses to Yen Fu, only Chang Ping-lin's was haughtily disrespectful. He did grant that the greatest threat to the Revolutionaries came not from Yen Fu himself but from the "politicians [who] can make use of his theories to hoodwink the nation," but he cursed Yen Fu for providing the politicians with such theories in the first place. He tried to discredit Yen Fu's scholarship and his knowledge of history, which he said he had totally misused. And, in McCarthy-like fashion, he tried to cast doubt on Yen Fu's "loyalty," claiming his study abroad had turned him against his own people:

Perhaps people do not know about Yen Fu. When he was young he went abroad and studied in the West, where he was so frightened and impressed by their race that he came to see yellow men as base and vile, and as badgers from the same mound, whether Manchu or Han. Therefore he is not interested in *either* constitutionalism or revolution.[40]

Chang Ping-lin was the only one of the Revolutionaries, however, to go beyond Yen Fu and his commentaries and systematically criticize Jenks's thesis itself, although actually what he criticized was not so much Jenks's thesis as its applicability to China. In what might have been good practice for a similar study of Marxism, Chang Ping-lin took great pains to argue that Jenks's categories for a universally fixed social evolution simply did not fit the realities of Chinese history. He also delivered the telling comment that sociology was still a very primitive and inexact science.[41] But that was a rare warning, heeded by next to no one.

It is no coincidence, however, that Chang Ping-lin should be the only one of the intellectual Republican Revolutionaries who dared to curse the dean of Chinese Darwinists (who, indeed, rewarded Yen Fu for his efforts at sociological prognostication with one of

China's ultimate insults: "Surely his face is too thick!"),[42] for Chang Ping-lin, of all the Revolutionaries, was the least impressed with evolution. That, perhaps, also has something to do with the fact that, on one last aspect of "the racial question," he, of all the Republican Revolutionaries, strangely appeared the most liberal.

The remaining racial question was what was to become of "racism" after the revolution. After the majority race had reasserted its right to self-rule, what was going to happen to the Manchus, Mongols, Moslems, and Tibetans—the remaining four of "the five great races of China"?

The Manchus, of course, would still be among "the remaining," even after the revolution. For both Sun Yat-sen and Chang Ping-lin swore that their "racism" would not lead to a racial blood bath. In repudiation of the bloody rhetoric of firebrands like Tsou Jung, Sun Yat-sen declared:

> I have heard people say that the racial revolution is to wipe out the Manchu race. Such talk is very wrong. . . . We do not hate Manchus; we hate Manchus who hurt Chinese. If, when we carry out the revolution, the Manchus do not try to stop us or harm us, we will definitely not seek to revenge ourselves upon them. When they first conquered the Han race, they would capture cities and slaughter for ten days before putting up their swords, but that is not something humans would do. We will certainly not act that way.[43]

Indeed, Sun Yat-sen insisted, despite his implication that the Manchus were not human, that his great principle of "racism" was not "racist" at all, but only political: "*Min-tsu chu-i* does not mean that when we encounter people of a different race we will drive them away; it means that we will not let people of another race seize our race's sovereignty."[44] "We who oppose the Manchus," added Chang Ping-lin, "do not just say, 'You are of the Aisengoro clan, and we are of the Chi clan or the Chiang clan, and we fear that you will befoul our blood line.' All we say is, 'Return our country and return our sovereign rights.'"[45]

But what then? Would *min-tsu chu-i* cease to be a reflection of evolutionary law after the Han race regained its sovereignty? What of the four major minority *min-tsu,* not to mention the dozens of minor minority *min-tsu*? Did not they too have an evolutionary duty to seek *min-tsu* autonomy? Did not they too have a right to *min-tsu chu-i*?

Only Chang Ping-lin came close to saying yes. He did not say that right was "evolutionary." Instead, in fuzzily metaphysical, "natural-right" fashion, he simply granted that, if the Han race had a right to racial sovereignty, than naturally the others did too. But clearly it was a right he hoped the others would not invoke. He granted it, in context, to prove that his party's *min-tsu chu-i* was not the kind of racism that would countenance the enslaving of other races, but he hoped the races in question, the Manchus, Mongols, Moslems, and Tibetans, realizing that fact, would forever choose to stay with China. Coming dangerously close to his opponent, Liang Ch'i-ch'ao's, contention that the minority races were already virtually Chinese, he assured those races and all people skeptical of his *min-tsu chu-i* that Han Chinese would never treat Manchus, Mongols, Moslems, and Tibetans as white Americans treated blacks.[46]

Sadly, however, he acknowledged that those races might, nonetheless, want independence, and, if they did, if they insisted, then China, he said, should let them go:

> If the Moslem chiefs' hatred for the Manchus has so penetrated them to the bone that they extend their grudge to us, and fiercely desire independence to restore the domains of their Turkic ancestors, then we should give in to their desires, knowing they but look on us as we look on the Manchus. If there is no other way, it would even be all right to give them the land west of Tun-huang and enter into a Holy Alliance with them, to break the right arm of Russia.[47]

Chang Ping-lin's *min-tsu chu-i* made him at least a little sympathetic with the aspirations of other *min-tsu*.

But, in this instance as in so many, Chang Ping-lin proved a rare bird. The espousal of *min-tsu chu-i* did not lead the majority of his confederates to any such sympathy. Instead, the more Darwinian of them won the day with the argument that the Han race, the fittest race, had the evolutionary right to rule not only itself but also "its" minorities. The Han race had the right to rule all the races in "Chinese" territory—even though it had been the Manchus, not the Chinese, who had forcibly brought Manchuria, Mongolia, Turkestan, and Tibet into that territory! Manchu territory, once Chinese, was to be inviolable.

The exasperated Revolutionary pamphleteer, Ch'en T'ien-hua, just before he drowned himself in Tokyo Bay, declared that "it is in accord with recognized political principles for majority, superior races to rule over minority, inferior races,"[48] and for many the matter was just that simple. But Wang Ching-wei added an important qualification, with the strange help of a rather obscure Western wise man, Johann Kaspar Bluntschli, whom Liang Ch'i-ch'ao had turned to two years earlier, in 1903, in his *anti*-anti-Manchu cause.[49]

Bluntschli, said Wang Ching-wei, believed that, when it was necessary for several races to exist in one state, "the greatest and strongest race should act as the foundation of that state,"[50] with which statement Wang Ching-wei most heartily agreed. And yet there were two ways, said Wang Ching-wei, in which such a principle could be followed: The many could be subordinated to the one; or the one could accept the many and assimilate them. The first policy, however, might "arouse resistance." And anyway, as it was "directly opposed to the principle of human equality," it was "definitely nothing a civilized country ought to have."[51] Therefore the Republican Revolutionaries were for the second. They were for assimilation (*t'ung-hua*):

> The *min-tsu chu-i* our party advocates wants to have all our races assimilated into one race, so that there will be one national people. We do not mean that, except for the Han race, no races will be allowed to live in China. We mean that all races should be assimilated into the Han race so that they will all be Chinese.[52]

And such assimiliation was not to be thought tragic. Bluntschli said, said Wang Ching-wei, that certain races simply did not have what it takes to stand alone. They needed to be protected or annexed.[53] "The feeble ability of today's Manchus, Mongols, Moslems, and Tibetans," said Wang Ching-wei, "fits perfectly with what [Bluntschli] says. Therefore, to cause them to be assimilated into the Han race is in utmost accord with principle.[54]

It was all so obvious: "The Han race created China. The Han race will restore China. The Han race will be able to make China turn from danger to safety and from weakness to strength. Therefore to absorb the other races is the Han race's right, not the Han race's duty, though perhaps, as the Han race *must* assimilate the other races it takes in, if it is not to leave the nation future worries, it is not only the Han race's right, but also the Han race's duty."[55]

Wang Ching-wei, therefore, bolstered by Bluntschli, whom he had stolen from Liang Ch'i-ch'ao's camp (by cleverly ignoring half of what Bluntschli said), could reply to the rhetorical question he asked himself, Did *min-tsu chu-i* mean that all of "China's" *min-tsu* should be free? with a confident and unequivocal "No."[56] And yet, although Liang Ch'i-ch'ao had first brought in Bluntschli to argue against both Republican and racial revolution, he would have answered Wang Ching-wei's rhetorical *min-tsu* autonomy question with an equally unequivocal "No," and he would have cited the very same Bluntschli passages to support his answer. For *racially* what he hoped would come out of a constitutional monarchy was exactly what Wang Ching-wei hoped would come out of a racial revolution. Both were "intoxicated" (as Liang Ch'i-ch'ao put it) with the vision of a China in which the Han, Manchu, Mongol, Moslem, and Tibetan races would all be united as one "national people" (*kuo-min*) or as one "great race" (*ta min-tsu*), with the Han race, of course, at the center,[57] one "great race" united in a spirit not of "tribalism" but of "nationalism," what Liang Ch'i-ch'ao called *ta min-tsu-chu-i* (great racism), to distinguish it from the *hsiao min-tsu-chu-i* (petty racism) of the Revolutionaries.[58]

The Revolutionaries were incensed by the epithet *petty,* convinced as they were that their "racism" was a prerequisite for a Chinese China, but their post-revolutionary ideal was indeed Liang Ch'i-ch'ao's *ta min-tsu-chu-i* and, after the revolution finally came, that is precisely what they advocated. They did not call it *ta min-tsu-chu-i*—but *min-tsu-chu-i* was redefined to mean "nationalism" (what Liang Ch'i-ch'ao had meant by it in the beginning).

Happily, however, that nationalism did not lead to any policy of forced biological assimilation, despite Wang Ching-wei's original contention that assimilation was the only civilized way to accord with evolution. To symbolize, instead, a happy unity that stressed neither racial assimilation nor subordination (and definitely not annihilation), the first flag of the Republic of China was composed of five stripes of different colors for the five great races now of China, stripes equal in size, of course, only on the flag, but which nevertheless were colorfully separate.

In the People's Republic, the stripes have become stars, one great one and four lesser ones, all of a color. But the major minorities and even the minor minorities are still separate and are even, indeed, protected (somewhat) in "autonomous regions" (the Outer Mongols are now out, but the Inner Mongols are still in). Whether such "protection" is in accord with or in defiance of the laws of evolution (or civilization), "Han chauvinism" (*ta Han-tsu-chu-i*) has clearly, in theory, been repudiated. But just as clearly, as both the Darwinian proponents and the Darwinian opponents of racial revolution so fervently hoped, the Manchus, (Inner) Mongols, Moslems, and Tibetans are still both on the flag and under it.

DEMOCRACY, LEAPS FORWARD, AND EVOLUTION BY RICKSHAW. Racial revolution, or Han self-reassertion, was not, of course, enough. Ming T'ai Tsu, founding father of the Ming dynasty, had accomplished that much, said Sun Yat-sen, when he seized China back from the Mongols. But Ming T'ai Tsu had not done anything else. He had not revolutionized Chinese government (though he had made it more revolting) and therefore (two hundred and seventy-six years later!) foreigners—the Manchus—had been able to come in and conquer China all over again. If now, then, China was to be not only *again* Chinese but *forever* Chinese, it would have to have a political revolution to go with its racial revolution. If China was to be Chinese once "and for all," it would have to have "democracy" (*min-ch'üan-chu-i*)—the second of the Three People's Principles.[59]

"Democracy," said Sun Yat-sen, "is the basis of the political revolution."[60] He said that in 1905. But he probably believed it in 1895, when he launched his first abortive coup from Canton.[61] As we have already seen, however, by the time he and his confederates finally started propagating their beliefs, no longer secretly but in their own Chinese journals, the "best" arguments they could muster in favor of democracy were quite as unoriginal as the "best" arguments they could muster against the Manchus. For, again, Yen Fu and Liang Ch'i-ch'ao had voiced those arguments years before.

Actually, "all" the "best" arguments came down to only two, both of which were Darwinian. Democracy was the government of the fit—it was what made the fit the fit—and as such it was the wave of the future—it was an evolutionary stage that every fit people must reach. Yen Fu had said that in his first essays in 1895; Liang Ch'i-ch'ao had said it over and over again a year later in the *Shih-wu pao.*

Sun Yat-sen did not say it, clearly, until 1905. But, when he did, he tied not only democracy but all three of his Three People's Principles to the theory of evolution. He made his "ism" a Darwinian "ism." He made the first Chinese doctrine that almost filled the doctrinal vacuum in China the theory of evolution had helped create an evolutionary doctrine. For he introduced his Three People's Principles, in the founding editorial of the *Min pao,* as principles extrapolated from the evolutionary history of the West:

> I have traced the evolutionary progress of Europe and America to the three great principles: nationalism [or "racism"], democracy, and socialism. At the fall of Rome, nationalism flourished and the various countries of Europe and America [*sic*] became independent. But, when they became monarchical and overbearingly authoritarian, so that those of low estate suffered beyond endurance, the idea of democracy arose. At the end of the eighteenth and the beginning of the nineteenth century, despotism collapsed, and constitutional government grew up in its stead. The world was opened up. Human knowledge soared. Material things developed. More was done in a hundred years than in a thousand years before. But after economic and political problems, a socialist movement began, and the twentieth century is now a period that must be dominated by socialism.[62]

The Three People's Principles were not just three good ideas; they were reflections of three stages of social evolution. The strata of social evolution had been laid bare in the West, and *mirabile dictu* those strata showed the way to the future, for the West had moved on ahead. Sun Yat-sen leapt, therefore, to a faith he would maintain all his life: "All the countries of the world progress first from nationalism to democracy and then from democracy to socialism."[63] And that meant that democracy *would* come: ". . . the imperial system just cannot compete with republicanism. . . . Those prosper who flow with the tide of evolution. The imperial system will never be able to survive."[64]

The Three People's Principles, therefore, each had one foot in evolutionary historical determinism. As early as 1903 (although still eight years after Yen Fu's essays) another Republican Revolutionary, who styled himself "Ching-an," wrote an article entitled "The Evolution of Forms of Government," in which he promised China that "in the twentieth century, a complete and flawless racial republic must appear."[65] Evolution itself, he said, would see to it:

> "Heaven selects, things compete." "The fittest survive." So it is for all things, and especially for forms of government. . . . Even the brave cannot resist Heaven. At times of great change, scholars exhort and excite people, popular bands brave hardship and scorn, heroes arise to launch and direct things. Great change seems to come because such men love action. Ha! Since when do such people really love action? They are simply being driven by the force of evolution. They are forced by the situation and cannot do otherwise.[66]

This was the kind of evolutionary determinism that would help "determine" Karl Marx's Chinese victory. Scholars, party members, and heroes would bring democracy to China, but like those who Alexis de Tocqueville said had brought democracy to America, they would all be "blind instruments in the hands"[67]—not of God, but of evolution. Evolution, like Hegel's "Spirit," would wield "the infinite mass of [the masses'] wills, interests, and activities" as "the tools and means . . . to accomplish *its* end."[68]

Sun Yat-sen knew little if anything of Hegel or de Tocqueville, and despite his professed Christianity he rarely spoke to Chinese audiences of God or any spirit. And yet he would have agreed with the spirit, if not the letter, of de Tocqueville's statement that "it is not necessary that God himself should speak in order that we may discover the unquestionable signs of his will. It is enough to ascertain what is the habitual course of nature and the constant tendency of events."[69] For, like de Tocqueville, he believed men *could* ascertain the "course of nature," and that he *had* ascertained it—in American history. The course of nature ran towards democracy. Democracy was "predestined" by evolution.

That strange belief Sun Yat-sen shared with Yen Fu, Liang Ch'i-ch'ao, and even K'ang Yu-wei. All agreed that democracy was an evolutionary stage. What they were not agreed on was whether or not it was the *next* stage.

The Republican Revolutionaries said it was; the Reformers said it was not. The Reformers said—at least in the beginning—that the next stage was constitutional monarchy, and that it would be evolutionary suicide to try to skip that stage and go from monarchy to democracy without going through constitutional monarchy. For "progress," said Liang Ch'i-ch'ao, "has steps, which you cannot take two at a time."[70] And woe to those who tried: "If a people's intellectual and moral level has not yet reached the point at which they are capable of democracy, and yet they set one up regardless, it is a most dangerous thing."[71] Their democracy would fail, monarchy would reassert itself in worse form, and their country in the whole chaotic process would be perilously weakened.

Yen Fu had said in 1895, "Our people are not yet ready to rule themselves,"[72] and both he and Liang Ch'i-ch'ao were still saying the same thing a decade later. In January 1906, however, freshly goaded by the Revolutionaries to argue again against great leaps forward, Liang Ch'i-ch'ao took a sudden step backward. He professed a new doctrine of awesome significance for modern China. He argued that it was not just true that "China today cannot possibly institute a republican constitutional system. . . . China today cannot yet institute a monarchical constitutional system."[73] For "the Chinese people," he said, "do not yet have the ability to practice parliamentary government.[74] . . . Constitutional monarchy is indeed our party's stated political goal, and we will not stop until we reach it, but. . . . today we still cannot yet put it into practice. . . . [for] the level of our people is still not up to it."[75]

Constitutional monarchy was *not* the next stage. There was another stage that would have to come first. Any country, said Liang Ch'i-ch'ao, before it could get from monarchy to any form of constitutionalism, would have to go through a stage of *k'ai-ming chuan-chih,* "enlightened dictatorship."[76]

Enlightened dictatorship was for the ruled, not the ruler.[77] "Enlightened dictatorship," said Liang Ch'i-ch'ao, "serves actually as a transition to constitutionalism and as a preparation for constitutionalism." It was natural and necessary, being one step in the natural "order of a state's progress."[78] And therefore, like constitutional monarchy itself, now pushed one step further into the distance, it was a step that could not be skipped.

If all countries needed enlightened dictatorship at the same point in their evolution, however, they did not all need it with the same urgency or for the same length of time. For the intensity of the struggle for existence varied. Countries that needed enlightened dictatorship the most, said Liang Ch'i-ch'ao, and, he added ominously, the longest, were those that had undergone long periods of "barbaric dictatorship" and then "incomplete dictatorship" (that is, dictatorship too inefficient to keep order), those in which there were many races and in which the masses were intellectually immature,[79] and those engaged in fierce foreign competition.[80] Those, alas, were the conditions, said Liang Ch'i-ch'ao, of China, and enlightened dictatorship, therefore, was most fittingly China's next stage.

Strange as it may seem, this espousal by Liang Ch'i-ch'ao of a necessary stage of enlightened dictatorship marked for him a Darwinian step back towards Spencer, or at least back towards the Spencerian side of Yen Fu's *T'ien-yen lun.* Actually it marked two such steps.

First, Liang Ch'i-ch'ao openly sought Spencer's help when he claimed that enlightened dictatorship was the "fittest" form of government for China (for the time being). For it was Spencer, he said, who had realized that fitness and fitness alone was what counted. "Years ago," he wrote, "Darwin expressed his law of biology with the words *yu sheng lieh pai* (The superior win, the inferior lose). But Spencer changed it to *shih-che sheng-ts'un* (The fit survive), meaning that, if something is fit, even if it is inferior it will be superior, and, if something is unfit, even if it is superior it will be inferior. Therefore, when we discuss things, we should only seek the 'fittest,' not the 'best'."[81]

Liang Ch'i-ch'ao was either mixed up or less than honest. He was at least forgetful. For he himself had told Chinese readers five years earlier that *yu sheng lieh pai* and *shih-che sheng-ts'un* were different translations of one sociological phrase.[82] Now he blamed "Darwin" for the former, which was the Japanese translation of Spencer's "survival of the fittest," (which Liang Ch'i-ch'ao had himself introduced into Chinese from the Japanese), and he lauded Spencer for the latter, which was Yen Fu's translation of the same thing. Granted, Yen Fu's translation was better, but in neither case was Spencer's original phrase a revision of Darwin's theory, for that phrase preceded Darwin's theory, and, although Darwin, it is true, never used it in his *Origin,* he later said he wished he had, because Spencer's phrase, he felt, so fit his theory.[83]

Be that as it may, Liang Ch'i-ch'ao's understanding of Spencerian wisdom gave him an even more refined argument against democratic revolution. Of course, he could say, democracy was better than dictatorship,[84] but, at this stage of China's evolution, dictatorship was fitter than democracy. Democracy was not fit for China, because the Chinese were not yet fit for it.

That was why "skipping stages" (*lieh teng*) would never work. If a country tried to skip a stage, it would end up in a stage it was not fit for. If it decked itself out in trappings that did not fit, it could not *be* fit—and would not survive.

Arguing further against stage-skipping, Liang Ch'i-ch'ao took his second step back towards Spencer. To attempt to skip a stage, he insisted, was as foolish as to attempt to graft a foreign body onto a social organism that had no natural affinity for it. Both attempts would cause unfitness because both would interfere with the process of nature. After all that he had said in praise of Huxleian "spirit" and "human action" (*jen wei*), he now came back to argue with Spencer that men should let "the process" (*t'ien hsing*) alone. Forcing change that had no roots in nature would be like "breaking short a stork's legs or lengthening a duck's." And "insisting on skipping a stage" would be like "pulling up rice shoots to help them grow."[85] Interference led to disaster–and revolution was interference.

But how could dictatorship not be, enlightened or no? And how of all things could dictatorship be Spencerian? Enlightened dictatorship, admitted Liang Ch'i-ch'ao, would be interference, but it would be interference on behalf of the process, in defense of the process. It would interfere only with anti-process interferers. An enlightened dictator would nurture the people. He would raise them (as would Yen Fu's ideal Sage)[86] until they grew up enough to rule themselves. In that sense he would still be a *fu-mu kuan* (a parent official) as were officials of old, except that he would raise his charges to be independent. He would not do this, however, by going against their natures. He would sow seeds and water roots (more was mixed up than the metaphor but, in this sense, at least, he was still a Huxleian cultivator). But most important, throughout the dangerously long process of raising, cultivating, and implanting, he would "be able to reconcile conflict within and insure progress in the conflict without,"[87] so that China could survive and win time to grow up. For growing up, as Spencer had said, was the only way up.

All these arguments were designed to counter the idea of the "naturalness" of revolution. Through them, Darwin, "revised" by Spencer, became a champion against a revolution as well as one for it. It must, therefore, have been with a wickedly gleeful sense of triumph that Liang Ch'i-ch'ao claimed that the inspiration for his long essay, "On Enlightened Dictatorship," was "a statement from the last testament of the [revolutionary] martyr, Ch'en T'ien-hua [the one who had drowned himself in Tokyo Bay]: 'We must use enlightened dictatorship, if we wish to save China.'"[88]

Liang Ch'i-ch'ao's counter-revolutionary usurpation of a "revolutionary" idea, however, was fraught with more ironies than anyone could see at the time. For if Liang Ch'i-ch'ao, as Chang Hao has pointed out, was soon to divorce himself from "his" idea of enlightened dictatorship, and go back to campaigning for a constitutional monarchy,[89] the real Revolutionaries were to wed themselves to it. Sun Yat-sen's famous "period of political tutelage" (*hsun-cheng*), a "necessary, transitional stage from monarchy to republicanism,"[90] was supposed to be, whether he called it so or not, a period of "enlightened dictatorship," and so was the much later period, equally "necessary," of the "dictatorship of the proletariat." Liang Ch'i-ch'ao had once again provided his opponents with their very "best" arguments of the future.

When Liang Ch'i-ch'ao's essay first appeared, however, it enraged the Revolutionaries and made Sun Yat-sen argue more excitedly than ever that evolutionary stages could be skipped. He argued that China could *lieh teng,* and in so doing hit at last upon an argument that was almost, at least, original. It was an argument that in time would raise its own flurry of ironies and "contradictions," but it was nonetheless of vital importance to Chinese Darwinism and China's future, for it was an evolutionary argument that threatened to undermine the "authority" of evolution.

Actually, Sun Yat-sen thought up his favorite "proof" of the naturalness of stage-skipping at least two years before Liang Ch'i-ch'ao's essay on enlightened dictatorship came out, when China still had to skip only one stage, not two. His favorite proof was his famous railroad simile, which he was to use so often that, despite the tragedy of his forced abdication of the presidency to Yuan Shih-k'ai, there would be a certain poetic justice in his subsequent brief appointment as Director of Railways.

At any rate, he first used his railroad argument, as far as I know, in 1904, in a Honolulu Chinese newspaper in which he rebutted an anti-Republican article by the chief writer of another Honolulu paper, the *Pao huang pao* (Preserve the Emperor news):

> According to what he [Ch'en I-k'an, the offending author, who was obviously up on his Liang Ch'i-ch'ao] says, to be in accord with the order of evolution, it is necessary to go through constitutional monarchy before it is possible to establish constitutional democracy. He does not

> know that, when something is new under the sun, it must evolve like that, but, when it already exists and has already been used to great effect, then we can imitate it, so that "the last can be first." Just consider the fact that China, which has never had railroads and which has only recently started to build them, uses nothing but the latest models. If we were to do as he says, then China, being today in only the earliest stages of railroad development, should use only the old English and American equipment of several decades ago, which it could then exchange gradually for newer stuff, until as a final result China would be able to use the new style trains of today. For only thus would we be "in accord with evolution's order." . . . But is there any such logic as this on this earth? Are there any people so stupid?[91]

No railroad builder in his right mind would choose the first train invented over a modern model. "Even women and children," said Sun Yat-sen, "could tell which is better."[92] And, therefore, no country in its right mind should choose constitutional monarchy over democracy: "Therefore it takes no wise man to know that constitutional monarchy is not fit for China."[93] For, democracy having "evolved," constitutional monarchy was "outmoded."

The great alleged fact of life that made that true, that lent real promise to the ancient idiom "The last shall be first" (*hou lai chü shang*),[94] and hence lent "backward" China real hope that it might "catch up" with the "forward" West, was that the fruits of evolution were there for the picking. The modern day could be plucked of its best: "Other people's good things . . . we can choose and use."[95] And best of all, those good things could be picked up at a fraction of their original cost. "For their material civilization," said Sun Yat-sen, "the Western countries spent two or three hundred years' effort before getting today's results. But we can simply pick up those results and use them. I ask you, gentlemen, is that cheap or is it not cheap?"[96]

If that sentiment was only "almost original," it was because Yen Fu, as we have seen, had already promised a full decade earlier, as he too surveyed the Westerners' fearsome progress, that "what for them was difficult will for us be easy."[97] But Yen Fu's sense of ease was not quite the same as Sun Yat-sen's. Yen Fu should have appreciated Sun Yat-sen's railroad lesson. He himself, after all, had learned to navigate modern steamships without, presumably, becoming first a master junkman, but he did not apply that experience to the ship of state, nor conclude that China's ability to buy and operate modern ships proved that it could with equal ease adopt "modern government." If he had thought of democracy as just another modern tool, he would still have argued that all Chinese would have to be trained for a long time before they could use it. But he did not seem to think of democracy as a tool. It was more like a state of mind that a people would have to grow up to. That is why he stuck to his metaphor of the social organism.

But Sun Yat-sen, with his railroad train, switched his metaphors from the organic to the mechanical, and that put him on a very different track. When he said that "the future of our China is like the building of a railroad,"[98] he left agriculture and went into industry, indeed, into the construction business (or into the reconstruction business). "The republics of the world," he said, "can be divided into two kinds: one comes about through natural evolution; the other is built by man power (*jen li*)."[99] And China would have the latter.

What had happened? Here at last was an original idea—but with a vengeance. After all that the Republican Revolutionaries had said about evolution being for democracy and about revolution being natural, Sun Yat-sen now suddenly turned on natural evolution, looked it in the face, and said, "No thank you." He seemed to take his stand with Huxley, in defiance of the cosmic process. "We must definitely not follow evolutionary change," he said; "we must have change through human effort, for only then will our progress be quick."[100]

But in that last clause he showed that he was still not really with Huxley. For he rejected the cosmic process, not because it was too cruel, but because it was too slow. Evolution was still going in the right direction, and left to its own devices it would still bear China to democracy. The trouble was that it would bear it there only in its own sweet time, and China could not wait. At least it did not want to. The West had reached democracy in the natural way, but that way now seemed the hard way, the long way. The West had taken the high road, and China would take the low road, and China would get to wherever people were going before them. It was that happy vision that allowed Sun Yat-sen to say disparagingly of the fearsome Westerners: "They have only had natural progress; We shall have man-powered progress (*jen li te chin-pu*)."[101] And that would make all the difference.

But what *was* the difference? What in Heaven's name, or in Darwin's name, was the difference between "man-powered progress" and "natural progress"? How could one become two? Liang Ch'i-ch'ao said, "We must progress the natural way. We cannot jump stages." Sun Yat-sen said, "No! We must progress the human way, and we can jump stages." But where in evolution was there room for two ways? Did Sun Yat-sen, who claimed Darwin as his foremost Western mentor, want China to act unnaturally? Even if he did not want China to stand in Huxleian (or King Knutian?) manner *against* evolution, did he now want China to rise above it? And was that not just as super-natural a position as that commanded by Huxley's defiant "microcosm"?

Sun Yat-sen introduced his train to try to railroad China to democracy, by stage-jumping coach, but his train was really a jinrickshaw. It was a *jen-li-ch'e,* a "man-power cart," an odd conveyance indeed for a social organism.

One can easily see why Sun Yat-sen *wanted* a jinrickshaw, why, indeed, he had to have one: there was no other way to catch up to the West. If *jen li* was not a factor in evolution, then nobody could do anything about anything, and, if the West had evolved beyond China, and kept evolving, then China would always be behind. If *jen li* could not at least speed up the process, then China had no hope—unless, of course, the West got sick, from "internal contradictions," so sick that it would slow down or even stop and let China surpass it. That was, indeed, a second hope, which Yen Fu, Sun Yat-sen, and Mao Tse-tung all seized upon, but, if Chinese wanted to *do* anything, and of course they did, they had to cling first and foremost to the first hope, that even in a world governed by evolution, *jen li* (man power) or *jen wei* (human action) could make a difference.

Liang Ch'i-ch'ao and even Yen Fu, the most Spencerian of China's Social Darwinists, were also passionate apostles of *jen wei.* Despite their disagreement with Sun Yat-sen over stage-skipping, they believed no less than he that *jen wei* was vital to China's survival and progress. They too really believed in evolution by jinrickshaw. They just did not believe that a jinrickshaw could take short cuts. They believed that those who knew the way could pull China forward by pointing out the way, by educating more people to pull, and by educating everyone to stop dragging their feet.

Liang Ch'i-ch'ao and Yen Fu also mixed their metaphors. They too wavered between Taoist Darwinism and Confucian Darwinism, between evolutionary determinism and evolutionary determinationism, between Spencer and Huxley. If we can be permitted one last time to switch our own metaphors in midstream, abandoning our jinrickshaw, Yen Fu, Liang Ch'i-ch'ao, and Sun Yat-sen too seemed to believe that all peoples were on great galleys drifting down one river. But they could row, so they could increase their speed and catch up with and pass each other. They could also get stuck, drop anchor, foolishly try to row upstream, rock their own boat, or sink. But if they could learn to row together, in the right direction, they could move much faster than the river itself and spurt to the head of the fleet. It was when the river took meandering loops, however, that Sun Yat-sen suddenly cried, "We can rush to the head of the fleet with a portage!" But Yen Fu and Liang Ch'i-ch'ao said, "No, that would be unnatural. It would leave us high and dry." What they did not admit, however, was that their rowing too was unnatural, if nature was the river.[102]

Yen Fu, Liang Ch'i-ch'ao, and Sun Yat-sen, therefore, really differed only in degrees in their belief in progress by man power. But Yen Fu and Liang Ch'i-ch'ao never came as close as Sun Yat-sen to declaring and accepting the fact that there was a world of difference between man power and nature, between civilization and evolution, between technology and biology. Sun Yat-sen alone, in his common-sense distinction between "evolutionary change" and "change through human effort," strode unwittingly to the brink of Alfred Russel Wallace's common-sense "discovery" (backed, however, in his case by years of scientific study) that, with the evolution of the human hand and brain, the law of natural selection, for man, had been virtually repealed, that, as Loren Eiseley put it, thenceforth "the brain and hand alone" could "order the environment that once ordered them."[103]

Sun Yat-sen probably had not the foggiest notion of the Wallace-Darwin debate about the brain, and he did not really hit upon Wallace's theory. He only almost did. His construction metaphors almost led him to see that the house that Jack built was not built as Jack was. They almost led him to make a distinction between technological progress and natural evolution, a distinction sorely needed in a China which had so thoroughly confused the two. They almost led him clearly to state that man had not pulled himself forward, from ape to man-ape to ape-man to man, even though he had, perhaps, pulled his civilization forward, at least from bleak to less bleak to less bleak still. But Sun Yat-sen did not really make a clear distinction between biological evolution and social evolution, partly because he was not concerned with biological evolution—he was concerned with revolutionary action—and partly becaue the distinction that he almost made was not really distinct in his own mind.

He did not say that all human civilization was man-made. He did not even say that all democracies were man-made, for he said that, "although the establishment of America as a country, through seven years of bloody warfare, would seem to belong to the realm of human action, the American people's democratic nature was completely the product of natural evolution."[104] He was trying to distinguish human action from natural action, but he was actually only labeling certain human action "human" and other human action "natural." For surely the only real difference between the establishment of American democracy and the hoped-for establishment of Chinese democracy was the difference between certainly difficult invention and hopefully easy adoption (the crux, of course, of the train argument). Both were in the same realm, even if one insisted that invention was "naturally" mothered by necessity, for how could adoption, in this case at least, be less so? Sun Yat-sen himself had insisted that China had to adopt democracy to survive. But he chose, nonetheless, to view that concession to necessity as a matter of choice, and therefore human.

Alfred Russel Wallace would have made no such distinction between the American way and the Chinese way, for his point was that all human handiwork, regardless of its degree of difficulty, was unnatural compared to the flesh-and-blood handiwork of nature itself. For there was an element of choice, even if only among limited alternatives, in all human action not present in the "action" through which human beings had evolved. Man had not chosen either his head or his hands, but he had chosen to do all sorts of odd things with them, multitudes of which, thought Wallace, could have been of no conceivable advantage to him in his struggle for survival. Wallace was struck by the "extravagances" of the human head, so much so that he re-entertained, to Darwin's dismay, suspicions of divine manipulation in the pre-human process. He found something so *un*natural in the selection of a brain already capable of a thousand tricks unnecessary to its first caveman possessor's survival that he was moved to wonder whether or not "some higher intelligence may have directed the process by which the human race was developed."[105] But that was only a suspicion. The point he insisted on was that, with or without the direction of a higher intelligence, once man *was* developed he was let into a world that no other species selected by evolution had ever been in before. Wallace re-established Homo sapiens as *wan wu chih ling,* "the highest of the ten-thousand things," for Homo sapiens was the only creature on earth with a mind "sufficiently developed to remove his body from the modifying influence of external conditions and the cumulative action of 'natural selection'."[106] Homo sapiens was the only creature that evolution had set free.

For Wallace, this "freedom" was the result of an historic breakthrough that had taken place once and for all. But Sun Yat-sen spoke as if such a breakthrough was a matter of present, individual choice. The Chinese people could break ahead of the natural process if they wanted to. It was "humanly possible" but had not yet happened. Sun Yat-sen offered mankind a choice to act humanly or naturally that Wallace had not offered at all. Wallace had said that, once men had minds, all their action was mindful, stupid though it might be. Sun Yat-sen said no. Men could act mindfully or they could act mindlessly, and, if they chose to act mindlessly (Wallace would have said, "Aha, that too is a *human* choice"), they would be right back where their forefathers started—as pawns of natural law. "Unconsciousness," said Sun Yat-sen, "is the way of cosmic evolution, and that is natural. Consciousness is the way of man-made evolution, and that is not natural. The former is slow evolution, the latter is fast evolution. Our responsibility is to stimulate the slow way with the fast."[107]

Sun Yat-sen did not come to his conclusions after years of scientific study. His concern and sense of responsibility for China's place in the world were too strong to allow him to sit back as a scientist might and ponder what Huxley called "the question of questions for mankind—the problem which underlies all others, and is more deeply interesting than any other—the ascertainment of the place which man occupies in nature." He did ponder that question, and Huxley's other questions, "Whence our race has come, what are the limits of our power over nature, and of nature's power over us? to what goal are we tending?"[108] but he jumped to his conclusions. Therefore his distinction between the human and the natural, as a Darwinian distinction, was undistinguished.

But Sun Yat-sen was grasping at a common-sense distinction that neither Huxley, Wallace, nor Darwin himself had ever honestly faced, the distinction between having a head and using it, between rolling up your sleeves and dirtying your hands and "sleeving your hands and looking on" (*hsiu shou p'ang kuan*), between being able and being willing. Perhaps, as Wallace said, man as a species was man, and that was that. Perhaps it made no sense to say that man could choose to be either natural or human. But as the Taoists, who had made the same claim but the opposite choice, had always said, man could at least choose to be "active" or "passive," and that was all Sun Yat-sen really meant by "human" and "natural."

All Sun Yat-sen was really doing was decrying Taoist passivity and insisting on Confucian action. And, in good Confucian fashion, he grounded his faith in action on human will power. He was convinced that China could make social progress if it wanted to, as Confucius and Mencius had been convinced that individuals could make moral progress if they wanted to. So he, too, as had Yen Fu and Liang Ch'i-ch'ao before him, echoed Confucius's words, "If I want goodness, then goodness will appear,"[109] when he said, "If China today wants wealth and power, then it shall have wealth and power."[110] And he too echoed Mencius's words, "It is that you do not, not that you cannot,"[111] when he said, "The question is no longer whether China can or cannot know what to do, or whether it can or cannot do it. It is simply whether it wills or wills not to do it."[112] If China had the will it would have the way, said Sun Yat-sen, using the Chinese version of that proverb, which Mao Tse-tung would later echo.[113] And, using one more proverb, which Mao Tse-tung again would echo, he said, "If we believe in our hearts that we can do a thing, then, be it as difficult as moving mountains or as filling in the seas, the day shall finally come when it shall be accomplished."[114]

This was a fine, traditional, Chinese affirmation of "the human spirit," not a contribution to Darwinian science. Indeed it was an affirmation that raised thorny questions for Darwinian science in two related but different ways. For, when Sun Yat-sen said that will power could move mountains, his metaphor took him not only beyond nature but against it.

His metaphor took him beyond its traditional self to "modern man's" faith in the "conquest of nature." And in that way it took him back once again towards Wallace and Huxley. For, although his first fuzzy notion, that man could leave the natural way and take the human way, to break away from the natural way to break the way for the natural way, was not in accord with the thinking of either Wallace or Huxley, his less mystical (or mixed up) faith in man's moving mountains was. Both Wallace and Huxley were convinced that man did indeed have great "power over nature." Wallace said man had the power to remove himself bodily from the lists of natural selection, or at least to "remove his body from the modifying influences of external [natural] conditions." And Huxley said man had the power "with every step of [his] progress in civilization. . . . [to] become more and more independent of the state of nature,"[115] indeed "to subdue nature to his higher ends,"[116] to bend "the Titan to his will."[117]

Of course, when Huxley said that, he was not just thinking of bending mountains and seas to man's will. He wanted to bend the ruthless, creative force behind both man and "man's" natural environment to man's will. More important still, he wanted to bend man's own "human nature" to man's will, to man's "good will." But it was man's power, limited though it was, over "the rest of nature" that first gave Huxley hope man might also have power over himself.

Huxley never believed man had the power to win a complete victory over either his environment or his "original sin,"[118] or even to win a lasting incomplete victory. With a "Chinese" sense of cycles (albeit stretched by geological time), which oddly enough went unappreciated by his newly optimistic Chinese readers, he offered in the end no more lasting hope than this:

> That which lies before the human race is a constant struggle to maintain and improve, in opposition to the State of Nature, the State of Art of an organized policy; in which, and by which, man may develop a worthy civilization, capable of maintaining and constantly improving itself, until the evolution of our globe shall have entered so far upon its downward course that the cosmic process resumes its sway; and once more, the State of Nature prevails over the surface of our planet.[119]

But Huxley at least believed man could conquer nature for a while. So did Wallace. And both believed that that was what man should do. Neither was a Taoist.

Huxley, of course, was one of those who had been most wildly excited by Darwin's (and Wallace's) discovery that man was natural, and in enthusiastic response to Darwin, in his own first book, *Man's Place in Nature,* he thought he had put man in his place.

But, in *Evolution and Ethics,* he let him out again, not as far as Wallace had, with his renewed intimations of the miraculous, which had made Darwin fear for natural man's very life ("I hope you have not murdered too completely your own and my child"),[120] but far enough. Both at least were convinced that, however "natural" man's beginnings were, what counted now was not man's place in nature but that place in nature that man carved out for himself.

Probably both Huxley and Wallace were concerned first and foremost with man's moral rather than physical conquest of nature, but both did believe that man could conquer much of physical nature, and that belief was one of the most exciting "morals" Chinese readers drew from Huxley's part of *T'ien-yen lun.* Certainly it was an idea that excited Sun Yat-sen, who said, "Today, with the flourishing of science, we finally know that man can conquer Heaven."[121]

Sun Yat-sen said "conquer Heaven" (*sheng t'ien*) not "conquer nature," because he chose to use a literary allusion, one indeed that Yen Fu had already used in translating from Huxley.[122] But Sun Yat-sen really meant "conquer nature," and in a way that was actually more down to earth than Huxley's. For Sun Yat-sen was no more antagonistic to "Heaven," or "nature," as the creative "cosmic process" than Yen Fu was. He wanted, perhaps, to "do better than the cosmic process," another possible translation of *sheng t'ien,* by speeding man on along *its* way, but, as far as real conquering was concerned, he only wanted to conquer nature in accord with nature. By knowing, through science, creative nature's laws, man could conquer all the rest of nature's creatures.

Perhaps, therefore, Sun Yat-sen might have done better to have availed himself of the ancient Chinese triad of Heaven, Earth, and Man (*t'ien ti jen*), for then he could have said that man by knowing Heaven could conquer earth. At any rate he was *not* declaring war on Heaven; he was greeting with rejoicing the new scientific creed, imported from the West, that man could indeed "subdue" the earth and win at last his promised "dominion" over all of it.

But the question remains, just where did Sun Yat-sen's belief in the conquest of nature and the surpassing of nature leave him as a Social Darwinist? Was not his championing of *jen wei* only a less bitter declaration of independence from evolution than Huxley's? Was not Huxley telling evolution, "Let us alone. We will not go your way any longer," while Sun Yat-sen was saying, "Thanks, but we can take it from here"?

That was almost what Sun Yat-sen was saying, and perhaps on those grounds he should have made a formal declaration of independence. But before the Revolution of 1911 he could never quite bring himself to do it.

Some of his colleagues came closer to making such a declaration. Hu Han-min, in his aforementioned "defense" of Yen Fu, openly warned against taking too seriously the "Darwinian" dogma of the "social organism," claiming that it was a useful theory for pulling the Chinese *ch'ün* together, but that, if it led people to believe their future was materially and not mentally controlled, they would submit to "natural necessity" and lose faith in human action. And, if that happened, the fortifying idea of the "social organism" would only make China's weak *ch'ün* weaker.[123] Hu Han-min did not repudiate the "social organism" altogether, but, by making it again a metaphor, he knocked the wind out of its sails (or cells).

An anonymous author of an article on "The Progressive Evolution of Foreign Relations," published in The *Wai chiao pao* (Journal of foreign relations), revived, somewhat, the social organism, but only to declare that human social organisms were free from the laws that ruled all others. In an elaborate multi-stepped scheme of evolution through the three great realms of the inorganic, the organic, and the conscious, he described a human group (*jen ch'ün*) as being "like" other organisms, but not exactly. For other organisms were caught up in the endless cycle of the conservation of matter, changing round and round from the organic to the inorganic and back again. But "a human group," he said, "is different. Its rise and fall and life and death are all controlled by human action, the general rule of which is progress without cycles and change without stagnation. . . . Material evolution is controlled by natural forces, and so must follow a fixed progression. But a human group moves in the boundless." And then he added, in support of Sun Yat-sen, "Its elements [individuals] are now different from primitive human beings, and it has long since freed its form from the organic realm, so it is not at all inconceivable for it to advance by skipping stages."[124] The rest of nature went round and round, while man alone went upwards, free to advance "under his own steam" and "by leaps and bounds."

Another champion of *jen wei* over *t'ien hsing,* Chang Ping-lin, who, as we shall see, was eventually to call "nature" one of the "four delusions,"[125] dipped at one point into the proverb bag to come up with the very ultimate in traditional expressions of faith in man power: "Heaven must allow what man desires" (*jen chih suo yü t'ien pi ts'ung chih*). That statement put Heaven and nature in human bondage, but only proverbially, of course. It took another anonymous writer, this time writing in the *Kuo-min jih-jih pao* (The national people's daily) to pen a truly original declaration of human independence, in an article entitled "Ko t'ien" (Down with Heaven).

The author of "Ko t'ien," declaring himself, a bit more literally, in "revolt against Heaven," repudiated all "heavenly" theories, religious, philosophical, or scientific, that set up deterministic tyrannies over man. And he included on his black list *t'ien-yen lun,* the "theory of heavenly evolution."

The theory of heavenly evolution, he said, was much better than earlier heavenly theories, or heavenly myths, but still, by stressing Heaven, greatly crippled *jen wei.* It threatened to knock the spirit out of heroes:[126]

> In modern times in Europe there have been the so-called *t'ien-yen* scholars, . . . who hold that there is nothing that has not evolved from Heaven, but that human rule is necessary to save the world from the cruelties of Heaven's process. That is, granted, a new way of looking at Heaven, and it should suffice to awake from their dreams those people who have looked to Heaven for their happiness. But empty words are not the equal of empirical facts. Instead of "heavenly evolution," I would rather say "human evolution."[127]

He did not by "human evolution" mean the evolution *of* man; he meant evolution *by* man. He had obviously been most struck, when he read *T'ien-yen lun,* by the phenomenon of *jen tse* (*artificial* selection). For he said that it made no sense to talk of natural selection at every juncture when man could so obviously select what he wanted regardless of Heaven:

> The sacrificial animals in the Great Temple are better than other animals and should be favored by Heaven. Yet they are killed. And it is man who kills them. If you explain this as part of *t'ien-yen,* then Heaven must make men kill them, but if this second Heaven wins, then the first Heaven loses. If the second Heaven can undo the first, then Heaven is hurting Heaven.[128]

And what sense could that make? It was man who could kill Heaven's fit, and it was man who could preserve Heaven's unfit. So man was in control.

The rebel against Heaven had seized upon artificial selection (admittedly a remarkable business) to dismiss the idea of natural selection, although it had been artificial selection which had helped lead Darwin to his *discovery* of natural selection. Totally ignoring natural selection's contribution to man's origins, the anonymous rebel declared that nothing really counted but *jen wei*:

> The saying goes, "Human determination can conquer Heaven." Today I would change that to "Human determination can replace Heaven." It can replace Heaven because man with his own power can carry out his own will. There *is* no Heaven.[129]

Perhaps Sun Yat-sen should have cried out, "Amen!" But he did not. In the end, having urged his countrymen to abandon the evolutionary way and take the human way, he refused to send them out on it all alone. He held on to the *t'ien* of *t'ien-yen* and refused to break completely with the more Spencerian evolutionaries. He never openly repudiated his purely deterministic colleague, who had said that even those "heroes" who most "loved action" were really "being driven by the forces of evolution."[130] And he did not give up his own claim that "the progress of civilization is brought about by nature and cannot be escaped."[131] He maintained in the very speech in which he cried, "We must not follow evolutionary change, we must change through human action," that "we may not say that China cannot be a republic for, if we do, we shall be contradicting evolution's laws."[132]

Contradicting his own faith in man power, Sun Yat-sen refused to declare man's independence from evolution because he wanted evolution's backing. Like Yen Fu and Liang Ch'i-ch'ao before him, he refused to give up either his fish or his bear's paws, his determinationism or determinism, because determinism was his only assurance that he would win. He wanted to believe, and he made himself believe, that the triumph of his democracy was preordained.

In the end, therefore, Sun Yat-sen seemed to echo the ancient adage *Fata volentem ducunt, nolentem trahunt* (The fates lead the willing and drag the unwilling), except that he added one promise: If one really went "willingly," one could stride ahead of the leaders, and get where the fates were going far sooner than the dragged.

Sun Yat-sen did not revolt against Heaven, nature, or evolution. He did not yet, as Huxley did, urge mankind into a kind of cosmic civil disobedience. And yet his mixed-up metaphors did undermine the authority of "Darwin's" laws, as they were then understood in China, for they pointed to a "human way" which was different from evolution's way, even though not opposed to it. And, if there was such a way, others, if not Sun Yat-sen himself, could wonder whether the laws of nature were fit laws for men. "What does Heaven say?" asked Confucius.[133] Perhaps Heaven only said "Do as I say, not as I do," and "said," in its open book of history, "Follow the Three People's Principles." Perhaps that was Heaven's only mandate. Perhaps, aside from that, "Nature red in tooth and claw" was not "a school of virtue," nor of anything much else.

These were questions that Sun Yat-sen did not ask; he only raised them by implication. But as the flamboyant response of the author of "Ko t'ien" made clear, the door was finally open for such questions.

DARWIN AND SOCIALISM: PRO AND CON. The third of the Three People's Principles opened the door for Darwinian doubts even further, at the same time that it led the Constitutional Monarchists and the Republican Revolutionaries into one last Darwinian debate. This was the principle of "the people's livelihood" (*min-sheng-chu-i*), perhaps the most original of the three, for China, but also, most assuredly, the fuzziest. Sun Yat-sen was clearly for it, but he never said clearly just what he was for, certainly, at least, not before the Revolution of 1911, and not really even after it. Nor, of course, was he ever able to define his third principle in action.

He did, however, equate it with socialism, and for simplicity's sake so may we, especially as the debate over *min-sheng-chu-i* became not only a debate over socialism à la Sun Yat-sen, but a debate over socialism in general, for which there already existed a separate Chinese term (another re-Sinified Japanese translation), *she-hui-chu-i,* the term Sun Yat-sen's opponents usually chose to use even when they were attacking his idea. Sun Yat-sen and his cohorts, therefore, were for socialism, although they described it in different ways. Liang Ch'i-ch'ao and his cohorts were for and against socialism, although they too described it in different ways. It was a bewildering debate, out of which no clear description of socialism ever emerged. Happily, however, both camps' descriptions of socialism have already been admirably described elsewhere, by both Chang Hao and Martin Bernal.[134] We can thus afford to slight the socialist content of China's first debate over socialism and concentrate solely on its Darwinian content.

The debate over socialism, the last great debate between Sun Yat-sen and Liang Ch'i-ch'ao before the Revolution of 1911, was Darwinian in two ways, first, because it took place almost entirely within the bounds of the Darwinian question: "How can China best struggle to survive?" The first and foremost question asked was not "Is socialism just, right, and good, and do our people cry out for it," for the debate over socialism was not first and foremost a moral or humanitarian debate. It did indeed promise to focus attention on the suffering of the Chinese masses, but it was not inspired by it. It was inspired by concern for the fate of China.

It is safe to say that both sides were still more immediately worried about their country than they were about their people. They were more immediately aware of nations' inhumanity to other nations than they were of Chinese inhumanity to other Chinese. Both camps were camps of exiles who encountered abroad one more foreign "ism," and both asked of it first and foremost, "Will it fit our country for survival?"

The debate was also Darwinian, however, because each side claimed Darwin was on its side. The question, therefore, soon became not "Are we for socialism?" but "Is Darwin for socialism?" or, phrased less personally, "Is socialism in accord with evolution?"

The Constitutional Monarchists and the Republican Revolutionaries disagreed on these questions. On this issue, as on all others, they saw themselves in two camps, violently opposed to each other; and they published thousands of words attacking each other's positions. But on the socialist front, at least, their struggle was not yet any clear "struggle between two lines," for lines were crossed, both within each camp and between them. There was, in fact, considerable agreement between the two camps, in retrospect almost more striking than the disagreement.

For one thing, with the exception of the group within the reformist camp, everyone seemed agreed not only that socialism was "good," but that it was a state towards which "evolution" was moving. The first time Liang Ch'i-ch'ao ever mentioned socialism, in his biography of K'ang Yu-wei, published in the December 21, 1901 issue of the *Ch'ing-i pao,* he awarded it his ultimate compliment by insisting that "my teacher's philosophy is socialist philosophy." Western socialism, he said, had been started by Plato and developed by St. Simon and Comte, but K'ang Yu-wei, "without ever reading their books," had come upon much the same ideal thanks to the inspiration of the "Li yun's ta t'ung."[134] Therefore, as the *Ta t'ung* would surely one day come, so would socialism. Indeed as one anonymous "reporter," probably Liang Ch'i-ch'ao himself, would later note in the *Hsin min ts'ung pao*: "Socialism is quite similar to the *Ta t'ung*-ism advocated by recent scholars in our country. Although it cannot yet be put into practice today [*ecce* a bone of contention], as the destiny of the world moves daily towards civilization, there can be no doubt that this ideology will be the final victor."[136]

Both camps were also agreed, however, that modern Western socialism had arisen in Western evolution as a cure for a Western disease, a disease that China, forewarned, could avoid, with a dose of timely preventive medicine. This was the as yet unlabeled "Western internal contradiction" argument that we saw bringing good cheer to Yen Fu as early as 1895. He was the first to say that the West was virtually doomed to "great rebellions" led by "parties seeking to level rich and poor," already arising because Western progress had pulled the rich and poor so far apart.[137] And he was the first to suggest that, while the West was thus "laid up," China could catch up if it could avoid the Westerners' sickness.

Liang Ch'i-ch'ao agreed:

> I believe the economic society of Europe and America has fallen to such a point that there is no way not to have a revolution. But China's economic society need only be bolstered a bit to make it follow the proper path of development and progress. The dangerous method of revolution need not be used.[138]

Sun Yat-sen also agreed, seemingly even with Liang Ch'i-ch'ao's last sentence. "Anyone with any vision," he said, "knows that Europe and America cannot avoid a social revolution." But China could. And China should. For, as he declared in an almost Burkeian passage:

> Revolution is something to be used only when there is absolutely no other way out. We cannot use it over and over again and wound our people's spirit. When we carry out our racial and political revolution, we must at the same time think of ways to reform our society's economic organization, to prevent a later social revolution. That is our greatest responsibility.[139]

Of course he also said, in the very same speech, "We do not want a rich minority to monopolize wealth, so we want a social revolution."[140] As always, he was a man of many tunes.

Sun Yat-sen wanted a revolution to prevent a revolution. His opponents had good reason to be confused. But the trouble was simply that he used the term *social revolution* to refer to two different things, to the violent, "natural" upheaval he foresaw for the West, and to the calm, deliberate "reform" effort he envisioned for China. The preventive social revolution he planned, therefore, was to be as peaceful as a reform movement: "In such countries as England and America, the social revolution may require military force. But, in China, the social revolution will *not* require military force."[141] Indeed, said he, echoing Yen Fu, "the social revolution in Europe will be difficult, but in China easy,"[142] for in China, "the social problem" faced by Europe had not yet arisen.

This, in retrospect, was the most peculiar point on which Sun Yat-sen, Liang Ch'i-ch'ao, and Yen Fu all agreed. They all thought China did not yet *have* any serious "social problem." The peasants' problems apparently were not thought "pressing." "The dangers of the social problem," said Sun Yat-sen, "lie in the future."[143]

Here was proof of the "bookish" nature of early Chinese interest in socialism. Chinese Reformers and Revolutionaries alike simply accepted the notion that "*the* social problem" was that between industrial capital and labor. China, not having much modern industry, ipso facto did not have much of a social problem. "The social problem in Europe and America," said Sun Yat-sen "has reached the point of no return, but in China it is still in its infancy."[144] And therein lay China's hope. Here once again Sun Yat-sen, as Yen Fu and Liang Ch'i-ch'ao before him, seized upon the logic of Mao Tse-tung's later "poor and blank" argument, to make "backwardness" an advantage: "Today it would be much easier for China than for Europe or America to institute socialism, because the social problem is brought about by the advance of civilization. Where the degree of civilization is not high, the social problem is not great."[145]

Of course he went on to warn that, if China did not institute his socialism, the social problem "will surely occur in the future sometime, and when it does, it will be too late to save the situation without having a big revolution."[146] Liang Ch'i-ch'ao supported him on that point: "Today, all the great problems of the other countries of the world are being transplanted without exception to our country. That is inevitable. Therefore, it should be perfectly clear that it will not do to look at this socialism problem as some 'disaster on the opposite shore.'"[147]

Liang Ch'i-ch'ao agreed with Sun Yat-sen that something would have to be done about the social problem before it got serious. And he probably shared Sun Yat-sen's faith that "before great disasters occur it is easy to prevent them."[148] Sun Yat-sen, moreover, agreed with *him* that in this case the disaster to be prevented was a violent popular upheaval and that it should be prevented by the government's reform. But Sun Yat-sen also contended that it would take a violent revolution to create a reforming government, and Liang Ch'i-ch'ao did not agree with that.

So Liang Ch'i-ch'ao still attacked Sun Yat-sen for advocating a social revolution. That attack was not quite fair, but Sun Yat-sen had on occasion said that he wanted a social revolution (so, of course, at an earlier time, had Liang Ch'i-ch'ao) and he did, of course, include socialism in his general revolutionary platform of the Three People's Principles, so it is no wonder that the Reformers should ignore his reformist language and attack him as a social revolutionary. That is why it was on the issue of social revolution that the debate over socialism began.

Liang Ch'i-ch'ao's argument against social revolution, however, need not be repeated in any detail, for it was exactly the same as his argument against racial revolution and political revolution. Any revolution, he said, that would cause internal disorder (*nei luan*) would invite the kind of external disaster (*wai huan*) that would lead to the nation's death (*kuo wang*). In the struggle for existence, any country weakened within would fall prey to imperialism without. That was his first argument against socialism, an old one, and, of course, Darwinian.

His second argument was also Darwinian, and an old one, but in new context it entered new territory. Socialism was good, said Liang Ch'i-ch'ao, and of course it would come, but China was not yet ready for it, and to institute it now would therefore be to break the evolutionary law against *lieh teng,* the "skipping of stages," which, as he had proved before, was in general a mistake and would in the end prove impossible. Specifically, it was a mistake because it would fit China with the wrong weapons for the present contest.

Liang Ch'i-ch'ao had come back from his visit to America convinced that the present contest was with capitalism. The West's most lethal weapon, capitalism, lay behind the new "fox method" of imperialism, a method far more insidious than the traditional "tiger method,"[149] and capitalism had just reached its ultimate form. It had produced "the towering spirit of the twentieth century. . . . the new, soaring demon king of economics"—the trust.[150] This was the creature Liang Ch'i-ch'ao discovered in America that most fascinated and frightened him. He was convinced that, within a hundred years, the whole world would fall under the political control of a few big nations and under the economic control of only a few score big companies.[151] If China did not do something, if it let foreign trusts move in, then in fifty years, he said, there would not be a single Chinese capitalist left in China. All would be eliminated.[152] And, when that happened, "when other countries' capital fills our country, then that will be the day our four hundred million countrymen become horses and oxen forever."[153]

But this "soaring demon king," the trust, was natural, said Liang Ch'i-ch'ao: "The fact that the political world must move towards imperialism and that the economic world must move towards trusts is all part of the natural destiny of the struggle for survival and natural selection. It could not be otherwise."[154] America had fought its trusts for twenty-five years, and still they existed: "Hence we can see that the natural power of evolution is not something human effort can resist."[155]

Trusts were a natural and unpreventable outgrowth of unbridled competition. And yet they had grown out of it in reaction against it (Liang Ch'i-ch'ao had his own sense of the dialectic), just as socialism itself would:[156] Indeed (and this was his cleverest point) trusts marked a *transition* to socialism. For trusts conquered their competitors without destroying them. They did not insist that they "die at their horses' feet." They let those that could not lick them join them.[157] They incorporated their competitors and hence grew bigger and bigger and, as they did so, they came to have less and less competition and moved the world towards socialism.

But, for the present, the world was not socialism's, but the trusts', and they were "socializing" it at the expense of others' independence. Therefore, if China wanted to preserve its independence, if it someday wanted its own socialism, not the West's, then it would have to fight the trusts. It would have to fight foreign trusts with its own trusts, foreign capitalism with Chinese capitalism. For in the struggle for existence, one had to fight fire with fire (an adage *not* found in *The Origin of Species*).

Therefore China now needed trusts and capitalists, not socialists:

> Today what China most urgently ought to study is the production problem, not the distribution problem. Why? Because the production problem is a problem of international competition. The distribution problem is a problem of internal competition. . . . And our survival depends on the former. If we cannot solve that, then we will *have* no distribution problem.[158] . . . We need big capitalists to keep out the foreigners.[159] . . . My economic policy would make the encouragement and protection of capitalists, as a united force against the foreigners, primary and everything else secondary.[160] . . . We should encourage capitalists first and protect laborers second.[161]

Liang Ch'i-ch'ao's argument, not against socialism but against "socialism now," met the Revolutionaries' challenge with a double Darwinian defense. The laws of evolution, he said, forbade stage-skipping and decreed that the present struggle for existence be a capitalistic struggle. In effect, therefore, the laws of evolution decreed that China must go through a capitalistic stage before it could possibly reach socialism. But that argument made the debate over socialism not only Darwinian but proto-Marxist—and almost *really* Marxist. For Marx was present. This was the first Chinese debate in which he had a chance to take part. The Revolutionaries had him in their camp. They mentioned his name in almost every article they wrote on socialism. And yet, strangely, they kept him on the sidelines and never really used him.

Perhaps they did not yet know him well enough. Perhaps they

saw him only as an authority on a kind of capitalism that did not yet exist in China. Perhaps they realized that he did not offer much support for their cherished hope that the last could be first. At any rate, although they anticipated many of the questions raised in later Marxist debates, and indeed many of Mao Tse-tung's answers, the Republican Revolutionaries never switched this first debate over socialism to Marx's level. They continued to meet Liang Ch'i-ch'ao's challenge on Darwin's level—and yet still without trying to rebut Darwin.

They had rebutted the "falacious doctrine" (*miu lun*) of unskippable stages, of course, before, in the democracy debate. But this time "the train defense" did not quite work. Democracy might be a modern device, ready and waiting to be adopted like the latest train, but socialism was still on the drawing board. As the reformers were wont to exclaim, "No one has yet been able to put actual socialism into practice even in the civilized countries of Europe and America. How could we do it in China?"[162] But this challenge was easily met with the "backwardness-is-an-advantage" argument. Sun Yat-sen again simply abandoned his train for a cart, so that he could get off the Western track and avail himself of some Chinese proverbial wisdom: "The cart before can give warning" (*ch'ien chü k'o chien*). The West had gone wrong. Capitalism was a rut, and China could avoid it and pass the West by.

This was a subtly different kind of stage-skipping, however, and founded on a different evaluation of the stages to be skipped. The Republican Revolutionaries had been against Liang Ch'i-ch'ao's constitutional-monarchy stage because it was outmoded and unnecessary, but they were against his capitalistic stage because it was "bad." An element of moral indignation was finally making its way into the argument over socialism, even though it was a strangely academic indignation.

China's first socialists showed their first social concern by fulminating against the evils of social inequality not in China but in the West. They railed not as Lu Hsun would, a decade later, against a "man-eating" Chinese society, but against a dog-eat-dog Western one, so dominated by "a minority of capitalists" that the whole worth of Western civilization suddenly seemed thrown into question. "The truth or falsity of civilization in the world," said Hu Han-min, with an accusing finger leveled at the West, "is determined by how much autonomy and freedom people can enjoy. If the vast majority of the people must grovel in submission to a minority of capitalists, is it civilization or barbarism, progress or regress?"[163] And Sun Yat-sen added, in what was at once a startling qualification of his original faith in progress and a proto-Leninist implication that Western capitalism was behind the ills of the whole world: "Civilization has good fruits and bad fruits. We must pick the good and avoid the bad. In the countries of Europe and America, all the good fruits are enjoyed by the rich, and the bad fruits are eaten by the poor. It is because a minority has seized all the benefits of civilization that we have such a world of inequality."[164]

Without, perhaps, quite realizing what they were doing, Sun Yat-sen and Hu Han-min had begun to question "evolution's wisdom." They both looked evolution's latest gift in the teeth and said, "Oh what a revolting development this is."

Of course, the rejected fruit of capitalism was something of a sour grape. It was a newly alleged source of Western power that seemed depressingly out of reach, and Chinese leapt, therefore, at any theory that found it tainted. That was socialism's first appeal. Before Chinese ever saw socialism as a cure for Chinese ills, they liked it because it diagnosed Western ills. They liked Western socialists, met in Japanese translation, because they were Westerners soured on the West who painted capitalism, the source of Western power, as an evil and as a weakness. They liked Western socialists because they confirmed Yen Fu's earliest report: that the West had wealth and power—and inequality, and was hence a "hot-house plant."

Socialism offered China hope, whether or not it spoke to China's ills. And even as unpracticed theory its existence won China some revenge. That could be seen in Hu Han-min's triumphant question, "Is this civilization or barbarism, progress or regress?" And yet the obvious psychological satisfaction Hu Han-min, Sun Yat-sen, and others found in socialism does not prove that their dislike for capitalistic inequality was insincere. Nor does it prove that their own socialism remained devoid of social concern. For even sour-grape socialism could inspire social concern. China's social problems might not have led Chinese to look at socialism, but socialism did lead Chinese to look at China's social problems.

Granted, in the beginning many were so happy to find explosive inequality in the West that they quite overlooked inequality in China (except between Manchu and Han). Indeed, they even denied it. Sun Yat-sen exulted in the fact that "in China, at present, capitalists have not yet appeared."[165] And Liang Ch'i-ch'ao, without incurring the immediate ridicule of China's "social revolutionaries," insisted that "in China the two classes of the extremely rich and the extremely poor do not exist."[166] But Hu Han-min, encouraged though he too was at China's subsequent "advantage," was inspired to look closer. And he found inequality: "Today Chinese capitalists have not yet appeared. Mr. Sun has mentioned that fact in his speech, and even Mr. Liang admits it. But although there are no capitalists there *are* already landlords."[167] And landlords, he implied, were just about as bad. Indeed they were protocapitalists, responsible for a "land problem" that would have to be solved *before* one could solve the "social problem."[168] For it was precisely because the West had failed to solve its land problem that it had gotten into its social problem. Therefore Hu Han-min became the first to call for "the nationalization of the land."

No one knew it, but the "Sinification of Marxism" was getting under way, as men like Hu Han-min and Sung Chiao-jen, who conveniently translated Marx's new terms, *bourgeois* and *proletaruns* (*sic*), as "rich gentry" (*fu shen*) and "the common people" (*p'ing min*),[169] made "the land problem Chinese socialism's first order of business. Sun Yat-sen, too, associated "the land problem" with "the social problem," and landlords with capitalists,[170] and he too finally advocated nationalization of the land:

> If, as in foreign countries, all land rights fall into the hands of a minority group of capitalists, we will inevitably end up with a capitalist

> dictatorship that will be worse than monarchic dictatorship.[171] . . . We are against a minority seizing economic power and monopolizing society's sources of wealth. . . . If land and big business are both nationalized, then the profit that comes from them can go to the people as a whole. . . . *Min-sheng-chu-i,* therefore, will get rid of the capitalist minority, so that the people can all enjoy freedom of production. Thus *min-sheng-chu-i* is state socialism.[172]

Capitalists were evil, landlords were evil, and trusts, as superhuman capitalists and superhuman landlords, were naturally far more evil still. And they were in complete opposition to socialism. They were not the proper transition to it. Feng Mao-lung (Feng Tzu-yu—"Freedom Feng") therefore denounced Liang Ch'i-ch'ao for "trying to introduce a monster" into China: "The trust in the future will be a great poison in China, and the only antidote will be the institution of socialism."[173]

Liang Ch'i-ch'ao might be right in his analysis of the West. There, perhaps, capitalism and trusts had naturally evolved, as a stage to socialism. But such a capitalistic stage was a hard way and the wrong way to socialism, for capitalism could never evolve peacefully into socialism; it could only lead to a situation, said Sun Yat-sen, demanding "killing and the shedding of blood."[174] And that was not the kind of social revolution or evolution China wanted. On that score, Sun Yat-sen and Liang Ch'i-ch'ao, without admitting it, agreed. They both were engaged in a proto-Marxian debate, and they both recognized a natural dialectic of evolutionary progress in the West. But they rejected it for China. Without yet labeling the idea Marxist, they rejected the Marxist idea that things had to get bad before they could get better. On the social level, if not the political level, they agreed that one stage did not have to explode into the next.

Liang Ch'i-ch'ao thought the capitalistic stage could be controlled. Sun Yat-sen thought it could and should be avoided, however natural it had been in the West. For Sun Yat-sen had decided that all that was natural was not good. "In general," he said "social phenomena cannot be completely allowed to go their natural way.

They are like trees. If you let them grow naturally, their branches will go in all directions."[175]

Here, in a simple simile, in a casual bit of rhetoric in the middle of a political speech, was a remarkable revision of what had been one of the most dominant tenets of Chinese Social Darwinian faith since 1895. Before this, Sun Yat-sen had cried out as loudly as anyone that nature was moving inexorably towards the good. Indeed, he had reiterated only sentences before the axiom that "the progress of civilization is brought about by nature and cannot be escaped."[176] But now he said that nature did not always take the best way to the good. On its way up it could branch in all directions. And left to its own devices it would. So man could not leave it to its own devices. What was true of "civilization" was true of nature and evolution as well. Evolution itself had "good fruits and bad fruits," and it was up to man to "take the good and avoid the bad." It was up to man to prune evolution—and to prune his own evolution. Without saying so, and probably without realizing it, Sun Yat-sen had met Liang Ch'i-ch'ao's first Darwinian challenge, or Spencerian challenge, with a Huxleian defense.

But what of the second challenge, Liang Ch'i-ch'ao's contention that capitalism could only be fought with capitalism? In the struggle for existence, which surely did exist, could China *afford* to skip capitalism? Could it afford a "moral way"? Real Marxists would later argue over such questions in China, but the Republican Revolutionaries and Constitutional Monarchists did so first. And the Republican Revolutionaries answered the Constitutional Monarchists' second challenge this time simply, forthrightly, and prophetically, with a description of "state socialism" more honest than any that would ever be given again. Said Hu Han-min: "If the country becomes a great landlord and a great capitalist, then foreign capital will be no worry."[177] "We are not against capital," added Sun Yat-sen; "we are against capitalists."[178]

So the Revolutionaries did not deny that the outside world was in a stage of capitalistic struggle. They argued, instead, that state socialism could hold its own by being state socialism on the inside and state capitalism and state landlordism on the outside. They argued with yet another variation of the Chinese Darwinian formula of unity within for struggle without.

There was one more Darwinian challenge to socialism from the reformist camp, however, potentially more troublesome—the charge, picked up from Western opponents of socialism, that socialism would kill progress.

Liang Ch'i-ch'ao himself rarely pressed this charge, although he had introduced it in his introduction of the thought of Benjamin Kidd. He preferred to praise socialism as a distant, *Ta t'ung* goal. But other Reformers pressed the charge with no mercy. Liu Hsien-chih, writing not for Liang Ch'i-ch'ao's journal but for Yang Tu's *Chung-kuo hsin pao* (New China news) gave a clear statement of it, prefaced with a classical appeal to Darwinian science. "Our party's aim," he wrote, "has been to study the original principles of human evolution and use Darwin's two great laws of 'the survival of the fittest' and 'adaptation to the environment' to deduce the way by which our country and our people can change to meet our environment and thus prove fit to survive."[179] One *seeming* way, he said, was socialism, a nice-sounding theory, but one with a fatal drawback:

> If we look only at the facts, those who talk today about socialism have as their aim the destruction of present society's outer organization. Now the organization of society is surely not beyond improvement, but, if we must abolish contests of ability and bring all people to one level, the result will be to put the superior and the inferior in one class, and the mechanism of progressive evolution will gradually break down. If we must take all private property and let the state do business with it, the result will be that everyone will rely on the state and not value independence and self-government. And if everyone fails to value independence and self-government, then society will have even less chance of progressing. The aim of our ideology is to promote society's progress. Therefore we do not want socialism.[180]

Chiang Chih-yu echoed these sentiments in Liang Ch'i-ch'ao's *Hsin min ts'ung pao.* Equality, he said, was a noble idea, but one to be wary of, for it denied the "natural distinctions" between "the smart and the stupid and the diligent and the lazy." If one were to force equality, "one would have to seize what the smart and the diligent have and give it to the stupid and the lazy, and the result of that would be to make everyone content to be stupid and lazy, and as a consequence the world would regress."[181]

Here at last was Social Darwinism to warm the heart of William Graham Sumner, who had cheered his fellow American proponents of free enterprise, in the 1870s and 1880s, by calling socialism "a plan for nourishing the unfittest":[182]

> Let it be understood that we cannot go outside this alternative, liberty, inequality, survival of the fittest; not-liberty, equality, survival of the unfittest. The former carries society forward and favors all its best members; the latter carries society downwards and favors all its worst members.[183]

We do not know whether or not Chinese capitalists warmed to the words of Chiang Chih-yu as American capitalists did to the words of Sumner, for Chinese capitalists have been far more silent and self-effacing than their American counterparts. Yet it is at least safe to say that, as far as political and philosophical theoreticians were concerned, Sumnerian Social Darwinism, before the Revolution of 1911, was never popular in China. The examples quoted from Chiang Chih-yu and Liu Hsien-chih are rare examples.

Certainly the Republican Revolutionaries were not bothered by Sumnerian Social Darwinism. "Freedom Feng" dismissed it with scorn:

> When socialism was just getting started, there was a theory among European and American scholars that attacked it, saying that, if socialism were put into practice, humanity would no longer make any distinction between rich and poor or noble and common, and therefore the desire to compete would die out and the world would regress. This theory in the latter half of the nineteenth century was worthy of discussion, but it has by now become a worn-out doctrine of the past that anyone with any knowledge at all can refute.[184]

Unfortunately, he himself chose not to refute it any further, except to say that socialism was the goal of progress, that it *was* progress, and that it was ludicrous, therefore, to say that socialism would prevent progress. His final defense was nothing more than a reaffirmation of Sun Yat-sen's faith that evolution was on socialism's side. And yet he reaffirmed that faith at the same time that he relegated progress through competition to the ranks of worn-out nineteenth-century notions. Progress backed by nature was taken out of nature's hands. Evolution and progress were now so thoroughly confused that progress could be sent off on its own without the evolutionary mechanism that had supposedly set it going.

On the eve of the Republican Revolution, Sun Yat-sen and his colleagues still claimed to be Darwinists. But so strong was their faith in themselves and in human action, so strong their conviction that their way was right, that they were almost ready to get ahead over evolution's dead body.

DARWIN AND THE ANARCHISTS

The Anarchists, too, did violence to the theory of evolution, even though most of them were avid, if not rabid, evolutionists. Their "violence," however, was partially, at least, intentional, for they were the first group of avowed Darwinian revisionists. And as *avowed* revisionists, they ushered in the dawn of a new Darwinian day.

The Anarchists were the madly brilliant first generation of Chinese anarchists, who first discovered Western anarchism, in the first decade of the twentieth century, who first discovered in the writings of Mikhail Bakunin and Prince Peter Alekseyevich Kropotkin, and in the deeds of all the Russian nihilists, the kind of "ismic" answer to China's problems so many Chinese were looking for.

This first group to swear allegiance, or at least qualified allegiance, to *wu-cheng-fu-chu-i,* or "no-governmentism," which is what "anarchism" became in Chinese, was initially divided into three groups, two of ideologues and one of activists. The ideologues were the Tokyo group, led by Chang Ping-lin, Chang Chi,

Liu Shih-p'ei, and his wife, Ho Chen, and the Paris group, led by Wu Chih-hui and Li Shih-tseng. The activists were all the Anarchists in the field, those in China, led by no one, but best typified by Ch'iu Chin, Hsu Hsi-lin, and Wu Yueh. They were the courageous, if woefully inaccurate, bomb-throwers and the plotters of romantic, if needlessly suicidal, uprisings, who were not necessarily philosophical Anarchists, but who were Anarchists self-proclaimed at least in their espousal of assassination.[185]

Of course, almost all of the Anarchists, ideologues and activists, had toyed with the idea of assassination, and as some of the activists became ideologues, so did some of the ideologues become activists. Ch'iu Chin edited her journals and Wang Ching-wei threw his bomb. Shih-fu (Liu Ssu-fu) blew off the fingers of his left hand only to write bombastic articles with his right. But, in general, those who spread the most ideas stayed far from the fray, while those in the midst of the fray spread only one idea—which they underlined with their lives—that violence for right was right.

In any case, it is difficult, if not foolhardy, to talk about "the Anarchists" as a group, for they were terribly poor at grouping. The leading ideologues were *all* rarae aves, and they found it hard to flock together despite their anarchic feather. Chang Ping-lin (who must also be numbered among the Anarchists, even though we have already numbered him among the Republicans, and yet who was really too rare to be numbered with anyone) and Wu Chih-hui despised each other. Liu Shih-p'ei hated Wu Chih-hui and liked Chang Ping-lin, but soon had a falling out with Chang Ping-lin. Chang Chi left Tokyo altogether and joined the Paris group. The Tokyo "group," therefore, which had only "gotten together" in 1907, had fallen to pieces by 1908.[186]

The Paris group was more cohesive personally, but not much more so intellectually. Its principal members managed to cooperate on various projects up to and, for a time, even after the Revolution of 1911, but certainly never as a political or even apolitical party of the *i hsin i te* (one mind and virtue) variety. They remained of several minds, and even of several virtues, as their Chin te hui (Progressive Virtue Society), the closest thing they ever had to an organization, allowed its members, in Buddhistic fashion, different levels of moral vows.[187] The leaders of the Paris group, therefore, never tightly organized either themselves or their thoughts, and their famous and infamous journal, *Hsin shih-chi* (A new century), like its short lived Tokyo counterpart, *T'ien i pao* (Natural right), offered its readers an Anarchist platform that was anarchic indeed.

And yet we still have a right to consider the Anarchists as a group—because their readers did. Their readers considered them a very special group, however disparate and eccentric, introducing a new, foreign creed to a country that needed a new creed. Despite all their conflicts and contradictions, the foreignness of their creed made them a group to be listened to.

Moreover, the anarchy in their thought only increased the final influence of Anarchist ideas. For, by presenting no clear "party line," they made it very easy for their readers to accept some of their ideas without accepting others, to accept *many* Anarchist ideas without accepting the idea that gave the Anarchists their name—that people would be best off with no government.

Most of the people who read *Hsin shih-chi* or *T'ien i pao* and favored revolution probably took their stand in the end with the Republicans, not the Anarchists, making the same distinction that Hu Han-min made in the *Min pao* in 1906: "The difference between us and the No-Government Party is that we believe that the Chinese people, at their present level, have to have a government."[188] But having said that the Anarchists' ideal was too utopian for now—"at the present level"—they felt free to accept other Anarchist ideas, and did—especially Darwinian ideas.

The Anarchists' Darwinism proved to have tremendous appeal for many more people than were ever converted to anarchism itself. Indeed, the Anarchists' contribution of Kropotkin's great revision seemed to be precisely what all Chinese Social Darwinists had been groping for. And most could happily accept it without giving up their belief in the need for government.

Before looking at the Anarchists' Kropotkinism, however, we should look at the scientific level of the Anarchists' Darwinism in general, for they laid great claims to being "scientific." The first word of *The New Century* was "science," and the Anarchists claimed their whole ideology was based on science—just as the Marxist-Leninists would after them. And the science above all others that their ideology was supposedly based on was the science of evolution.

SCIENCE. At least two of the leading Anarchists had had a chance to learn more of scientific Darwinism than almost any other Chinese writers since Yen Fu himself. Li Shih-tseng had formally studied biology, zoology, and botany in France and had supposedly read the writings of both Lamarck and Darwin;[189] and Wu Chih-hui had studied evolution on his own for several months in England. He had not, as Daniel Kwok suggests, "become acquainted" with the theory of evolution in England, but surely he became better acquainted with it there, better acquainted with it by far than most of his contemporaries.[190] Like his contemporaries, however, he had first become acquainted with the theory of evolution in the early writings of Yen Fu. He had even taught Yen Fu's article, "On Strength," with its introduction to the theory of evolution, to his students at the Nan-yang hsueh-t'ang (Nanyang Academy) in 1899.[191] It was not until a decade later that he seriously studied evolution from Western sources. He *may* have studied evolution during his first stay in England, from 1903 to 1906, but, by his own account, it was in 1909 (which he wrongly called "the year before the first year of the Republic") or, more probably, in 1910, after "it was no longer possible to publish *The New Century*," that he "retreated with [his] family from Paris to the western suburbs of London" and delved into evolutionary studies, to edify himself, his children, and his countrymen—and to support himself:

> At that time I had no way to earn a living. But near where we lived there was a relatively well-stocked library. So I went there every day to borrow books and the latest illustrated magazines on such subjects as the theory of evolution, astronomy, geology, anthropology, natural history, even chemistry and physics. Everyday I brought some back, and when I had read them I explained them to my children. And I also

> wrote out [Chinese] versions which I sent to the Civilization Book Company [in Shanghai] for publication, so I could get some money.[192]

It was at that time that he translated Dennis Hird's *A Picture Book of Evolution* and Joseph McCabe's *La Paleontologie* and wrote his "novel" about evolution, *Shang-hsia ku-chin t'an* (Talks of Heaven and earth and past and present).[193]

The works on evolution that Wu Chih-hui read and translated in England may have been of widely varying degrees of sophistication, but all, at least, were concerned with real science and "real Darwinism," not Social Darwinism. And yet Wu Chih-hui read them all only "when it was no longer possible to publish *The New Century*," only after his greatest days as an Anarchist essayist were over. The readers of *The New Century*, therefore, reaped no benefits from his scientific study. Nor, although this time there was no such simple explanation of the fact, did they seem to reap any benefits from Li Shih-tseng's study of biology. If Li Shih-tseng and Wu Chih-hui really knew much more than their contemporaries about "real Darwinism" during their days as editors of *The New Century*, they seemed reluctant to "let on."

The mostly wildly unscientific articles on evolution to appear in *The New Century*, however, were not written by either Li Shih-tseng or Wu Chih-hui, but by outside contributors. And yet the published comments of Li Shih-tseng and Wu Chih-hui on such articles, critical though they were, were rarely as devastatingly critical as they should have been. Comments and articles alike, therefore, reflect the fact that, in China's first Darwinian decade and a half, scientific understanding of Darwinism had not progressed by leaps and bounds.

One of the weirdest bits of Darwiniana to be published in *The New Century* was "On the Origin of Man," sent in by a contributor from Holland who came to the same kind of Darwinian grief, only more so, that Liang Ch'i-ch'ao had come to a full decade before, when he tried to explain evolution to his students in Hunan.[194] The contributor from Holland made a disastrous attempt to mix

The Origin of Species with *The Book of Mountains and Seas (Shan hai ching),* and other ancient texts.

He began by acknowledging Darwin's proof that men had evolved from apes, although he added that China's own *Li chi* (Book of rites) offered a proof of the same thing that was two thousand years older. But he then offered his own "advanced" thesis, that men had "not only evolved from apes, but from many other kinds of animals as well." New species, he said, could only be born when different species mated. Thus, horses and donkeys produced mules, and turtles and snakes produced tadpoles. So too, "when a tigress bears three sons, one may be a leopard." The leopard got his spots because his tigress mother mated first with a tiger and then with some other beast. European pigeon-breeding, he said, substantiated all this (so Darwin himself bred misconceptions). It was obvious that men too must have come from mixed parentage. And the proof lay in *The Book of Mountains and Seas.*[195]

All the races described in that book must once, he said, have existed. There were some "with ox heads and human bodies" and others "with human heads and snake bodies," just as in European paintings there were human beings with lion heads and elephant heads. "In those days," he said, "there really must have been people like that. Such kinds of people must originally have been produced by all sorts of animals mating together." For example, "Apes are closest in form to men, and their nature is the most licentious. Therefore those with ox heads and human bodies must be men born of the union of apes and oxen."[196]

Such a theory led to the view that different races might have very different ancestors, although many seemed to be related as canine cousins. Discussing some of China's ancient neighbors, the Chinese Hollander said that "the *Shuo wen* quotes ancient sources as saying that the Ti were a breed of dogs and the Ch'iang and Chie were breeds of sheep," and such accounts, he added, were probably true. China's Moslems, or perhaps Moslems in general, were also of dubious ancestry: "It cannot be for no reason at all that Moslems are popularly said not to eat dogs or pigs because their grandfathers were boars and their grandmothers bitches. (Some think that Moslems cannot bear to eat dogs or pigs because they think them unclean, but that is not the case. That might explain their not eating them, but why would they respect them? Surely the Moslems do take pigs and dogs to be their ancestors.)"[197]

He did admit that the Chinese race did not know "racial purity" either. "The old books say," said the contributor from Holland, "that the Yellow Emperor's daughter was married to P'an-hu (P'an-hu was the Yellow Emperor's hound)."[98] But then he went on to a second evolutionary problem which would have given Darwin himself pause for thought. He asked which had evolved first, man or woman:

> When the human race was born, someone had to come first. Suppose we say that man was born before woman. No man has a womb. None could give birth. If we say that man and woman were born together at the same time, it would be like a religious myth, based only on imagination with no real proof. Having studied the matter, I would venture to say that woman came before man, and I do have plenty of proof.[199]

His proof was of two kinds, the first "historical": "The ancient histories say that Lady Wa made men," which proves that "in those days it was a woman's world." (So much for Adam's rib.) But his second proof was linguistic. He found "fossilized" evidence of evolution in his own etymology of Chinese characters: "All those who created characters must have had a basis for their creations. Our Chinese word *shih* [beginning] is composed of the woman radical and the element *t'ai* [roughly homonymous with *t'ai* (womb)], thus signifying that the human race began in the womb of woman. The character *hsien* [first] is composed of *niu* [ox] and *jen* [man] signifying that the first human beings had ox heads and human bodies."[200]

All the above, despite seeming implications to the contrary, somehow went to prove that men and women, and all races, were related and equal, and that all mankind (and womankind) should therefore live together in harmony and peace. The contributor from Holland ended with a plea to his readers to realize that fact, and with a bravo to the Anarchists for already looking in the right direction.

Perhaps it was thanks to the bravo that Wu Chih-hui treated the contributor from Holland as politely as he did, although perhaps, too, there was sarcasm in the opening compliment he paid him: "In our very immature Chinese community of evolutionary scholars, he may truly be reckoned as a towering authority."[201] He did attempt to criticize the contributor from Holland scientifically, however, and in so doing proved that he knew much more about Darwinism than the "towering authority" did. Wu Chih-hui pointed out that Darwin's theory was not based on ancient rumors but on scientific study of things as real as bones. The bull-headed theories of the contributor from Holland, he said, were the products of his own and others' imagination, and he warned him not to Darwinize old books. "Old books," he said, "are all great obstructions, and there is nothing more harmful to China's modern society." Indeed, if Chinese were going to treat old books as science, he sighed, perhaps (as someone else had suggested) Chinese *should* swear off Chinese books for a decade.[202]

And yet, as Wu Chih-hui exposed the Darwinian "immaturity" of the contributor from Holland, he also revealed his own—just as interesting and probably more important, however much less flamboyant.

Having helpfully explained, for example, that the process of evolution had taken eons and eons and that no monkey had ever given birth to a man, Wu Chih-hui gave the following explanation of just how the countless species between amoeba and man *had* evolved:

> Throughout the course of a trillion years, each and every one of this multitude of living things, pursuing the principles of evolution, evolved and advanced according to its ability. . . . Those of little ability are perhaps still simple organisms. But those whose ability was most suited to evolving gradually became the various species of monkeys, and ten trillion monkeys then ceaselessly went on evolving and advancing, again each following the principles of evolution. A part of them, some sooner,

> some later, all struggling to get ahead, evolved into wild men and then gradually turned into the civilized men of today. Another part of them only evolved into gorillas and baboons, and then gradually into the savages of today. And still another part of them evolved so slowly up to now that their descendants are the kind of man-apes that still exist in Africa and such places. As for the ordinary monkeys of today, at the time our ancestors evolved into men, their ancestors were probably still among the lowest forms of living things.[203]

This was not "mature Darwinism" even in 1908. The implication that modern races represented different stages of evolution (which did not square with the Anarchists' frequent decrials of racism), the implication that certain races, relatively speaking, were "low life," and that savages had "evolved" into civilized men as apes had evolved into savages, was not necessarily a "backward" idea, for many respected Western scientists still thought Darwin had proved the same thing, Darwin himself having spoken all too often of "the highest races" and "the lowest savages"[204]; but the idea that all species were trying to evolve up one and the same ladder towards manhood (or beyond), and that their varying degrees of progress, their "advancement" or "backwardness," reflected varying degrees of ancestral "ability"—that was not Darwinism at all. Granted, certain of Darwin's statements might easily have led one to leap to such a conclusion. He had, for example, said that "any animal whatever endowed with well-marked social instincts, the parental and filial affections being here included, would inevitably acquire a moral sense or conscience, as soon as its intellectual powers had become as well, or nearly as well, developed as in man."[205] But even such a statement should logically have led, at its wildest, to a vision of a world "peopled" by moral animals, by lions who would lie down with lambs, or be conscience-stricken if they did not, not to the vision of a world peopled by people who *had been* animals, animals who had all finally "made it" through anthropomorphosis.

There was a discrepancy between Wu Chih-hui's commitment to science and his knowledge of it, which he further demonstrated in his comments on another Darwinian article he chose to publish in *The New Century,* "Nan-nü tsa-chiao shuo" (The theory of promiscuous copulation between men and women), which, needless to say, gained a far greater notoriety than the article "On the Origin of Man."

The author of "Promiscuous Copulation" called himself Chü P'u, and from the state of his scientific knowledge, and his style, one can identify him fairly safely as "the contributor from Holland" himself.[206] Chü P'u was a non-misogynous misogamist, who was convinced that the institution of marriage was close to the root of all evil and the greatest obstacle to the continuation of evolutionary progress. For marriage monopolized love and cramped its power. Love was the force that had made the world and that made the world go round. And love was the force behind evolution. But to keep *man* evolving, love had to be free, "public," and universal—and marriage had made it private. Therefore, said Chü P'u, "If today we wish love to spread and our people to progress, we must begin with the abolition of marriage; we must begin with promiscuous copulation."[207]

By *promiscuous copulation* Chü P'u seemed to mean both "free love" and miscegenation, both of which he deemed necessary for the progressive evolution of the human race. For "without promiscuous copulation," he said, "a race will not advance."[208] He therefore proposed a social revolution for eugenicist reasons, based again (if we can assume that he *was* the contributor from Holland) on an undeniably weird mixture of ill-understood Darwinism and ill-gathered "facts":

> The world has long accepted the fact that man evolved from various species of animals. Now it has ever been a true principle that, as the different species of animals were able to progressively evolve by mating together, so would the different races of man, if they mated together, be able to evolve even further. But why, then, over the last ten million years has the human race on this earth not made much evolutionary progress? And why have the Chinese people not only not progressed but even regressed? I cannot help but bitterly hate the people who invented marriage![209]

Chü P'u never explained the biological mechanics of the matter, but he argued that it was an evolutionary law that odd bedfellows would bear outstanding offspring, especially if they *loved* each other:

> In the early days of the human race, it was truly a woman's world. There were men, but not many. And, as things that are scarce have value, women at that time fought for husbands, but men did not fight for wives. . . . Some of the women who did not get to mate with men, however, mated with animals. Women, therefore, copulated promiscuously, and evolution advanced at its fastest, and the people that were born were of the very best sort. . . . But later, as the number of women fighting for husbands became much greater, and as there were still not enough men to go around, marriage had to be instituted to avoid strife. Even so, however, although marriage rites at that time did exist, they did not forbid women to copulate promiscuously, and as many still did, the evolution of mankind was still speedy.[210]

Chü P'u had discovered a marvelous way to believe both in evolution and in the superiority of the ancients. The ancients had been better because the moderns had regressed. Already in the Chou dynasty marriage rites were getting worse, but women were still, he said, allowed to choose their own mates, and elopements were not forbidden. "It is obvious, therefore," he concluded, "that at that time marriages could be free. And as women were thus able to follow those they loved, the sons they bore were still virtuous and wise. Licentiousness was at its greatest in the Spring and Autumn period, but then too the most men of talent were produced. That proves the truth of what I say."[211]

Thereafter, through the T'ang dynasty, people did not progress, but neither did they regress, for, although marriages were not free, chastity was not yet a virtue. It was the Sung Confucians who were responsible for the regression of the Chinese race: "It was their theory that it is a small matter to starve but a great matter to lose one's chastity that cast women into the eighteenth level of hell" and so weakened and stultified the Chinese race that foreign races could invade and rule.[212]

Racial intelligence and racial strength, therefore, could only be regained by a return to promiscuity—at home and abroad. Chinese in Malaya had proved that it worked abroad. Miscegenation with Malays, so Chü P'u claimed, had produced children stronger than Chinese, even though pure Malays were weaker. And the Japanese had proved the theory at home: "Japan in the past was called the Dwarf Land, because the Japanese race was so short, but ever since prostitutes ran rampant their race has been improved."[213]

Lest there be any doubt as to the eugenic value of promiscuity, Chü P'u said, one had only to look at the Sages. Jesus, Confucius, Lao Tzu—all were "illegitimate," as were the heroes Ch'in Shih Huang and Han Wu Ti. "Hence we know," said he, "that all men who have been at all outstanding have been bastards."[214]

Darwin (among others) was being abused, and Wu Chih-hui knew it—in part. He upbraided Chü P'u for his unscientific reliance on ancient books as authoritative sources for either history or natural history, and he especially decried Chü P'u's conclusion, based on such books, and, said Wu Chih-hui, on a unique form of Chinese superstition, that the ancients had evolved to a level far higher than that of their degenerate descendants. He especially disliked that conclusion because anarchism, he said, was based on the principle that "in the world there is progressive evolution and no devolution."[215]

Finally, Wu Chih-hui ridiculed as quite without basis Chü P'u's contention that Japanese prostitutes had added to the stature of their race.[216] Still Wu Chih-hui commended Chü P'u for showing sharpness of thought and courage of conviction, and said of him, with no *obvious* sarcasm, "I would certainly predict that with such knowledge and ambition his future contribution to the world of evolutionary science will not be slight." Moreover, in showing sympathy for Chü P'u's advocacy of free love and the abolition of marriage and even, although to a less extent, of "promiscuous copulation," he let two of Chü P'u's weakest "Darwinian" propositions go unchallenged. He cited, in seeming defense of promiscuous copulation, the "fixed scientific principle that, when different kinds (*i lei*) mate, the progeny is good and, when the same kind mate, the progeny is not good."[217] He seemed to have heard of the dangers of "inbreeding," but gave no clearer definition than Chü P'u himself of what a breed was. He failed, therefore to dispute the idea that evolution had "progressed" thanks to the getting together of dog and cat, donkey and horse, beauty and beast.

Wu Chih-hui seconded Chü P'u in his most wildly romantic contribution to the science of eugenics, his insistence that love itself was eugenic. In advocating promiscuous copulation, Chü P'u had not advocated ravishing orgies of self-gratification. He had advocated love matches, for they, he insisted, would produce the best seed. And Wu Chih-hui agreed: "Women who mate, become pregnant, and give birth because of love will definitely have better sons than those who mate because they are forced to."[218]

But Darwin had never said anything of the sort. He had said, one must admit again, in his most famous Lamarckian lapses, that "virtuous tendencies" might be inherited "after long practice" over "many generations,"[219] and so he might be accused of saying that "good behavior" could be inherited, but, even if it could, it took *time*. Darwin never promised that there would be any "definitely" inheritable betterment in the "good breeding" of children born of "happy homes." And he certainly never promised that any kind of good behavior could be inherited overnight, thanks simply to a lovely conception. However mysterious he found mutations, he had never attributed them to love. Nor had he relied on love in his breeding experiments. He had bred highbred pigeons, not lovebirds—through arranged marriages.

It would be ridiculous to ridicule Wu Chih-hui for not understanding love's place in evolution—for no one does. It would also be most unfair to ridicule him for not understanding the latest Mendelian genetic theories of his day, for Mendel's theories had only recently been rediscovered.[220] But one can say, nevertheless, that Wu Chih-hui's comments on Chü P'u's theories revealed a very poor understanding of the "mechanics" of evolution even as Darwin himself had first presented them.

Those comments, however, were all written before Wu Chih-hui took his second "sabbatical" in England, before he spent months borrowing books on evolution from a British library. He wrote and published only one important work before the Revolution of 1911 written after his evolutionary study, and as a result of it, and that was his aforementioned "novel," *Talks on Heaven and Earth and Past and Present.* And yet that work, too, although devoted almost entirely to evolution, was still quite un-Darwinian.

Wu Chih-hui's *Talks on Heaven and Earth and Past and Present* told the story of evolution in a most novel way indeed, although the book was not really a novel. It did have a cast of characters and a plot, but the plot was finished in the first chapter, as soon as it had gathered together the cast of characters on a ship so they could talk. "Their" talks, lively and humorous, gave Wu Chih-hui *his* chance to talk to China's youth.

Wu Chih-hui wanted to popularize "common knowledge" about "man's place in nature" that was not yet common knowledge in China. He wanted to tell his young compatriots, or at least all who could read, the true story of the sun, the moon, and the stars, and of the earth and its formation, and of its elements and its seasons, and of its first living things, and of their evolution into fish, birds, and beasts, and finally men. But he wanted to tell of all those things not just as an introduction to science, but as a background to "social science." He did not want to spread scientific knowledge just for the fun of it, or even, first and foremost, for the scientific or technological use of it. As he made perfectly clear in his introduction, he wanted to increase China's scientific knowledge so as to decrease China's resistance to change.

He wanted to drive from Chinese thinking the stultifying notion that things "have always been like this (*hsiang-lai ju tz'u*),[221] and that things should therefore be left alone, that things were sacrosanct because "they have existed since ancient times" (*ts'ung ku suo yu*).[222] These two phrases had caused great harm, and their warped corollary, that "everything the ancients had was better than that of today," had caused even greater harm, especially as it had "greatly obstructed the theory of evolution."[223] So Wu Chih-hui set out to tell the story of evolution to prove that "actually, in the universe things are similar every instant but not the same; there is not a single thing that has existed since ancient times."[224]

And therefore there was not a single thing that could not, would not, or should not change.

Wu Chih-hui was still using the theory of evolution as Liang Ch'i-ch'ao had first used it back in the *Shih-wu pao* days of his famous argument, "We must change or be changed." Wu Chih-hui, therefore, showed himself much more interested in the *fact* of evolution than in the *manner* of it, and that, perhaps, was why he failed to demonstrate much understanding of the manner of it.

Another reason, however, was, ironically, that he was trying, again for the aforementioned purpose, to make evolution "understandable,"—and interesting, and entertaining, and thus acceptable. And he did. His *Talks* were "conversational" and extremely witty—too witty. They made the story of evolution sound like the history of the survival of the wittiest, of those who used their wits to change their spots.

Wu Chih-hui, with justification, prided himself on being a humorist, but his humor was so pervasive that it is now next to impossible to tell whether his evolutionary "mistakes" were mistakes or merely jokes. He should have paid more careful attention to the mechanics of evolutionary change, for he set himself up as one who claimed to know scientifically how China should change now; but he did not pay careful attention to the mechanics of evolutionary change, and the explanation of evolution he gave, therefore, even after his London study, was quite unscientific.

Perhaps, as we have said of Lu Hsun and his "unfilial" amoebae, Wu Chih-hui knew better. But he too, if he did, did not let on, and his young readers, and most of his old ones, certainly did not know better. Wu Chih-hui's unscientific explanation of evolution should have weakened his authority as an apostle of a "scientific ideology," but it did not, because his readers had no way of knowing his explanation was unscientific.

His fanciful story of evolution, however, fit very well with his scientific ideology. Moreover, it was quite compelling—precisely because it was not completely fanciful. Wu Chih-hui had so much fun with the sun, the moon, and the stars that, even when he came down to earth, he did not get to the biological evolution of man until the *Talks'* last two chapters; but when he did get there, he provided a host of interesting and quite accurate evolutionary information. He knew the names of geological ages and of the categories of living things. He could talk about invertebrates and vertebrates, amphibians and mammals, marsupials and placentals, and he knew the order in which such types of "creatures" had evolved. He was not the first to tell of such things in Chinese (although I still do not know who was). Lu Hsun, for one, had used most of the technical terms Wu Chih-hui used, three years earlier in his article "Jen ti li-shih" (The history of man), but he had told that history very summarily and in terse classical style.[225] Wu Chih-hui's vernacular account made it come alive. He talked of kangaroos and anteaters as well as of fossils and dinosaur bones. And he even made dinosaurs come alive as "huge wall-tigers," that is, geckos.[226]

He also talked about comparative bone structure between man and beast to prove that there was more to the theory that men and monkeys were relatives than that they were look-alikes. In his "novel" one can still sense his own sense of excitement at discovering skeletal "proof" of evolution in British museums:

> Huang Hsing said, "Ah, I wish I had met you sooner, Miss Wang. When I once visited a museum in London, I remember that just inside the entrance there was a big glass case with the skeleton of a horse in it, with the skeleton of a man standing alongside. We said that it was probably a great race horse that had died and been put there along with his trainer for posterity. But now I see that we were totally wrong. They were probably comparing human bones and horse bones." Chi-ing said, "Right! Right! What you saw was put there by Mr. Darwin!"[227]

His whole book conveyed his excitement at discovering "the basic story" of natural history. He lamented that he could not show his readers—his characters' listeners—all the illustrations in a book called *Jen-lei chin-hua hsueh* (Human evolution),[228] perhaps one of the books he later translated, but he painted an exciting enough picture of human evolution in his lively prose.

He came to scientific grief only when he tried to tell how the fascinating related classes, orders, and families of beasts he described had actually evolved out of each other. It was here that he made evolution from the very beginning a matter of wits and will power. Consider, for example, the tale of the geckos who evolved into birds:

> If we want to talk about our real ancestors, they were the geckos who evolved at that time in the coal forests. They were pitiful indeed, scurrying in all directions seeking cover. They naturally couldn't count on prospering. But how had they ever gotten into such a pathetic predicament? It was all because certain of their upright elders were well-meaning but stupid. They had concluded that the monstrous behavior of their elder [dinosaur] brothers was no way to last for long, so they had decided to make their own bodies very small. But they never guessed that being small was no good at all. For they were now just right for their elder brothers' breakfasts and dinners. They hadn't a chance for survival. So all the little brothers got together for a conference, at which several voiced their anger that their parents, not being able to get them born with more feet to flee for their lives with, hadn't at least given them some wings so they could fly up higher than thirty feet, where the long-necked monsters couldn't get them. But afterwards, lo and behold, Heaven granted their desires. Whatever they wanted in their hearts, if they wanted it badly enough, they were able to produce. So they really did change their two front feet into two great wings. And it was their group, flourishing in the Mesozoic era, that became the birds of today.[229]

Wu Chih-hui was obviously having fun when he used the idiom *t'ien sui jen yuan* (Heaven grants men's wishes—usually used not as a "rule," but as a wish in itself). "Men's wishes" was a fair description of what he was attributing to his geckos, but he surely was not thinking of a "Heaven" that was a Santa Claus God. And yet this idiom can stand very well as a statement of Wu Chih-hui's theory of evolution as he presented it. For everything he said in his *Talks* supported the idea that *t'ien* as nature did produce mutations in response to will.

He told of clever Australian egg-laying mammals (mammae having already been "thought up," so that mothers would not have to leave newborns unprotected) who had struck upon the idea that it would be even more "convenient" if they could keep their eggs inside them until they were ready to hatch, and "sure enough, once they put their minds to it, in no more than a few generations their eggs became placentas and they could give birth to their children straight away."[230]

And so the story went on: "There was a breed of small kangaroos who practiced with all their might until they learned how to raise children who were so hardy that they didn't have to be kept in pouches anymore."[231] Once that happened, mammals were let out of the bag. They were ready to take off, and they did. Soon an anteater was born who sired, in proto-Noah fashion, a family of four great branches, one of which "peopled" the earth. It produced squirrels and then monkeys, and the great tailless apes. There were five ape brothers, progenitors of todays gibbons, orang-utans, gorillas, chimpanzees—and us. But only our ancestor gave up trying to learn the skills of his father and brothers. He alone "wracked his brains trying to dream up new tricks." And he succeeded. He thought up the idea of walking on his back two legs so that he could free his front two legs to do something else. That was the first step forward, the first great step towards manhood. His descendants kept working on his idea, using their heads, and their hands, for a million years—and turned themselves into men, thus proving, said Wu Chih-hui, that "the proverb has it right: 'If only you don't give in, you can grind a pestle to a pin.'"[232]

The "mechanics" of evolution, therefore, were will, wit, and incredible determination. The story of evolution according to Wu Chih-hui was not the story of natural selection at work among creatures blindly struggling to survive. It was the story of intelligent selection among creatures consciously struggling to better themselves, to gain a higher standard of living. Wu Chih-hui was obviously aware of the survival value of the newly evolved parts that were to lead to such higher standards—mammary glands, placentas, and human heads and hands—but he never said such things were *naturally selected.* He said, instead, that they were chosen and refined through practice by their clever beneficiaries. Indeed, even when he let one of his characters admit she was speaking loosely and in a joking manner when she used the words "practice and refinement" (*hsiu-lien*) to describe creatures' efforts to evolve, and even when he had her admit that over tens of thousands of generations species might change "without realizing it," he still let her immediately go on to say that species really changed "in accord with their environment and *the ambitions of their hearts.*"[233] "Profitable variations," therefore, had become "good ideas" or one's "heart's desires."

Wu Chih-hui put design into evolution literally. He made design an inside job—which was as un-Darwinian as creation. He did just what Huxley had ridiculed. He made all species their own Sir John Sebrights.[234] He made the evolved responsible for their evolution—and for their *future* evolution. That was the clue to what he had really done. Had he not read back into the story of man's evolution his fervent desire for man to be in control of his "fate"? He wanted mankind to keep going, so he wanted his readers to look at evolution and take heart. He wanted them to say, "If the duck-billed platypuses could do it, so can we."—"As long as you think. . . ," said Wu Chih-hui.[235]

That kind of thinking, of course, made the transition in his story from organic evolution to social evolution very easy, because there was no great divide. Minds, even if they had been little minds, had been at work since the beginning. Amphibians had made up their minds to leave the sea, placentals had made up their minds to stop laying eggs, and "Lao Wu," Fifth Ape Brother, had made up *his* mind (in the words of his descendant, Liu Pang) to *tou chih pu tou li* (compete with brain instead of brawn).[236] That decision, followed by millennia of perfecting practice, had made the brains of Lao Wu's progeny "grow better day by day,"[237] until they were man-sized—or wild-man-sized—and were ready to think of tribe, speech, fire, and civilization. It was all one process, progress won by thought. Evolution was, had been, and always would be the story of "enlightenment."

Now, lest we laugh too quickly, we should sit up and remember that the absolutely awe-inspiring result of evolution *has* been enlightenment. Somehow light has come out of darkness. Somehow sounds have come out of silence. Somehow consciousness has come out of unconsciousness. But the end was not the means, at least not for Darwin. He wanted no part of Wallace's belief that "some higher intelligence may have directed the process by which the human race was developed,"[238] but neither would he have accepted the notion that any "lower intelligences" had done so. That is why he tried his best to explain the process in terms of an *unconscious, natural* selection.

It is true that Darwin could say that "a little dose, as Pierre Huber expresses it, of judgment or reason, often comes into play, even in animals very low in the scale of nature,"[239] but that dose came into play, if it did, only in the pursuit of life, liberty, and happiness, not in the pursuit of evolution. And, at any rate, it was a little dose. The dosage Wu Chih-hui prescribed was ridiculous. He so personified the denizens of his primeval seas and forests that he had at least "the best of them" trying to personify themselves—exactly like Emerson's worm.[240] The difference, he said, between those who used wombs and those who laid eggs was the difference between "the smart and the stupid."[241] Those apes who did not become men were those (like Lu Hsun's monkeys) who "were unwilling to study in earnest."[242]

Again, we must hastily say that Wu Chih-hui may have known better. Surely he was having fun with evolution, trying to "spice up" his account. He said, after all, of one of his storytellers (who was telling his story for him) that she wanted at one point to "throw in a few sentences of oil, salt, soy sauce, and vinegar." But her purpose in so doing, he said, was to make the principle of evolution "even clearer in everyone's mind."[243] And that could not have been the result.

Wu Chih-hui's story, as he told it, was pseudo-science. It was presented between exciting scientific covers, and illustrated, though only in words, with exciting scientific illustrations, but it was pseudo-science nonetheless, and yet marvelously suited to the pseudo-science of anarchism.

THE GREAT REVISION. Wu Chih-hui's account of evolution quite ignored "the law of mutual struggle." Peter Kropotkin's revision of the theory of evolution challenged that law's authority. And Kropotkin's counter-law, "the law of mutual aid"—the great revision—formed the "scientific basis" of Chinese anarchism. That basis was beautiful, inspiring, and shaky. And we must look at its shakiness before looking at what Wu Chih-hui and Company built on top of it.

Kropotkin's great revision was most effectively a revision of *Social* Darwinism, but its "legitimacy" rested on a supposed revision of Darwinism proper. In the 1860s, just after he had read *The Origin of Species,* Kropotkin, as a young officer of "the Mounted Cossacks of the Amur,"[244] had been allowed to roam the wastes of eastern Siberia and northern Manchuria making geographical explorations. In the course of his wanderings, however, he had come to have "serious doubts. . . . as to the reality of that fearful competition for food and life within each species, which was an article of faith with most Darwinists, and, consequently, as to the dominant part which this sort of competition was supposed to play in the evolution of new species."[245] Evidently already thinking of himself as a Darwinist in general, he had "eagerly looked" for that "fearful competition," but had failed to find it, "even in those few spots where animal life teemed in abundance."[246] Instead, he said, "In all these scenes of animal life which passed before my eyes, I saw Mutual Aid and Mutual Support carried on to an extent which made me suspect in it a feature of the greatest importance for the maintenance of life, the preservation of each species, and its further evolution."[247]

It was only years later, however, in 1883, when he was already a confirmed Anarchist, indeed a confined Anarchist in the French prison of Clairvaux, that he found just the touch of academic support for his suspicions necessary to crystallize his thinking. "I found," he said, "in a lecture by a Russian zoologist, Professor Kessler, a true expression of the law of struggle for life. 'Mutual aid,' he said in that lecture, 'is as much a law of nature as mutual struggle; but for the progressive evolution of the species the former is far more important than the latter.' Those few words . . . contained for me the key of the whole problem."[248] And those few words inspired him to gather the materials that would eventually make up his great revisionist work, *Mutual Aid.*

Kropotkin realized from the beginning, of course, that Professor Kessler's suggestion "was, in reality, nothing but a further development of the ideas expressed by Darwin himself in *The Descent of Man.*"[249] But he felt that Darwin had not done his own ideas justice, that he had not seen clearly enough that his remarks about man's "social instincts"[250] did indeed necessitate a revision of his original theory of *The Origin of Species.* And of that book Kropotkin was openly critical. "The idea which permeates Darwin's work," he wrote, "is certainly one of real competition going on within each animal group for food, safety, and possibility of leaving an offspring. . . . But when we look in his work for real proofs of that competition, we must confess that we do not find them sufficiently convincing. If we refer to the paragraph entitled 'Struggle for Life most severe between Individuals and Varieties of the same Species,' we find in it none of that wealth of proofs and illustrations which we are accustomed to find in whatever Darwin wrote."[251] Consequently, he concluded, "to admit a pitiless inner war for life within each species, and to see in that war a condition of progress, was to admit something which not only had not yet been proved, but also lacked confirmation from direct observation"[252]—his own observation. And yet, it was precisely that unproved and unconfirmed "admission" that had given rise to the worst tenets of Social Darwinism.

It was Social Darwinism, in all its forms to date, that Kropotkin really detested. "When my attention," he wrote, "was drawn, later on, to the relations between Darwinism and Sociology, I could agree with none of the works and pamphlets that had been written upon this important subject. They all endeavored to prove . . . that the struggle for the means of existence, of every animal against all its congeners, and of every man against all other men, was 'a law of Nature.'"[253] And that was the "law," said Kropotkin, that everywhere led to despicable license:

> It is known to what conclusions Darwin's formula, the "struggle for existence," had been developed by his followers generally, even the most intelligent of them, such as Huxley. There is no infamy in civilized society, or in the relations of the whites towards the so-called lower

> races, or of the strong towards the weak, which would not have found its excuse in this formula.[254]

"It is horrible," echoed Kropotkin's English friend, W. Bates, "what 'they' have made of Darwin."[255] But Kropotkin did not absolve Darwin so completely. "Even during my stay at Clairvaux," he proclaimed, "I saw the necessity of completely revising the formula itself and its application to human affairs."[256]

In revising the formula itself, he still sought to be a scientist. But, in reapplying it to human affairs, he became, of course, himself a Social Darwinist, even if he was a revisionist Social Darwinist. For he too turned to nature and its evolution for "lessons," even as he decried the lessons others learned from it. In that regard, he was much more a Social Darwinist than Huxley, even though he had singled out Huxley as his chief opponent among those who had dabbled in "the relations between Darwinism and Sociology," even though, indeed, it had been Huxley's "atrocious article, 'The Struggle for Existence; a Program,'" of 1888, that finally impelled Kropotkin to put his "objections" in "a readable form."[257]

But why was Huxley his chief opponent? Before Kropotkin ever finished *Mutual Aid,* Huxley had already delivered his own revisionist lectures on *Evolution and Ethics,* and Kropotkin had read them.[258] Here again, "we confront the enormous paradox. Huxley's lectures are decidedly not an exposition of Social Darwinism. They actually represent an attack on Social Darwinism."[259] Ethically, Kropotkin and Huxley stood opposed to the same enemies, the "fanatical individualists" who used Darwin's "law" to excuse their every "infamy," to those who would use the "gladiatorial theory of existence" as a model for mankind. Why, then, should they be at each other's throats?

Kropotkin was at odds with Huxley in much the same way Yen Fu was, only more so, for Kropotkin had scientific sensibilities as well as "religious proclivities" to be offended by Huxley's pessimistic nay-saying to the cosmic process.[260] Kropotkin had absolute faith that the way of nature was for the good, because he was convinced that it was producing the good. He saw good in nature and good coming out of nature. Therefore he and Huxley repudiated "the gladiatorial theory of existence" in very different ways. They were at odds because Huxley *accepted* "the gladiatorial theory of existence" as the reality of the way of nature and *then* repudiated it, while Kropotkin repudiated it because he was convinced that it was *not* the way of nature.

Huxley in his "atrocious article" had said that, "from the point of view of the moralist, the animal world is on about the same level as a gladiator's show."[261] But Kropotkin said, "No, it is not!" He catalogued the teeming herds of different "sociable animals," most of whom he pictured as happy herbivores, who were most *un*gladiatorial, and then exclaimed:

> How trifling, in comparison with them, are the numbers of the carnivores! And how false, therefore, is the view of those who speak of the animal world as if nothing were to be seen in it but lions and hyenas plunging their bleeding teeth into the flesh of their victims! One might as well imagine that the whole of human life is nothing but a succession of war massacres.[262]

The *scientific* truth of the matter, said Kropotkin, was that "association and mutual aid are the rule with mammals. We find social habits even among the carnivores"[263]—and among even less likely candidates. Kropotkin found mutual aid at work up and down the length and breadth of the whole animal kingdom, down indeed "to the lowest stages of the animal world."[264] In a Chuangtzian flourish he even pronounced that mutual aid had "been noticed among the dung beetles."[265]

So Kropotkin protested that cosmic nature *was* a school of virtue. He saw a great moral lesson in cosmic nature almost everywhere he looked: "Don't compete!' . . . That is the *tendency* of nature. . . . That is the watchword which comes to us from the bush, the forest, the river, the ocean. 'Therefore combine . . . practice mutual aid!' . . . That is what Nature teaches us."[266]

Nature might be a harsh school of virtue, and it most assuredly was a utilitarian school of virtue. "Life in societies," said Kropotkin, "is the most powerful weapon in the struggle for life."[267]

The social virtues had to be learned or else. For "the unsociable species," he said, ". . . are doomed to decay."[268] But like all utilitarians, Kropotkin saw no reason to think any less of virtue because it worked.

There was something odd, of course, in the ease with which Kropotkin made nature into a school of virtue. He turned a description into a prescription, he turned "normal behavior" into a norm for behavior, but he did so with no obvious authority. When he first said that, "happily enough, competition is not the rule either in the animal world or in mankind,"[269] and when he said that "association and mutual aid are the rule with mammals,"[270] he meant by "the rule" only what normally happened. But, before he was done, what happened "as a rule" had become what good men should follow as a "rule."

In his attempt to make morality natural, therefore, Kropotkin, like Liang Ch'i-ch'ao and so many others, still left man in an unnatural position. Nature might be a boarding school, but one still had to be outside it to go to it. Moreover, Kropotkin gave man the unnatural ability to follow its rules or not "as he saw fit." But so did Huxley. Huxley agreed with Kropotkin that nature could be a school, that it could be a utilitarian school of hard knocks. But its hard knocks taught not virtue, he thought, but its opposite. He disagreed with Kropotkin, therefore, only on the nature of the school, and on whether or not mankind should attend it.

But this disagreement over the nature of nature led to a further disagreement over the nature of human nature. Huxley and Kropotkin were agreed that human nature was in large part an inheritance from pre-human ancestors, but they were in such violent disagreement over the nature of those ancestors that they were, naturally, in violent disagreement over the nature of the inheritance.

Huxley was convinced that "men's inheritance from the ancestors who have fought a good fight in the state of nature" was "their dose of original sin,"[271] that "their inheritance . . . from the long series of ancestors, human, and semi-human and brutal" was an "innate tendency to self-assertion,"[272] a primordial "instinct of unlimited self-assertion,"[273] the "condition of victory in the struggle for existence,"[274] which "every child born into the world will still bring with him."[275] That inheritance left men still ill-natured, and made it sadly true that "with all their enormous differences in natural endowment men agree in one thing, and that is their innate desire to enjoy the pleasures and to escape the pains of life; and, in short, to do nothing, but that which it pleases them to do without the least reference to the welfare of the society into which they are born."[276]

This was a low opinion of human nature based on a low opinion of animal nature, and it infuriated Kropotkin (if such a "generous personality"[277] can ever be said to have been infuriated), for Kropotkin *liked* animals, and respected them. Those "men of science," he said, who were so "anxious to prove the animal origin of man" that they "began to charge the savage [and his civilized cousins as well] with all imaginable 'bestial' features," were simply "not conversant with the social aspects of animal life."[278] They were not conversant with animals' social—that is, ethical—instincts.

Seeing mutual aid at work even "among the dung beetles," Kropotkin refused, of course, to subscribe to Huxley's "general consensus that the ape and tiger methods of the struggle for existence are not reconcilable with sound ethical principles."[279] With the possible exception of the gorillas and orang-utans, he said, "all the remainder of the monkey tribe—the chimpanzees, the sajous, the sakis, the mandrills, the baboons, and so on—are sociable in the highest degree."[280] And, if tigers were not, they were among the rarest of exceptions: "We can only name the cat tribe (lions, tigers, leopards, etc.) as a division the members of which decidedly prefer isolation to society. . . . And yet even among lions 'there is a very common practice to hunt in company.'"[281] So at least the mandrill and lion methods of the struggle for existence were reconcilable with sound ethical principles. For within those methods was mutual aid, "the positive and undoubted origin of our ethical conceptions."[282]

The very best part of our nature, then, was "bestial" before it was human. There was "a pre-human origin of moral instincts."[283] Beasts had bequeathed us their social instincts.

Up to that point, Kropotkin's and Huxley's argument over the nature of human nature was uniquely modern, Western, and Darwinian. They were talking, after all, about the evolution of human nature. But, from that point on, once they started talking about the nature of human nature as it *had evolved,* their argument once more became classically Chinese. It became exquisitely Hsun-tzian and Mencian—even before any Chinese took it up.

"Darwin," Wu Chih-hui once said, "was a Western Confucius."[284] That was a rare compliment from one so critical in so many ways of China's intellectual tradition, but it was a comparison that made little sense, except in the sense in which he probably meant it, that Darwin and Confucius were both men of tremendous intellectual influence. It facilitates another comparison, however, that does make sense, a comparison between Confucius and his "disciples" and Darwin and *his* "disciples," which was, indeed, of classical proportions. For Kropotkin and Huxley were to Darwin as Mencius and Hsun Tzu were to Confucius, and Kropotkin was to Mencius as Huxley was to Hsun Tzu, and Huxley, Kropotkin, and Darwin were to the tough Social Darwinists as Hsun Tzu, Mencius, and Confucius were to the equally tough Legalists—all, of course, not exactly, but to a remarkable degree.

As moralists against amoralists, Hsun Tzu, Mencius, Confucius, Huxley, Kropotkin, and even Darwin were all on the same side, united against any and all Social Darwinists or Legalists who tried to foist upon the world the notion that the law of self-preservation was the *only* law in "the Hobbesian War of each against all," that is, in "the normal state of existence."[285] Moral law, they insisted, should take precedence over the law of self-preservation. That was why Confucius, thinking of the individual, had said, "Good men and men of honor will not seek life at the expense of goodness; they may give up life for the sake of goodness."[286] And that was why Darwin, thinking of the species, had said, in agreement with Huxley's later dictum that the practice of goodness must be directed "to the fitting of as many as possible to survive,"[287] "We must. . . . bear the undoubtedly bad effects of the weak surviving and propogating their kind," even though "no one who has attended to the breeding of domestic animals will doubt that this must be highly injurious to the race of man."[288]

Darwin, Confucius, Hsun Tzu, Mencius, Huxley, and Kropotkin were all for mutual aid. They were divided only in their opinions as to what it was in our make-up that would or could motivate us towards mutual aid. But in that division, sides were drawn up across both cultures and centuries. Mencius and Kropotkin stood together against Hsun Tzu and Huxley, with Confucius and Darwin somewhere in between—if only because they were less "committal" than either pair of disciples.

Confucius and Darwin seemed all too painfully aware of the mixed nature of our nature, so that it is nearly impossible to pin down either of them on the question that so absorbed their followers, whether human nature was "basically good" or "basically evil." It was precisely because their sayings or writings could give encouragement to proponents of either view that their followers split so fiercely over the issue. Confucius could sound almost as pessimistic as Hsun Tzu: "I have never seen anyone who loves virtue as he loves sex,"[289] but he could also sound almost as optimistic as Mencius: "Is goodness far away? If I want goodness, then it is at hand."[290] And he could also say, "By the time I was seventy I could follow the desires of my own heart without transgressing the laws of morality."[291] Did that mean that the habits of a lifetime had at last "become instinctual"? Or did it mean that he had finally rid himself of the socially learned habits of selfishness that had too long kept his "natural goodwill" from fully expressing itself?

Darwin too was hard to pin down. He could sound like Kropotkin, when he said that "the aid which we feel impelled to give to the helpless is mainly an incidental result of the instinct of sympathy, which was originally acquired as part of the social instincts," and when he went on to say, "nor could we check our sympathy even at the urging of hard reason, without deterioration in the noblest part of our nature."[292] But he elsewhere admitted that "it is untenable, that in man the social instincts (including the love of praise and fear of blame) possess greater strength, or have, through long habit, acquired greater strength than the instincts of self-preservation, hunger, lust, vengeance, etc."[293] He therefore recognized "a struggle in man between his social instincts, with their derived virtues, and his lower, though momentarily stronger impulses or desires."[294] And he *seemed* to agree with Huxley that "the lower nature of man" was not working "for righteousness, but against it."[295]

True, Darwin did not see in man's "lower nature" any proof of Huxley's contention that "ethical nature may count upon having to reckon with a tenacious and powerful enemy as long as the world lasts."[296] He granted that man's lower nature was only "momentarily stronger" and predicted that "looking to future generations . . . virtue will be triumphant."[297] But that future triumph, of course, was not strictly speaking to be a triumph for human nature. It was to be a triumph for some super-human nature. And it would be a triumph that owed less to man's primordial social instincts than to his "recent" advances in "intellectual power" and "knowledge."[298] Darwin implied that, as human beings evolved, they had been and would be guided less and less by instinct and more and more by reason,[299] and, when he implied that, he sounded much more like Huxley than like Kropotkin.

For Huxley and Kropotkin and Hsun Tzu and Mencius were really divided over the issue of whether man's goodwill came from his head or his heart. Were good acts the results of good ideas or good feelings? That was the question behind the question of whether the good in man was "natural" or "artificial."

Huxley and Hsun Tzu, as we have seen, placed all their hope for goodwill in man's intellect, in his reason and good sense, not in his good instincts or good senses. In the "nurture-nature" argument over the nature of "conscience," they stood for nurture. Hsun Tzu insisted that ethical principles were learned: "All in one's nature is given of Heaven. It cannot be learned; it cannot be cultivated.

Propriety and righteousness, however, are born of the Sages. Men become capable of them through study. They are perfected through cultivation."[300] The remaining question, of course, was *Whence* were ethical principles learned? And Hsun Tzu seemed to beg the issue when he simply replied, "from the Sages." But he did make it clear that ethical principles represented the pooled wisdom of the Sages. The Sages themselves had not come upon them naturally. Their ethical principles had been thought up. They were still good ideas.

And so they were for Huxley. He may have tried to preserve his faith in a perfectly natural universe by claiming continued allegiance to his own axiom that "man, physical, intellectual, and moral, is as much a part of nature, as purely a product of the cosmic process, as the humblest weed,"[301] but he was never able to say how—morally—that was so. It was "in virtue of his intelligence"[302] that man was moral, but it was his intelligence that alienated him from the rest of nature, and from his own "lower nature" which was still his "basic nature." Huxley may have occasionally spoken of man's "ethical nature," but when he did so he still had man thinking up his ethical principles off the top of his head, which really meant pulling them out of the sky.

Kropotkin and Mencius, of course, did not. They grounded man's ethical principles in *feelings* that arose from deep within us. Kropotkin grounded them in earth itself, or at least in the lowest forms of dust that breathed. Kropotkin and Mencius placed their faith in feelings and were suspicious of the intellect. "Unless men are maddened in the battlefield," said Kropotkin, "they 'cannot stand it' to hear appeals for help, and not to respond to them. The hero goes; and what the hero does, *all* feel that they ought to have done that as well. The sophisms of the brain cannot resist the mutual-aid feeling, because this feeling has been nurtured by thousands of years of human social life and hundreds of thousands of years of pre-human life in societies."[303]

Mencius would not have understood Kropotkin's last sentence, but he would have recognized the first as his own. For he too had insisted that men "cannot stand (*pu jen*) the suffering of others.[304] And he too had insisted that in that inability lay the root of human goodness.

The trouble, therefore, was in our calculating minds. "The sophisms of the brain" might not be able to prevent "the mutual aid feeling" from arising, but they could prevent the practice of mutual aid to which that feeling should naturally lead. Our second thoughts could thwart our first impulses. We could be too clever for our own good. And the sophisms of some could hurt the feelings of many. "The understanding and the feeling of the millions," said Kropotkin, "are vitiated by teachings worked out in the interest of the few."[305]

Still it was more natural to be a good Samaritan than a bad one. "Some training," said Kropotkin, "is required in a lady of the richer classes to render her able to pass by a shivering and hungry child in the street without noticing it. But the mothers of the poorer classes have not that training. They cannot stand the sight of a hungry child, they *must* feed it, and so they do."[306] And so *could* all people. For "the training" and "the teachings" of a calculating society could be undone and discontinued. "Everyman," said Mencius, "can become a Yao or Shun."[307] For everyman, echoed Kropotkin, is well motivated from the wellsprings of his being.

So Mencius and Kropotkin said that "human nature is basically good," and Hsun Tzu and Huxley said that "human nature is basically bad," and still there is no knowing which side was right. And yet, as Chinese Anarchists were convinced absolutely that Kropotkin was right, it is important to see that Kropotkin's side of the argument was not watertight, even on its own terms. Kropotkin was most scornful of other Social Darwinists he saw engaged in getting "the millions . . . to accept the teachings of a war of each against all, which are offered to them under the title of science, but are no science at all,"[308] but his own counter-teachings, also offered under the title of science, were not much of a science either. He ridiculed other Social Darwinists for clinging to "the law of mutual contest" as "an article of faith" but his own counter-law of mutual aid was equally an article of faith. The reviewer of his book who praised him for "demonstrating unconquerable faith in the innate goodness of humanity,"[309] was perhaps more correct than he realized. But, when all was said and done, at the end of *Mutual Aid,* it was not clear whence Kropotkin's faith had come.

It must hastily be admitted, however, that Kropotkin anticipated the main objection that can be made to his book, and that he tried to answer it. But in the end he forgot his answer:

> It may be objected to this book that both animals and men are represented in it under too favorable an aspect; that their sociable qualities are insisted upon, while their anti-social and self-asserting instincts are hardly touched upon. This was, however, unavoidable. We have heard so much lately on the "harsh, pitiless struggle for life," which was said to be carried on by every animal against all other animals, every "savage" against all other "savages," and every civilized man against all his co-citizens—and these assertions have so much become an article of faith—that it was necessary, first of all, to oppose to them a wide series of facts showing animal and human life under a quite different aspect.[310]

He wanted to redress the balance. He wanted originally only to prove "that Huxley's view of nature had as little claim to be taken as scientific deduction as the opposite view of Rousseau, who saw in nature but love, peace, and harmony destroyed by the accession of man." For Rousseau, he said, "had committed the error of excluding the beak-and-claw fight from his thoughts; and Huxley committed the opposite error; but neither Rousseau's optimism nor Huxley's pessimism can be accepted as an impartial interpretation of nature."[311]

From that promising promise of an impartial interpretation of nature, however, he descended rapidly to an imbalance redressed only with another imbalance. For what was to be the other side became the only side—and in only seven pages. He set out on page 8 to write "a book on the law of Mutual Aid, viewed at [*sic*] as one of the chief factors of evolution—not on *all* factors of evolution and their respective values."[312] But on page 15, he described mutual aid, in his italics, as "*a law of Nature and the chief factor of evolution*"[313]—and especially of "progressive evolution."[314]

That, of course, was his strongest contention—his great revision—but it was his weakest argument. For it was based on two most ill-supported assertions: (1) that the thitherto alleged "chief factor" of evolution, the "fearful competition for food and life within each species,"[315] could not lead to "progressive evolution," and (2) that competition within each species was not very fearful anyway.

He asserted first that, "when animals have to struggle against scarcity of food . . . the whole of that portion of the species which is affected by the calamity, comes out of the ordeal so much impoverished in vigor and health, that *no progressive evolution of the species can be based upon such periods of keen competition.*"[316] He even contended that, "if natural selection were limited in its action to periods of exceptional drought or sudden changes of temperature, or inundations, retrogression would be the rule in the animal world."[317]

What Kropotkin said of Darwin, however, could now be said of himself: "When we look in his work for real proofs . . . we must confess that we do not find them sufficiently convincing."[318] For Kropotkin supported his first assertion only with another: "Those who survive a famine, or a severe epidemic of cholera, or smallpox, or diphtheria, such as we see them in uncivilized countries, are neither the strongest, nor the healthiest, nor the most intelligent. No progress could be based on those survivals—the less so as all survivors usually come out of the ordeal with an impaired health, like the Transbaikalian horses just mentioned."[319]

It is difficult to imagine how Kropotkin could have imagined that the survivors of an epidemic could be less healthy than the dead. Perhaps the dead had been immune to all diseases but that which killed them, while the survivors were immune only to that which they survived. But if so, then there still would soon be no survivors, and one would have to conclude that, given the bad luck of the one disease, none of the race were fit.

Kropotkin's more serious contention, however, one seemingly based on a belief in the inheritability of acquired characteristics, was that no "fitter breed" could come out of such calamitous natural selection because the survivors' "impaired health" would ill affect their issue. True, if the survivors' health was so impaired that they could not multiply at all, then again in the end there would be no survivors, but, if the survivors could recover enough to multiply, then evolution's only absolute standard would have been met. Kropotkin could not have been expected to know whether or not exhaustion could "weaken the genes," but he should have been able to find in his experience examples of once exhausted parents who recovered enough to bear healthy children; and he did, at least, admit that such survivors, however bedraggled, would be able to transmit unimpaired the toughness that had led them to survive.

"All that natural selection can do in times of calamities," he said, "is to spare the individuals endowed with the greatest endurance for privations of all kinds."[320] But what else could one expect evolution to do? Darwin had certainly never promised that it could do anything else. What Kropotkin promised, or wanted, of evolution was something that in true Darwinian terms was quite natural. Kropotkin wanted unnatural selection to select as man selects.

This can be seen from his example of "the Transbaikalian horses just mentioned," one of his few concrete "proofs" of the failure of "calamities" to produce progress. Look at "the Siberian horses and cattle," said Kropotkin: "They *are* enduring; they can feed upon the Polar birch in case of need; they resist cold and hunger. But no Siberian horse is capable of carrying half the weight which a European horse carries with ease; no Siberian cow gives half the amount of milk given by a Jersey cow."[321] But why in Heaven's name should they? Why should a Siberian horse, or any horse, carry any weight at all? Why should any cow give more milk than its calf or herd could use? Kropotkin was judging animal "progress" by human standards, reflecting human wants. He had forgotten Darwin's warning that "what natural selection cannot do, is to modify the structure of one species, without giving it any advantage, for the good of another species."[322] And he had forgotten Chuang Tzu's warning that, even were it possible, such good would be no good at all. Had Chuang Tzu gone to Europe on the trans-Siberian, he would have said that European horses were only fit to be misused. Wild Siberian horses more befit the *Tao,* although they too were nowhere near useless enough. To find a truly Taoist horse, Chuang Tzu would have had to go on to Africa and discover the zebra.

Kropotkin did not understand the fitness of the unfit. For his standards of progressive fitness were human standards that made no sense unless evolution was for us, which surely it was not—if it was "natural." And yet, even if one were to substitute "evolutionary change" for Kropotkin's "evolutionary progress," his contention that such change was not dependent on calamity and competition was still unfairly optimistic. For Kropotkin had actually challenged (and not altogether successfully) only one kind of calamity and competition. He was fooling himself (and many of his readers) when he spoke as if in so doing he had disproved the vital importance to natural selection of calamity and competition altogether.

Kropotkin fooled himself and many of his readers because he was so upset by an exaggeration, the undeniable exaggeration of Huxley's gladiatorial metaphor, that he rushed too blindly into the arenas to confront it, and so spent his energies belaboring the exaggeration without ever touching what was exaggerated.

Huxley's gladiatorial metaphor, said Kropotkin, was "a very incorrect representation of the facts of Nature," as indeed, *if taken literally,* were many of the metaphors Darwin himself had used. But Darwin had realized it. Darwin had stated clearly, said Kropotkin, that *struggle for existence* was to be taken "metaphorically."[323] And "the same remark which Darwin made concerning his expression: 'struggle for existence,'" Kropotkin went on, "evidently applies to the word 'extermination' as well. It can by no means be understood in its direct sense, but must be taken 'in its metaphoric sense.'[324] . . . As to 'competition,' this expression, too, is continually used by Darwin . . . as an image, or as a way of speaking, rather than with the intention of conveying the idea of a real competition between two portions of the same species for the means of existence."[325] If one really understood Darwin, therefore, and really knew nature, one would know that the truth of the matter was that "the word 'extermination' does not mean real extermination,"[326] and that "what is described as competition may be no competition at all."[327] The "fearful competition for food and life within each species," therefore, could not be the major factor in the evolution of new species, and could not even be very fearful, because it was the exception, not the rule.

But was that the truth of the matter? What Kropotkin meant by *real competition* and *real extermination* was obviously conscious competition and conscious extermination, real fights to the death. He meant, therefore, that in nature there were actually very few real gladiatorial contests, at least between members of the same species. There were few challenges to mortal combat over the last mango, few shoot-outs even at the water hole, and even fewer consciously carried out extermination campaigns aimed at cousinly genocide.

That, of course, was a good point, one with which most modern evolutionists would agree. G. Ledyard Stebbins, for one, has said that "physical combat, which results in the death of the less successful individuals, is one of the least common ways in which 'the struggle for existence' takes place."[328] But that in itself does not make the world a gentle place. The absence of murder does not mean the absence of death. Kropotkin could say that "one species succumbs not because it is exterminated or starved out by the other species, but because it does not well accommodate itself to new conditions, which the other does."[329] But, in his joy at the absence of extermination, he seemed totally to overlook the sadness of the succumbing.

Kropotkin's ability to overlook such succumbings, with elevated gaze, was quite remarkable, as can nowhere be better seen than in his following curt dismissal of "Malthusian competitors"[330]—and with them of the whole "Malthusian problem":

> In reality, the chief argument in favor of a keen competition for the means of existence continually going on within every animal species is—to use Professor Gedde's expression—the "arithmetical argument" borrowed from Malthus.
>
> But this argument does not prove it at all. We might as well take a number of villages in South-East Russia, the inhabitants of which enjoy plenty of food, but have no sanitary accommodation of any kind; and seeing that for the last eighty years the birthrate was sixty in the thousand, while the population is now what it was eighty years ago, we might conclude that there had been a terrible competition between the inhabitants. But the truth is that from year to year the population remained stationary, for the simple reason that one-third of the new-born died before reaching four years, and out of each hundred born, only seventeen or so reached the age of twenty. The new-comers went away before having grown to be competitors.[331]

The new-comers *went away!* They up and vanished, taking the problem of extermination with them. And "it is evident," said Kropotkin, "that if such is the case with men, it is still more the case with animals. . . . Each storm, each inundation . . . each sudden change of temperature, take away those competitors which appear so terrible in theory."[332]

But was there nothing terrible in the practice of that "taking away," and was not that taking away natural selection itself? Kropotkin could proudly state that "the importance of natural checks to overmultiplication, and especially their bearing upon the competition hypothesis, seems never to have been taken into due account," because he meant to prove that natural checks to overmultiplication also naturally checked competition—by removing the need for it. But that method of accounting had nature eliminating competition by eliminating the competition! How could it be worth it, when, as Kropotkin himself admitted, such beneficent natural checks eliminated—for example—"astounding numbers of winged ants," "incredible amounts" of pinemoths and "numberless quantities" of mice?[333] Was Kropotkin not like a medieval pacifist rejoicing in the fact that a great joust had been called off because of the plague?

And were not his competition-preventing "natural checks"

themselves great "competitions"? If after some natural check the kind of "fearful competition" he most detested did indeed seem reduced, was it not simply because the real competition had already been held? Kropotkin refused to admit that Huxley's metaphors, and Darwin's, could make some sense, even some literal sense, if only one took the word *competition,* for example, in a different sense. For if examinations could be called competitions, one could say that "natural checks," obvious or "invisible," did put all creatures "to the test."

Take as an unnatural analogy the alleged "cut-throat competition" among pre-med students taking medical boards. Although dirty tricks are not unknown, few students actually cut each other's throats, and yet the selection process is brutal, just as it was in the classic case of the Birmingham moths. In the industrially dirty vicinity of Birmingham, dark, mutant moths were able to take over as the majority group in a population thitherto dominated by pale moths, whereas in clean, non-industrialized areas the pale majority was more than able to hold its own, because on dirty trees the pale moths stood out, and on clean trees the dark moths stood out, and in each case the outstanding were the more readily eaten by birds.[334] In neither case did one type of moth exterminate the other, and yet in each case one type was (almost) exterminated. There was extermination, therefore, without Kropotkin's "real extermination." But from the point of view of the exterminated (if there is a mothly afterlife from which to have one) the difference between real and unreal extermination could seem only academic.

Multitudes of pre-med students are eliminated by their examiners, multitudes of moths are eliminated by their predators, and in neither case is there internecine competition; but a "competition" is held nonetheless, and in each case, natural or unnatural, the selection process is fearful. For in each case there looms the terrible seeming fact of life, which Kropotkin tried to "exclude from his thoughts,"[335] that "many are called but few are chosen."

Kropotkin did not like to admit that even this "lesser"

competition was necessary for evolutionary change. The Birmingham moths, although a modern example, would seem to be a dramatic proof of Darwin's theory that "new species" originate over the dead bodies of their parent species. Dark mutant moths did not, of course, actually owe their origin to the elimination of pale moths; they owed it to mutation. But the mutants only surfaced from their gene pool as their "normal" cousins sank. The color of the population changed from pale to mixed to dark only as the dark proved themselves fit and the pale proved themselves unfit. Kropotkin, however, seemed to want to believe that such changes, especially if they were "improvements," could take place without such eliminations.

Again, however, he was "incredibly" optimistic. "Competition," he said,"is limited among animals to exceptional periods, and natural selection finds better fields for its activity. Better conditions are created by the *elimination of competition* by means of mutual aid and mutual support"[336]–and by migration. Migration was for Kropotkin the great unrecognized Darwinian alternative to competition: "In the migrations of birds which took place at that time on a truly American scale along the Ussuri and especially in a migration of fallow deer which I witnessed on the Amur, and during which scores of thousands of these intelligent animals came together from an immense territory, flying before the coming deep snow, in order to cross the Amur where it is narrowest–in all these scenes of animal life which passed before my eyes, I saw Mutual Aid and Mutual Support."[337] Animals together sought greener pastures to "avoid competition,"[338] and, as they sought greener pastures, *mirabile dictu,* natural selection found its "better fields"!

Kropotkin found the secret of unfearful evolutionary change in Darwin's statement that "one of the most frequent modes in which Natural Selection acts is by adapting some individuals of a species to a somewhat different mode of life, whereby they are able to seize unappropriated places in Nature."[339] For "in all such cases," said Kropotkin, "there will be no extermination, even no competition—the new adaption being *a relief from competition, if it ever existed.*"[340]

What a nobly face-saving vision of evolution that was! Nature in the nick of time "adapted some individuals" to new lifestyles so that they would not have to compete with relatives still wedded to the old ones. Nature taught new sub-species to eat new foods so that they could leave plenty of the old for the old. Nature taught the new to move out so that the old could have room—to live in the style to which they were accustomed.

Evolution was a veritable exercise in filial piety. *Ch'ing* might *ch'u yü lan erh sheng yü lan* (indigo blue might come from blue and yet be bluer than blue), as the Chinese say in Hsun-tzian paraphrase,[341] but mutation, the very stuff of all such evolutions, worked for the good of the old as well as for the betterment of the new. Cactus finches became adept, through adaptation, at using cactus spears to dig out cactus bugs, and so left plenty of non-cactus bugs for their parent non-cactus finches.[342] Long-necked giraffes moved out and on to more etherial pastures, and so left the low-lying bushes to their parent short-necked giraffes. Nature for the good of all helped new creatures out of crowded "areas" by fitting them for "unappropriated" ones. Nature was wondrous. It made room for all.

But the embarrassing absence of short-necked giraffes suggested that something was wrong. Perhaps the real evolution of the present day giraffe had *not* been so gentlemanly. It is unlikely that long-necked giraffes ever had principles as high as their heads. It is unlikely that they ever held meetings and decided to move out as a group and leave the lower leaves for their squatter fellows. If they ever did migrate in search of greener pastures, it is probable that they migrated in mixed groups, and that in bad years the squattest starved. Moreover, it is almost certain that there were never giraffe populations made up of giraffes with very long necks and giraffes with very short necks. Throughout their long, drawn-out evolution long-necked giraffes probably ate usually from the same trees or bushes as their only slightly less fortunate fellows, getting, because they could take, and took, more than their "fair share." In bad years, therefore, they may even have engaged, on occasion, in a bit of "real competition," so that in the end their less fortunate fellows (as well, sometimes, as members of other species who had previously enjoyed their "unappropriated places") might literally have been "put out" by their growing uppitiness.

Even the famous Galapagos finches probably did not distinguish themselves by any gentlemen's agreement to divvy up their islands. It is only barely conceivable that various pairs of freaks (or couples one member of which had genes that were dominantly freakish) ever took refuge on various private islands and so produced subspecies in "one move." It is far more likely that descendants of similar "explorer finches"[343] (as Mr. Stebbins calls them) turned out different on different islands because the different "nature" of those different islands *selected* differently from *among* their descendants. Such selection, however, must have been as hard as any selection on those not selected. The fact that, much later, family reunions have been possible, that sympatric populations of all kinds of Galapagos finches have been able to live harmoniously on one and the same island, thanks to their specially adapted different appetites, does not prove that their evolution was harmonious.

Kropotkin saw altogether too much harmony in evolution. His great revision, therefore, in the end carried Darwin's own great whitewash of "the process" even further. Darwin had said, "We may console ourselves with the full belief, that the war of nature is not incessant, that no fear is felt, that death is generally prompt, and that the vigorous, the healthy, and the happy survive and multiply."[344] Kropotkin said that "the war of nature," within each species, was largely a myth: "Peace and mutual support are the rule within the tribe or the species; and . . . those species which best know how to combine, and to avoid competition, have the best chances of survival and of a further progressive development."[345] Trying as so many had before him to tell people that they *should* not fight each other, he told them that nature said that the fittest *did* not fight each other. He said that nature was a good teacher. But that good news from a naturalist was not, alas, unquestionably well-founded.

THE GREAT REVISION AND THE CHINESE ANARCHISTS. Well-founded or not, Kropotkin's great revision was a godsend to Chinese Social Darwinists, especially, of course, to the Chinese Darwinist-Anarchists, who first discovered him and introduced him to the rest. For Kropotkin's great revision confirmed Chinese Social Darwinists' deepest Darwinian "prejudices" and so allayed their deepest Darwinian fears.

"Proto-Kropotkinism" we have already seen, in the thought of Yen Fu, Liang Ch'i-ch'ao, and Sun Yat-sen, working its way into Chinese Darwinian thinking from the very beginning, and working precisely in every case to make Darwinism less frightening and less threatening to China, working to make it encouraging, promising, saving. But now the Anarchists had discovered "the real thing," Kropotkinism proper, which was *foreign* confirmation of Chinese Social Darwinism, at least to a large degree. They discovered Kropotkin's great revision and they leapt at it, because Chinese proto-Kropotkinists *were* still afraid of Darwinism, because they did not yet quite have the courage of their own convictions.

The frightening thing about Darwinism, of course, was its seemingly irrefutable demonstration that we live in an amoral and bloody universe, in which self-preservation is the only virtue. That idea, so immediately offensive to Christendom, was only a little less immediately offensive to traditional Chinese "religious proclivities." Certainly it offended Confucius's faith in a moral universe and Mencius's corollary faith that those who followed that universe's moral Way would survive, while those who did not would perish.[346] And it offended the faith of the *Chung yung* in a world in which "the ten thousand things live side by side and do not hurt each other."[347] It even offended the similar though allegedly amoral Taoist faith that "the way of Heaven is to benefit and not to harm."[348] The idea of an amoral universe was not only offensive to most Chinese; it was, indeed, frightening, for as Chinese faced, so they thought, their darkest hour, an hour that marked not the dawn of a new dark age, which would pass, but the dawn of an evolutionary battle that might be their last, this revelation of an amoral universe meant that their "moral superiority," which they believed had always and would always preserve them come what may, was actually of no account—even if it existed. Darwin took away Ah Q's last graspable straw.

Kropotkin restored faith both in the universe and in the earthly triumph of virtue, but still he upheld Darwinian evolution, and without recourse to any deity. His, therefore, was the perfect revision for China, for it made the universe good and morality vital and yet made it so naturally.

Chinese Darwinists, of course, even though they had often described the dangers of an amoral Darwinian universe to scare their people into struggle, had inconsistently stood up for the goodness of the universe all the while, with their invariable confusion of evolution and progress; and Liang Ch'i-ch'ao, at least, had done his best to salvage the vitalness of morality—even, if need be, in an amoral universe—by basing it à la Bentham on self-interest. But Kropotkin did both things better. For he marshaled natural fact to demonstrate that evolution was moral evolution as well as physical, that the process had produced morality and put it in us, so that we could be good not by being calculating, like Bentham, but by being uncalculatingly responsive, like Mencius, to our good-natured impulses. And, if we were, then, as Mencius said, we would survive, because we would be infused with the impulse to mutual aid, which was evolution's virtue—and the Chinese evolutionists' virtue of the *ch'ün.*

Kropotkinism was perfect, and the Chinese Anarchists seized upon it. They did not appreciate the questionableness of Kropotkin's scientific argument (and they probably would not have appreciated it if it had been pointed out to them). They simply seized upon his oversimplification and simplified it further. They made, indeed, everything "perfectly" simple. "If something is good, it survives," said Li Shih-tseng, ". . . if it is not good, it dies."[349]

That was a formula that went all the way back to the *Shu ching.*

It was an ancient Chinese "prejudice" (if not a universal human wish) that had worked its way in different forms into Confucianism, Taoism, Buddhism, and even before Li Shih-tseng said it, into Chinese Social Darwinism. It was a formula Chinese Social Darwinists had been trying to work their way back to ever since they had first accepted Darwin's theory of evolution and the seemingly hopelessly opposite formula that theory supposedly prescribed, that only "the strong survive."[350] But none had yet stated it as boldly or as baldly as Li Shih-tseng, who made it clear that evolution was the evolution of the good.

Li Shih-tseng, by no coincidence, was one of the first to introduce Kropotkin to China. He was the first to mention Kropotkin in the *Hsin shih-chi,* in which he published his own translations of articles both by and about Kropotkin, and he was the first to translate and publish, also in the *Hsin shih-chi,* at least the first two chapters of *Mutual Aid,* thus giving Chinese the potent term *hu chu.*[351] The first time he quoted Kropotkin, however, or as he put it "Bakunin and Kropotkin," Li Shih-tseng took his stand with them in support of their dictum that "humanity's evolution is from the morally imperfect to the morally more nearly perfect."[352] He said that "social evolution and biological evolution and the evolution of human wisdom are all one principle.[353] . . . Evolution . . . is nothing other than the pursuit of goodness.[354] . . . Only the good survive."[355]

That faith was a very pillar of the revolutionary Chinese Anarchist faith. And it was to be a pillar of the revolutionary Chinese Marxist faith to follow, and of the Thought of Mao Tse-tung. But neither Chinese Anarchists, nor Chinese Marxists, nor especially Mao Tse-tung would ever have admitted that it was a faith that first and foremost put China back into a Confucian world in which virtue triumphed and moral superiority again meant all.

Perhaps it is not surprising, therefore, that the one Chinese Anarchist who did *not* accept it was a Buddhist. The irascible eccentric, Chang Ping-lin, Republican, Anarchist, *and* Buddhist revolutionary, did not believe it, and he attacked it, in *tan chiang p'i ma* fashion (a single lance and a single steed) even though that set him against friend and foe alike. He attacked a pillar of Chinese Social Darwinian faith as no one had before him.

Chang Ping-lin, Anarchist though he insisted he was, did not accept "Kropotkin's and Bakunin's" dictum that evolution was on its way towards moral perfection. He did not deny evolution altogether (as he would later deny the oracle bones), but he denied that evolution would be our salvation. For he denied that evolution meant meaningful progress.

He blamed Hegel for inventing "progress." Darwin and Spencer had merely tried to substantiate the idea, with biological and sociological demonstrations, while Huxley had opposed them, as Schopenhauer had opposed Hegel (Chang Ping-lin had gotten part of Huxley's point out of *T'ien-yen lun*). "I do not say, however," said Chang Ping-lin, "that the theory of progress is wrong, for even what Schopenhauer pursued could be called progress. Nonetheless, if you say that progress [or evolution] must in the end be capable of reaching a realm of perfect beauty and pure goodness, I can find reason to quarrel with *that* anywhere you look."[356] For "from the smallest organism to mankind," he explained, "progressive evolution exists only in knowledge; morality has daily gone backwards."[357] Or at best, he said, in his ultimate analysis, "If one speaks of morality, then good progresses, and evil also progresses. If one speaks of material life, then pleasure progresses and pain progresses too."[358] Therefore "Evolutionary progress" (*chin-hua*) is one of "the four delusions," and "Those who look towards evolutionary progress [Yen Fu, Liang Ch'i-ch'ao, Wu Chih-hui, *and* Sun Yat-sen!] are as superstitious as those who pray to gods and spirits."[359]

Chang Ping-lin called his own belief, that good and evil and pleasure and pain evolve abreast, "The theory of dual evolution" ("Chü-fen chin-hua lun"),[360] a theory that led him on to his "Five no's theory" ("Wu wu lun"),[361] quite possibly the most radical cure for the human condition ever prescribed in history. But "the theory of dual evolution" itself was not really radical at all. Many of its pronouncements would be accepted as truisms in our day. The odd thing was that he alone thought to make them in his.

He tried to point out that, as far as life on this planet was concerned, evolution, biological, social, and technological, was a mixed blessing, if one measured either goodness or pleasure. Animals, for example, show only a little familial love, he said, and it neither lasts long nor is widely extendable. "Animals," he concluded, "have only a little goodness. Human goodness is somewhat greater. However, is human evil less than animals', or is it greater also?" Tigers and leopards, he said, kill men, but only as men kill cattle, so men are as evil as tigers and leopards. But tigers and leopards do not "slaughter their own kind." Men have done that from the very beginning, and their wars have gotten worse, not better. The progressive evolution of their intellect and knowledge has led to a progressive evolution in their knowledge of how to kill each other, and they have used it. They kill many more with firearms than they ever did with sticks and stones. War lust, said Chang Ping-lin was "something that tigers and leopards do not have; it is possessed by men alone." That is how "we can tell that, from the lowest organisms to mammals to man, good advances and so does evil."[362]

And so it was with pleasure and pain. Amoebas, he said, are easily pleased; men are not. But, the more men can enjoy, the more they desire, and the more pained they are if their desires are not met.[363] The greater their hopes, the greater their disappointments. Indeed, the greater their enjoyment of life, the greater their fear of death. Thus, death is harder for the rich than for the poor, and harder for the poor than for horses and cattle, and harder for horses and cattle than for fish and turtles.[364]

Not everything in that argument was a "truism," but there was something to it, even though not, perhaps, enough to lead to this conclusion:

> In ancient times, good and evil were slight. Today, good and evil are great. In ancient times, pain and pleasure were slight. Today, pain and pleasure are great. But, if that is so, should those whose goal is goodness and pleasure look on evolution as the greatest good fortune or as the

> greatest misfortune? The reality of evolution cannot be denied, but its usefulness? . . . There is none.[365]

Chang Ping-lin looked at evolution, asked if it was worth it, and responded with a resounding, Buddhist "No!" Evil kept pace with good, so evolution never got anywhere. Our ethical nature never got anywhere. Hsun Tzu and Huxley were right. We are all born with an ineradicable selfishness: "Infants barely beginning to speak will go at each other if they do not get their way."[366] Chang Ping-lin was convinced that "lust and love of killing" (*hao yin hao sha chih hsin*) and "love of winning" (*hao sheng chih hsin*) would be a part of human nature as long as there was any human nature. And evolution would not change it.[367] For these were not just selfish survival instincts, which might diminish if survival became less of a struggle. "How can one say," said Chang Ping-lin, "that men's killing of each other is limited to the struggle for existence or to competition for mates?" It was in our nature "to compete for no reason."[368] People fought for the fun of it, and always would.

So Chang Ping-lin was even more pessimistic than Hsun Tzu and Huxley, especially as he did not believe men's good sense could ever get the better, once and for all, of their evil urges. He did not believe that either *li* (rules of propriety) or laws could ever conquer human nature. For "*li* and laws," he said, "are made by evil men. And, even were there *li* and laws that were not evil, one could not carve officials out of wood. So those who uphold the *li* and laws would still be evil men, and evil men governing evil men would be like tigers ruling bears."[369]

The proof that evolution got nowhere lay in modern "civilization." Hsun Tzu, said Chang Ping-lin, thought *he* had reason to be pessimistic: "All Hsun Tzu saw was the struggle between seven rival states in China. They may have treated the people like dirt, but nowhere nearly as much as modern imperialists do." "The so-called civilized countries," he said, "in their killing of peoples of other continents and other colors are worse than Chie and Chou [the last tyrants of the Hsia and Shang dynasties]. For Chie and

Chou were only two individuals. Today, leaders and people together make tyranny."[370] Just look at the French in Vietnam: "If you live, there are taxes; if you die, there are taxes. If you beg for food, there are taxes. If you clean latrines, there are taxes. Those who criticize are killed; those who try to leave are killed; those who assemble are killed. The cruelty of such tyranny was never seen in antiquity. France is a country that eats people [the seed from which sprung "A Madman's Diary"?]. The Mongols and Moslems could never have held a candle to the French. But is this France not the France that invented 'freedom and equality'?"[371]

People were not getting any better, civilized or not. Chang Ping-lin allowed a Kropotkinite question: "Biological evolution is not yet complete. Granted, there is much robbing and killing among people today, but after a thousand generations might not people be perfectly moral?"[372] But he answered it in the negative. Chang Ping-lin, if *only* Chang Ping-lin, did agree with Huxley that "ethical nature may count upon having to reckon with a tenacious and powerful enemy as long as the world lasts."[373]

But Chang Ping-lin's *response* to that gloomy prognostication was not Huxleian at all. He *longed* for the end of the world! His "five no's" were "no government" (*wu cheng-fu*), "no territory" (*wu chü-luo*), "no human race" (*wu jen-lei*), "no life" (*wu chung-sheng*), and "no world" (*wu shih-chie*).[374] And they were a progression leading out, not up, the only kind of "progress" he believed in.

"No government," therefore, was no end; it was only the first great step, and republicanism was only a little step towards that. As the Manchu monarchy was going nowhere, China needed a republic to get things going, but even a republican government could not be clung to, for any government, if maintained, would turn into a "great brigand," as Chuang Tzu had said.[375] A republic would only be good if it could ultimately self-destruct. But "no government" was no good either as an ultimate solution. For it could not stop killing. It was better than government only because it could reduce the scale of killing. State power made possible great wars. "No government" might eventually reduce wars to murders. If Chie and Chou had not been kings their cruelty could never have reached so many.[376] But "no government" would not work if peoples could still have territories, for without governments the peoples of the world would still fight each other for the best territories, for the warm places, for places in the sun. "Why," asked Chang Ping-lin, "do the Russians dare talk of 'no government'?—Because their land is so bitterly cold that . . . they have no fear that other peoples would want to take it." Therefore "no government" would only work if the peoples of the world agreed to move every year, taking turns in the good places.[377]

But even that could never root out all evil. For the root of all evil was not in government, nor in property, but in us: "Even if over all the earth there were no governments and no territories, even if there were disarmament and communism, still wounding, killing, and destroying could not be stopped."[378] For men would still possess a murderous "sense of self" (*wo chien*). "Killing," said Chang Ping-lin, "is what we hate, but the root of killing cannot be broken. For if you wish to break the root of killing, you must first break one's sense of self, and, if you break that, you will also break off life."[379]

But there was the solution! Break off human life, not, of course, through genocide or geno-suicide, for killing was "what we hate," but through refusing to *transmit* life.[380] The human race could let itself die out. It could willingly become extinct. Extinction was nirvana!

All other forms of life would (somehow) have to cease as well, lest evolution repeat itself and bring man back. And the world would have to cease, lest it bring *life* back. But, as the wise already knew, the world really existed only in the eye of the beholder, and so would vanish when the eye did, although that, admitted Chang Ping-lin, "cannot be explained to ordinary people."[381] Still, even ordinary people should be able to conceive of stopping evolution by stopping life. And, if only they could, they could cause the death of evil.

It seemed so logical. And yet I suspect Buddhist masters of old would have been unimpressed. I fear they would only have shaken their heads and sighed that the cycle of transmigration was not so "easily" gotten out of. For they had never recognized death as an escape. Chang Ping-lin wanted out of our evil world, but his road to nirvana was this-worldly, for he looked to biological extinction. In a remarkable way he had secularized nirvana just as nineteenth-century Western utopian socialists (including Marx) had secularized Heaven. But he did not see that. He thought that in his "five no's" he had found a way out, for all mankind.

He did realize that his way would not be easy, that man, like men, would have to "work out his salvation with diligence." And he denied all possibility of "sudden enlightenment" or of a "great leap forward." He said, "Today, though we might wish to reach the 'five no's' in one great flying leap, it is impossible." Man could only advance like a "crippled donkey," and it was "to man's greater sorrow that he does not want to be a crippled donkey and yet has no way not to be."[382] And yet a crippled donkey could advance, and so could we. We could limp our way to the exit. All the world might be a stage but we could leave it, and the tragedy could end. "Exeunt omnes" could be our salvation.

That was Chang Ping-lin's anarchism: no government, no people, nirvana. But the cry "no people" was not popular. That was the last thing the rest of the Anarchists wanted. They wanted people, new people, good people to live happily ever after. So no one listened to Chang Ping-lin, at least no one listened to the last four of his "five no's," which was probably just as well, *except* that no one listened to the wisdom of his warnings either, to his warning that people could believe in "the religion of progress"[383] superstitiously, and that such superstition, like so many others, could lead to tyranny. No one listened when he said, "Alas, the stupid of old accused people of not accepting their fate; the fanatics of today will accuse people of not seeking progress." No one listened when he said, "The upholders of progress hate those who differ with them, and they will attack them for 'going against the laws of nature.'"[384]

None of the rest of the Anarchists listened to Chang Ping-lin, and so they did just what he feared, as would those who came after them. They saw no absurdity, no arrogance, no danger in setting themselves up as the arbiters of natural law, because they were too excited by the idea that nature's law was to make things better. "Evolution," they said, "means progress without end,"[385] and "Those who know the Way of evolution [should] follow it and help it."[386] And *they* knew.

They knew what Mao Tse-tung would soon know too, that "the world is progressing, and the future is bright. This basic movement of history no one can change."[387] So they knew, as Mao Tse-tung would also know, that "the darkness shall pass. The dawn is before us."[388] For the darkness was passing from human nature itself. Kropotkin had proved it. The rub in all human civilizations so far was being rubbed out. The rest of the Anarchists rejected Chang Ping-lin's pessimism, quoting Kropotkin to prove that mankind could be perfect.

When they did that, although they were in a disguise that fooled even themselves, they were Confucians rejecting Buddhism. The human race, on this earth, could be like Yao and Shun (which was *better* than being like the Buddha). And it *would* be. That was the new element, the historical determinism provided by "evolution," far though historical determinism had originally been from the mind of Darwin. But the faith in the possibility of human perfection was Confucian. Indeed, the way in which the Chinese Anarchists conceived of the process of that perfection was almost too Confucian to be perfectly Kropotkinian.

Wu Chih-hui said that "the human race is bent on progress thanks to its good moral sense."[389] For once, he fully agreed with his Tokyo rival, Liu Shih-p'ei, who proudly held that wise men, East and West—Confucius, Mencius, Kant, and Kropotkin—though using different words, all agreed that man had a good *hsin,* a good heart *or* mind (the Chinese word meaning either),[390] and that it was that that made possible human progress. Wu Chih-hui chose Mencius's term, *liang hsin,* for that good heart or mind, that moral sense, and described its struggle towards progress in Mencian terms as well: "Moral sense at every step wars with rampant human desires" (*jen yü*).[391] But, although Mencius's *liang hsin* could indeed, as we have seen, be made to sound Kropotkinian, Wu Chih-hui did not make it sound so, for he pitted it not only against our *jen yü* but also against our *i-ch'uan,* our biological inheritance, forgetting that Kropotkin's main point was that our *liang hsin* was *part* of our biological inheritance.[392]

When Wu Chih-hui pitted our moral consciousness against our bestial inheritance, therefore, he sounded more Huxleian than Kropotkinian, as he and Li Shih-tseng both did when they made *liang hsin* sound Hsun-tzian and intellectual instead of Mencian and emotional—or instinctual. "Man is an animal," said Li Shih-tseng: "He has evolved from apes to become man. His original bestial nature he has not yet shed, and so he knows only emotional love." But eventually he would know "intellectual love," which would be far better, because it would be extendable to all people.[393] "At the present stage of human evolution," added Wu Chih-hui, "moral consciousness is still no match for heredity."[394] But moral consciousness would win in the end. Indeed there were already encouraging signs: "The so-called superior races without exception are those that have a relatively large number of people whose moral consciousness is winning."[395] Moral consciousness would win because people were getting wiser.

Growth in wisdom, of course, could have been explained (up to a point) biologically. Better brains could be naturally selected. And Li Shih-tseng had indeed once listed wisdom as just another step in natural evolution. Evolution, he said, had started with gases (good old Neo-Confucian *ch'i*) and then gone to liquids and then to solids and then to organisms, from the simple to the complex, and then to man "from the stupid to the smart" and to society "from the benighted to the relatively right."[396] He admitted natural selection into his progression, admitting that the principle "The strong survive" was inviolable, but he said that with men that meant "The strong in knowledge survive,"[397] and both he and Wu Chih-hui spoke as if that meant strong in moral knowledge that could be learned by using our present heads, not waiting for better ones. Both Li Shih-tseng and Wu Chih-hui spoke, in Confucian fashion, as if we biologically already had everything we needed.

There was something even less biological, however, about Li Shih-tseng's claim that evolution was now moving "from the benighted to the relatively right." And other Anarchists used the same sort of language. Wu Chih-hui said that "truth and justice move every day towards progress."[398] And the very first paragraph in the *Hsin shih-chi* said that the goal of revolution—and evolution—was "reason," and that "everything not in accord with reason should be done away with," so that mankind might "progress ever closer towards what is right" [*yü chin yü kuei cheng-tang*].[399]

But such language was Hsun-tzian–Confucian. It brought to mind Confucius's dictum "Conquer yourself and return to propriety,"[400] and Hsun Tzu's insistence that self-conquest must be rational. It was at any rate strange Darwinism and Kropotkinism, for it seemed to give man not only innate feelings of solidarity with other creatures but an innate ability to know, or to learn, "the good." But that meant that man could know the good, could know the right, before he could get there, before he could evolve there, and that meant that there must be a good or a right "out there" to be known. Evolution's good existed before evolution. Mankind was becoming more and more what it "was meant to be." But, if that did not imply design again, then it implied a Platonic ideal or a Confucian *Tao* still embarrassingly transcendent. The idea that anything should be on earth as it is in Heaven was simply very hard to ground in biology—which was where Kropotkin had started.

Evolution had again turned into enlightenment, with no one admitting that that meant there was more to it than met the eye. And, turning into enlightenment, evolution very easily came back to the theory of natural rights. This can be seen in the very title of the foremost Tokyo Anarchist paper, the *T'ien i pao*, probably best translated as "The journal of natural rights," although more literally "The journal of *heavenly* rights." Both the Tokyo and Paris groups of Chinese Anarchists came to believe that the watchwords of the French Revolution, *Liberté, Egalité, Fraternité* (although Wu Chih-hui consistently listed them in the reverse order!)[401] were natural rights promised by evolution, a hard promise to prove.

Kropotkin and Darwin himself had done their best to prove that all men are brothers (even with all beasts), but they had not proven absolutely that brotherly love need follow.[402] The only freedom they had proven was the freedom of the free-for-all. Otherwise there was only non-freedom for the good of all. As for equality, they had proven nothing of the sort. Liu Shih-p'ei said that all men should be equal because they were all originally descended from a single tribe (although he then said they were not equal because they had evolved thereafter at different speeds),[403] but what Darwin had proved was that all evolution had depended absolutely on *inequality*. Kropotkin, trying to prove that animal populations were cut down without "competition," said that Transbaikalian horses in the winter "grow exhausted not because there is not enough food for all of them—the grass buried under a thin sheet of snow is everywhere in abundance—but because of the difficulty of getting it from beneath the snow, and this difficulty is the same for all horses alike."[404] But the Darwinian explanation was that the difficulty was *not* the same for all horses alike, for, although the snow was the same, the horses' "physical endowments" were not.

Natural-rights Darwinism was strange Darwinism, yet heartening Darwinism, for it dispelled even further the Darwinian fear that the Chinese were an inferior race that must perish. If evolution was on the side not only of virtue but of fraternity, equality, and liberty, then a Chinaman's chance was a good chance.

The Anarchists argued that the Chinese of course deserved and would have a place in the coming new world of brotherly love and mutual aid. The Chinese were not an inferior biological breed to be weeded out. The Anarchists insisted on that, even when they were most disgusted with their race. Wu Chih-hui, in a very discouraged article "celebrating" the one hundredth issue of the *Hsin shih-chi,* said that maybe the Chinese *were* a race of "sleep-loving pigs." Before, he said, he had only echoed such language to goad his countrymen into action; he had never believed it.

But now that description seemed more and more apt[405]—because his countrymen were not responding to *Hsin shih-chi.* "They have already soundly slept through a hundred weeks," he said. "I wonder how we shall celebrate the *Hsin shih-chi's* two hundredth issue!" But when he glumly concluded that "every people has its habitual nature, and it is not at all easy to change it,"[406] he was admitting that the trouble lay in bad habits. And habits could be changed. His people could wake up. When he sneeringly said in his essay "Pity the Pigtailed Chinaman" that "the Chinese are truly the most loyal and faithful of slave races,"[407] what he meant was, "but they need not be."

And that was the conviction that underlay even the most shocking Chinese anti-Chinese statement in *Hsin shih-chi,* a statement that on the face of it seemed to mark the lowest point in sixty years of expression of lessening Chinese self-respect. This was the statement of a wisely anonymous Chinese visitor, recently arrived in Europe, who sent *Hsin shih-chi* a report entitled "Thoughts on Viewing Prostitutes," which he began and concluded in this wise:

> During my recent stay in Paris, I went to see the public prostitutes, and suddenly realized the way the Westerners make progress. . . . I saw the public prostitutes of Paris and can only deeply sigh that the Chinese race is unfit and deserves defeat.[408]

Now how could any "Son of the Yellow Emperor" say that? Surely this was one of the most remarkable Chinese sentences that Darwinian vocabulary had ever gotten itself into—except that it did not really mean what it seemed to mean. The prostitute-viewer had not actually concluded that the Chinese race would be overwhelmed because white women, on close inspection, were overwhelming. He had, it is true, been intrigued by their physical attributes, but it was not their physical attributes that had led him to his conclusion.

Originally, he said, he and two Chinese friends had been "just looking" when they chose to step over the threshold of a Paris house of prostitution. So, when they were shown a circle of very scantily clad young females, they had taken a good look. And low and behold: "If one looked only at their upper parts, one saw billowing yellow hair and dark eyelashes, and when they opened their mouths, perfectly white teeth. Their aspect was exactly like the pictures on the boxes of foreign soap which we used to look at back home."[409]

But there was nothing *discouragingly* impressive about their looks, and all might still have been well, had it not been for "the rules of the house":

> We had planned to look, pay our money, and get out, but the house insisted that we could not. They said that there were spring-palace prescriptions for guests who did not carry through the rules of the house. So we did as we were bid, and accompanied two prostitutes to an inner chamber.

And there the author *was* overwhelmed:

> Surely nothing like that has ever been seen before in all history. Those who dream of "European civilization" simply can not imagine. We all felt embarrassed and most uneasy. We almost wanted to beat a retreat. For we felt that no fellow human beings should be so completely without a sense of shame. For three days we were unhappy about it, until we compared it with everything else we had seen and heard in Europe . . . and then suddenly we understood.[410]

The important thing, for success in anything, was to do away with "the enemies of one's occupation," and in *their* occupation the enemy was shame. "Europeans," he said, "all believe in doing things completely, in being absolutely thorough and totally professional in their work. . . . As I see it, Europeans in everything they do are as thorough and professional as these sellers of sex. . . . And it is precisely because of our own failure to be thorough and professional that we Orientals are so feeble!"[411]

The Paris prostitutes were professionals. They shared a pervasive European professional spirit that was the secret of Western progress. Orientals were feeble only because they failed to show similar spirit, not because they were physically or mentally unfit. And spirit, as good Confucians had known all along, was always at hand. It was like goodness. "I have never yet seen anyone whose strength was insufficient,"[412] said Confucius; "If I desire goodness, then goodness arrives."[413] "It is not that you cannot, but that you will not," said Mencius to King Hsuan.[414] And so it was with the Chinese people and "Western" spirit. They could show it if they wanted to. They could be equal if they wanted to. They did not have to wait for evolution to better their brains because they could make up their minds. Miraculously again, no Chinese entertained the truly dreadful thought that spirit or will power itself could be—or could not be—"in the genes."[415]

So back we come to the same old Confucian faith of Yen Fu, Liang Ch'i-ch'ao, and Sun Yat-sen. The only new thing was that the Anarchists had an even clearer sense, rightly or wrongly, that they had an evolutionary right to be equal, not just an evolutionary chance. For equality was natural.

That conviction, however, did more than just quiet Chinese fears of inferiority. Ill-founded though it might be on purely Darwinian grounds, it had the most salutary side effect of lessening Chinese racism. We have seen Chinese essayists from the very beginning of the Darwinian era in China madly scrambling to disassociate the Chinese from "the darker races" that they saw on all sides "going under." But now, ever so slowly, some of them, with Anarchists leading the way, began to argue as Liu Shih-p'ei did, that, "although the races of man are different, they are one in their humanity."[416]

Liu Shih-p'ei claimed that modern science had proved that man was a collection of elements, and that all men were collections of the same elements. "The materialists," he said, "can prove that men's bodies are the same," and "the idealists can prove that men's minds are the same."[417] So, either way, human nature was one. He did say that there were superior races and inferior races, that some were strong and some weak, some knowledgeable and some ignorant, and that those differences were indeed caused by the different conditions of different environments, but he was obviously thinking of those differences as cultural inequalities that could be erased.[418] For "it is in human nature," he said, "to return to equality."[419]

The Anarchists, therefore, believed, in theory at least, even though one can still find racial slurs in their publications,[420] that every race was basically fit and deserved to survive. So they became less afraid of confessing kinship with races that seemed "endangered species." They dared at last to take a stand with the "victim races" or, indeed, with "the *other* victim races," as few of their countrymen had dared before them.

The revolutionary writer who called himself "Ling Shih" had seemed all alone in 1904, when he cried out, "The plight of the blacks is ours also! All of us in the world who are under domination by the whites are blacks!" Moved by a translation of a book called *Hei nu hsu t'ien lu* (The black slaves' lament to Heaven) written by an American woman named Ssu-t'u-huo (Maria W. Stewart?), he had exclaimed, "Alas, who says the blacks are really an inferior race with evil natures and no souls!" He had cursed "mad whites" who claimed to be a higher race and who wanted to rule all others. "The whites," he had said, "in the name of civilization practice barbarism. Truly they are cruel and inhuman." And he had finally written, "With the tears I shed for the black people I weep for our yellow people."[421]

With those words he had sown the first seeds of Chinese interest in "Third-World solidarity," but the Anarchists now sowed many more. They too loved to write of "white barbarity,"[422] and with their faith in the evolutionary rightness of mutual aid they at last deemed it only natural for victimized races to stand together. And yet what was unique and all-important about the Anarchists' stand with other races was their insistence that their stand was not a *racial* stand. For the enemy, they insisted, was not the white race; it was the white race's governments, just as within China itself the enemy was not the Manchu race but the Manchu government.[423]

The Anarchists, therefore, were not just trying to rally one race or many races against another in an evolutionary racial war. They were trying ultimately to rally all peoples against all governments, for "government," they cried, "is the source of all evil"—whatever Chang Ping-lin might think.[424] And government was an affliction still suffered by all peoples. All peoples were in the same boat, because they were all in ships of state. Therefore Chinese could hope for solidarity with all peoples, with yellows, and blacks, and browns, and reds, and even with whites as well! Truly, China's Anarchists could promise long before Mao Tse-tung did that "we have friends the world over"[425]—and what a reassuring promise that was.

The Anarchists were the first to substitute an impersonal, "inhuman" enemy for the white peril, the racial enemy the Chinese had feared so long. And that substitution in itself did much to quell the decade-long Darwinian fear that modern world history—and natural history—was to be the story of a superior, biological breed of white men eliminating all lesser breeds of non-white men—including the Chinese. But that was not all. The new enemy, unlike the old enemy, was a patently unfit enemy, doomed itself to be eliminated by evolution. For governments were relics of the past. They, not the non-white races, were on their way to extinction. For they were anti-liberty, anti-equality, and anti-fraternity, and so they were anti-evolutionary and would die out—because "you can't stop progress"—or evolution. The new enemy, therefore, was a paper tiger. *All* governments were paper tigers. And China's government was a tissue-paper tiger!

That revelation was the third and crowning beauty of the Anarchists' argument; China's government was so pathetic that it would fall first, and so China at long last would again *be* first. As Ho Chen wrote, together with her husband, Liu Shih-p'ei:

> China's government favors passive rule and finds virtue in non-interference. Although a government in name, how far is it from no-government? Fools in their ignorance may say that China's government is irresponsible, and corrupt in the extreme, but they do not realize that the Chinese people are taking advantage of their government's corruption to escape the boundaries of human rule and so safeguard their boundless freedom. . . . But, from that point of view, China should be the easiest place in which to put anarchism into practice. Of all the countries in the world, China should be the first to have no government.[426]

China would be first, and the white Western nations last. For the white Western nations' governments were the strongest, and "the bigger they are," the harder they would be to make fall. The Western nations, not the rest of the world, would be backward.

So there it was again, paper-tiger-ism with its happy corollary, the-last-shall-be-first-ism, both of which we have seen from the beginning, in the thought of Yen Fu, K'ang Yu-wei, Sun Yat-sen, and even Liang Ch'i-ch'ao. Kropotkin's great revision had led the Anarchists in yet another way to the same great sigh of relief: "If Evolution be for us, who can stand against us? Victory shall be ours! Our very 'backwardness' shall make us fit!" That was the great revision's greatest message, which made it so inspiringly attractive.

But, from that point on, the great revision encountered a great contradiction. The Anarchists did not conclude that the Chinese should therefore just sit back, confident that China would come out ahead. They said, instead, that China must go ahead and get ahead through revolution—violent revolution sparked by assassinations. The good news that evolution was for brotherly love did not mean that China had to love its enemies, either at home or abroad. Unlike the poet who "dreaded talk of evolution," the Anarchists "dared hate" both the Manchu government and foreign imperialists, the minions of foreign governments, for both were anti-evolutionary.[427] Both were laying false claims to a Darwinian right to do their worst. Both were laying false claims to a "law of the jungle" Kropotkin's great revision had repealed.

Governments, said Li Shih-tseng, all kept power over their people through force alone. So, under all governments, "The weak are the meat of the strong."[428] That was the Chinese idiom first used to translate the law of the jungle, but Li Shih-tseng denounced it, and he specifically denounced the imperialists' use of it, calling on as exalted a "Darwinian" authority as Alfred Russel Wallace to denounce it with him. He triumphantly quoted from Wallace's "Militarism—the Curse of Civilization":

> I know that the historians of the future will stress over and over again in their world histories that we of our generation had not yet

> rid ourselves of our barbaric nature, and that we actually did not deserve to live in the civilized period of the nineteenth century. The nations of Europe in their actions are not worthy of the fair name of civilization, and surely today's governments, which use their strength to oppress the weak, are not the governments the nineteenth century ought to have.[429]

With Wallace's backing, Li Shih-tseng thus cursed all modern nationalists, who "insist on perpetrating barbaric killing, calling it 'the struggle for existence'."[430] For the Way of evolution, he said, was "universal love" (*po ai*).[431] And the Anarchists must therefore fight for universal love. But there was the contradiction. Anarchists, he said, must carry out assassinations of "enemies of humanity," that is, of men in government. "But is there not a contradiction between assassination and universal love?" he allowed a hypothetical questioner to ask. "Where!" he replied: "What sort of talk is that! They are enemies of the people . . . [to kill them] will be precisely what is meant by the saying 'to get rid of those who harm the group is to love the group.'"[432]

The Anarchists thus divorced their enemies from humanity. The impersonal nature of their primary enemy—the government—made those *in* government also impersonal, less than human, against the people and therefore "not people." They were also, of course, the very things that were arresting evolution. They were "reactionaries," and as reactionaries they were evolutionary hangers-on that ought to die out. The Anarchists were the first—though not the last—clearly to make their enemies reactionary, anti-revolutionary, anti-evolutionary, anti-nature, unnatural, inhuman beings—in short *yao mo kuei kuai,* "devils and monsters" that *kai ssu,* "deserved to die."

The Anarchists were ready to show anti-evolutionaries no more mercy than medieval Christians were prepared to show antichrists. And so, in yet another ironic parody of the religion they so scorned, the Anarchists displayed a "religious" ruthlessness, against evil—as they saw it. Decrying their enemies' appeals to the law of the jungle, they were ready, in the name of universal love, to turn on those enemies with a vengeance more fearsome than the law of the jungle. So the name of the Anarchist journal, the *T'ien i pao* (Natural right, or Heavenly right) came to resound with the most fearsome of its ancient connotations. For the Anarchists found in Darwinism a natural right, a Heavenly right, to wipe their enemies—"evolution's enemies"—off the face of this earth.

They did not, of course, do it. For the Anarchists never came to power (if Anarchists by definition ever can). A revolution came, and they helped it come, but it was not theirs; it was the Republicans'. And it was not even the Republicans' for very long. It became Yuan Shih-k'ai's and then the warlords'. And, although the warlords brought anarchy enough, it was not the kind the Anarchists had in mind.

Anarchism was rejected as a "viable Way." It did not seem practical enough—or nationalistic enough. When everyone else was shouting that, one way or another, nationalism was the only thing that could save China, the Chinese Anarchists decried nationalism, although clearly for nationalistic reasons. Wu Chih-hui insisted that opposing nationalism did not mean selling out the nation.[433] Instead, he protested, the Anarchists' program for revolution was actually "the perfect policy for saving [China] from a melon-carving."[434] And Li Shih-tseng added that "obstructing the revolution out of fear of melon-carving is like saying to someone about to die of asphyxiation, 'Don't open the window or you'll catch cold.'"[435] But most people were unconvinced. Even most Revolutionaries seemed to feel that a revolution that promised no government when it was over would not only court disaster but guarantee it. Most Chinese patriots longed for a government, a government that would do something, that would save them. And even the Anarchists themselves came to admit that no-government was an ideal many years away.[436] So most of them went into government.

But what, then, is important about the Anarchists? They helped further set things up for the Marxists. They helped further convert many people to certain beliefs from which it was very easy to be converted to Marxism. They helped many intellectual Chinese, therefore, to become Marxists. But surely not all those interested in anarchism became Marxists, and not all those converted to

Marxism were first Anarchists. Therefore, Chinese anarchism is not important just as an historical cause or an historical stage. Almost more important, to us, is the simple fact that the appeal of anarchism provides a "retrospective preview" of the appeal of Marxism. Marxism proved attractive to Chinese for the very same reasons that anarchism (initially) proved attractive—plus a very few more, which made a very great difference.

Both anarchism and Marxism spoke to the same Chinese hopes and fears, and both with only slight differences gave the same encouragement. Both offered a "Darwinian" social "science" to dispel the Darwinian fear that China would not survive. Both insisted that the *Tao* was for order, not disorder, that evolution was moral, and that China's enemies were evil. Both saved China from "backwardness" by declaring that the contest they seemed to be losing was not the contest. Both denied the natural rightness of a world of warring states and warring races. Both disarmed the seeming enemy—the white race—by declaring it not the enemy. Both substituted "a system" as an international enemy afflicting the white race more than any other. So both promised that the last could be first. Both said that evolution was for fraternity, equality, and liberty, but both put equality before liberty.[437] And both said that evolutionary struggle was a struggle between the forces of light and the forces of darkness, so that fraternity meant only fraternity with the forces of light. Both gave a natural right as right as any divine right to revolution, and in effect to vengeance. Both promised that evolutionary revolution must succeed. So both promised China certain victory. Both promised China perfect fitness. Both promised that China, once again, could be the fittest.

What Marxism offered that anarchism did not, besides, eventually, a Sinified landlord enemy closer to home than any government, and so an even greater source of explosive power born of hatred born of suffering, was an organization, a party, that seemed capable of carrying out evolution's work. The Marxists were "practical." The Anarchists were not. Moreover the Marxists soon seemed to have the pudding's proof, in the Russian Revolution.

The Anarchists had an ideal of great appeal which almost fit perfectly China's greatest hopes and fears, but which did not fit the need for action. Marxism did, and so proved fitter, and so survived.

When All Was Said and Done

On May 7, 1912, seven months after the Republican Revolution, ex-Provisional President Sun Yat-sen proclaimed to China that Darwin's day was over. "The idea that all must struggle against each other to survive," he said, "is already a theory out of date. Today in the evolution of the human race there is no way to survive except by helping one another."[1] The Anarchists must have been pleased. But one wonders whether Sun Yat-sen was thinking first of the human race, or of China, or of the Republicans.

Three months later, invited to Peking by President-fast-becoming-Dictator Yuan Shih-k'ai, he repeated the same theme:

> Before the twentieth century, the nations of Europe invented a new-fangled struggle-for-existence theory, which for a time influenced everything. Every nation assumed that "the survival of the fittest" and "the weak are the meat of the strong" were the vital laws on which to establish a state. They even went so far as to say that "might is the only right, there is no reason." This kind of theory in the early days of the evolution of European civilization had its uses. But, from the vantage point of today, it appears a barbaric form of learning.[2]

And three months later still, although no longer in Peking, he repeated it once again:

> Darwin's thesis that there was only "the right of might" in the world and no "reason" was widely parroted by later scholars, who took "the right of might" absolutely to be the world's only true principle.

> But, testing this against my conscience, I felt that I could not agree. For I sincerely thought that, although "the right of might" might indeed be in accord with the progress of natural evolution, still our naturally endowed consciences made reason difficult to deny. [I concluded], therefore, that evolutionary selection is the barbaric, material way of progressive evolution, but that reason and conscience are the way of progressive evolution for morality and civilization.[3]

Fairly or unfairly, Charles Darwin was being discredited. Evolution itself was now said to have passed him by. His own theory was passé, no longer fit, "barbaric." Darwin's day, said Sun Yat-sen, *ought* to be over.

But of course it was not. Only five months after he said he disagreed with Darwin, and seventeen days before Yuan Shih-k'ai had the leading Republican challenger in parliament, Sung Chiao-jen, assassinated, Sun Yat-sen already saw the calligraphy on the wall, and went *back* to Darwin, back to the first lesson Darwin ever taught the Chinese: "While this twentieth century is a world governed by the struggle for survival and the survival of the fittest, how can government, industry, or anything progress without struggle and competition?"[4] He meant, of course, parliamentary struggle, peaceful party struggle, but the assassination of Sung Chiao-jen ended all that.

The Revolution of 1911, as a Republican revolution, failed, and, when it did, Chinese had to ask all over again, "Now what?" And when they did that, struggle began all over again, and Darwin, all over again, was part of it, on all sides. So his day was not over. And yet his first and most important day was—the day of Darwinian revolution. His second day, from the Revolution of 1911 until the time his "ism" was eclipsed or superseded by Marxism, was in large part only a repeat of the first. It is true there were more Darwinian footprints in more essays than ever before, so that many have been fooled into thinking that Darwin's second day *was* his day. But very few of those footprints led in new directions. Most only made circles on old, well-trampled ground.

This was the era in which it was virtually impossible for children to get through school without having to write essays on the topic,

"Things compete, Heaven chooses; the superior wins, the inferior loses,"[5] essays probably now all lost. But, even if some could be found, although there might be much revolutionary talk in them, I doubt that one would find much that was revolutionary. For even the schoolchildren's mentors, even the editors and authors of *The New Youth,* the most famous and influential magazine of Darwin's second day, had very few ideas that were truly new (although many may have been independently thought up anew). It was the youth that was new, not their ideas. The whole May Fourth period, which *The New Youth* did so much to inspire, has been vastly overemphasized as a period of new thought. It was a period of the propagation of new thought. What was new was China's first "modern" student generation, ready at last to be affected by new thought. But almost all of the "new ideas," with only a few notable exceptions, already had been had—by Yen Fu, Liang Ch'i-ch'ao, Sun Yat-sen, and Wu Chih-hui. At least this was true of Darwinian ideas. Darwinian ideas can be found in essay after essay, on page after page. Every Darwinian theme that we have yet seen can be found—but very few new ones.

So we do have a right—if not, for humanitarian reasons, an obligation—to stop this study at the end of the first day, even though we must take one fleeting glance at the second day in order to ask our ultimate question: What, when all was said and done, did Darwin do to China?

During his second day, Darwin served first as the patron saint of the New Culture Movement. But the New Culture Movement was really only a resumption of the cultural revolution started by Liang Ch'i-ch'ao in 1902, but suspended during the "real revolution" of 1911, while people waited to see whether or not that revolution would solve everything—or anything. It did not. Yuan Shih-k'ai crushed the Second Revolution, and bequeathed the nation to a very motley crew of "warlords." When all that was clear, Ts'ai Yuan-p'ei, Ch'en Tu-hsiu, Hu Shih, Lu Hsun, and many others *resumed* the cultural revolution, partly because they despaired of safely being able to do anything else, and partly because they soberly concluded, for the time being, that a cultural revolution would have to precede a political revolution—or even reform—after all.

All the New Culture Movement's most famous cries, cries against Confucius, the ancestors, and the family, cries for "Mr. Science," "Mr. Democracy," and "The Literary Revolution," all had been raised more than a decade before. Liang Ch'i-ch'ao had called for a "literary revolution" in 1902.[6] The Anarchists, though no friends of his, had called for a "writing system revolution" in 1907.[7] Liang Ch'i-ch'ao, as we have seen, helped start the revolt against Confucius in 1902, and the Anarchists gleefully carried that on too in 1907 ("Confucius, in the barbaric period, could recognize a few more characters than other people.")[8] And in the same year they called for an "ancestor revolution" as the first step in a "family revolution" (and they were *not* trying to get their *ancestors* to rise up.)[9]

The New Culture Movement's cries were all cries Darwin had backed before, and he now backed them again in the same old way. He was the patron saint of the New Culture Movement for two simple reasons. First, his theory, so the New Culture Movement's leaders still insisted, "proved," as Liang Ch'i-ch'ao had said it had years before, that "the present surpasses the past and the future surpasses the present."[10] That was the faith behind the Anarchists' injunction to *tsun chin po ku* (respect the present and belittle the past)[11] and the Communists' later injunction to *hou chin po ku* (extol the present and belittle the past). But second, his theory "proved" that the modern was not just better than the old; the modern was fit, and the modern alone made fit. Darwinian fears were still very much alive, and the New Culture, therefore, was still to fit China for survival. "All that counts," wrote Lu Hsun, "is our preservation. We should ask only whether or not something has any power to preserve us, not whether or not it is of our 'national quintessence' [our cultural best]."[12] And Lu Hsun obviously thought China's old culture, her cultural best, had no such power: "Just look at our society; cannabalism, banditry, murder, traffic in human beings, phallic worship, black magic, polygyny, . . . everything in our 'national quintessence' is in perfect accord with the culture (?) of barbarians."[13]

Lu Hsun still saw the New Culture Movement as a Darwinian matter of life and death—China's life and death. That was clear, he said, even in the realm of literature: "Hereafter, we really do have only two roads. One is to embrace our ancient writing style and die. The other is to discard our ancient writing style and survive."[14]

But what of those who disagreed? What of those who defended the "national quintessence" and opposed the New Culture? It has been often said that those who did also opposed Darwin, that seeing the West's true face, so they thought, in World War I, they rejected the West and science—and Darwin's science first of all.

Perhaps some did. Lu Hsun later said that many Chinese cursed Darwin during World War I.[15] In 1930, Lu Hsun blamed misinterpretations of Darwin's theory at the time of World War I for the fact that the term *evolution* was in 1930 falling out of use.[16] And certainly Darwin's two greatest Chinese apostles were disenchanted enough with the West in World War I to cast doubts, at least, on the benefits of their Sage's legacy. Yen Fu muttered in 1918: "Having witnessed seven years of China's Republic and four years of Europe's bloodiest war in history, I feel that the European race's last three hundred years of evolutionary progress have all come down to nothing but four words: selfishness, slaughter, shamelessness, and corruption."[17] And Liang Ch'i-ch'ao in 1919 echoed that view in his famous *Ou yu hsin-ying lu chie lu* (Selections from a record of impressions on travels in Europe).

"The material progress of the last one hundred years," he said, "has been several times that of the preceding three thousand years, and yet, not only has our human race not gotten happiness out of it, it has brought us a host of disasters."[18] And those disasters he traced back to the theories of evolution and individualism! *The Origin of Species*' "thousands of words," said Liang Ch'i-ch'ao, "all came down to ten: 'the struggle for survival and the survival of the fittest'." And those ten words spelled trouble:

> Those two principles were put together with Mill's utilitarianism and Bentham's hedonism [all of which Liang Ch'i-ch'ao himself, of course, had most enthusiastically introduced to China] to form the core at that time of the English school. And at the same time, Max Stirner and Soren Kiergegand [*sic*] were loudly advocating individualism, which reached its worst in Nietzsche, who said that altruism was a morality for slaves, and who said that the extermination of the weak was the natural duty of the strong, and absolutely necessary for the world's destined progress. All these weird theories were based on Darwin's biology, and they perfectly fit the psychological temper of the time. So, for individuals, the worship of influence and money became the sacred rule, and, for countries, militarism and imperialism became the most stylish political policies. Surely it was out of this that came this great international World War.[19]

World War I, said Liang Ch'i-ch'ao, was a "retribution,"[20] and it seemed he meant a retribution for believing in Darwinism. But that was not what he really meant. Neither Liang Ch'i-ch'ao nor Yen Fu ever repudiated Darwin's theory of evolution. They repudiated misuse of Darwin's "theory of biology" (they finally tried to bring it back to that), and they repudiated misbelief in the universal fitness of the West, even though they had done more than anyone else in China to make Chinese believe in the fitness of the West, and more than anyone else in China to drag Darwinism into Social Darwinism.

When they came back to Chinese culture, however, when Yen Fu said, "When I look back again at the Way of Confucius and Mencius, truly it measures up to Heaven and earth and could benefit the whole world,"[21] and when Liang Ch'i-ch'ao cried, "Dear youth [of China]! Attention! Forward march! Millions of people on the far side of the ocean, distraught by the bankruptcy of material civilization, are sadly, desperately, shouting for help, waiting for you to come and uplift them,"[22] they still believed in Darwin. They had simply concluded that Chinese culture was the fittest after all.

That was the conclusion even of the most original of the "conservative" critics of the New Culture Movement, Liang Sou-ming. He was an even more ardent champion of Chinese culture and foe of Western culture than Liang Ch'i-ch'ao ever became, but he wrote his most famous work, *Tung Hsi wen-hua chi ch'i che-hsueh* (The cultures of East and West and their philosophies) in answer to the starkest of Darwinian questions, "Can Oriental culture in the end survive?"[23] and he found a perfectly ingenious Darwinian way to answer, "Yes."

The three greatest cultures of the world, he said, Western, Chinese, and Indian, represented three very different ways of life, the way of struggle, the way of harmony or accommodation, and the way of escape, and each was a natural product of evolution. Moreover, each was good or proper or fit in certain circumstances (there was a time and place for everything), but each was also a stage in a sequence, an evolutionary sequence that held true or would hold true for all societies. And, in that sequence, the Western way came first—which meant that it was the most primitive. The Chinese way was advanced. China was ahead. It was in danger only because it had leapt ahead so early in its history that its material civilization, and that alone, remained underdeveloped, which would have been of no import had China not been now so rudely dragged into contact with a still-barbaric West.

To withstand the affrontery of a first-stage culture, China would have to go back and re-engage in a little self-strengthening struggle of its own, and it would have to borrow certain things from the material civilization of the West, the fruits of a stunted spiritual growth. But China's basic position was correct, and advanced, and most fit. India's way was too advanced. It was the way of the distant future. And India's present position was proof enough that the way of the future was no way for the present. China's way was the middle way. The world would have to go that way. And when it did, China would again be China.

Here indeed was a way to have your Confucius and your Darwin too. Here was a new Darwinian defense of Chinese culture. But it was not very new. It was only the latest way—and the best to date?—of turning China's weakness into strength. And that was what every single major figure that we have looked at had tried to do, or would try to do, from K'ang Yu-wei to Mao Tse-tung! Liang

Sou-ming's thesis was only one more Chinese Darwinian attempt—and certainly a neat one—to contradict the wretched Westerners' allegation that China was dangerously unfit.

So, in the New Culture Movement, Darwin was still there on both sides. But Liang Sou-ming's side did not win. Most Chinese intellectuals were no longer, if they ever had been, content with harmony and accommodation. They *wanted* to struggle. They wanted to win back their self-respect and win back the respect of the world. So the conservatives lost. Hu Shih's "liberals," if for convenience's sake we may call them that, won on all fronts except the democratic. They even won, if one is content to decide the matter by counting converts, the great debate of 1923 over science and metaphysics. Whether as pragmatism or Marxism, "scientism," as Daniel Kwok has called it, won the day, and we need not describe how easily Darwin was made the patron saint of that. He had been the patron saint of scientism ever since Liang Ch'i-ch'ao's first statement of that creed in 1904.[24]

In their continued championing of "Mr. Science," however, and in their indignant denying of Liang Ch'i-ch'ao's alleged allegation that science was "bankrupt,"[25] the liberals still agreed with the conservatives on one important point—that science—and Darwinian science—could be abused, and that it had been abused by the warring powers of the West. It was perfectly true, said Hu Shih, that Darwin's theories "contained one very dangerous element," the notion that "the weak are the meat of the strong," a notion that was "the scourge of the present day."[26] But "for the weak to be the meat of the strong," he said, "is the way of birds and beasts; it is not the way of man." The trouble was that people "*took* the way of birds and beasts to be the way of man."[27] But that was pseudo-Darwinism: "Those who now uphold the theory of the right of might say that it is a law of evolution. But they do not know that above 'natural selection' there is still 'human selection.' 'Heaven and Earth are inhumane,' so 'the weak are the meat of the strong,' but human selection is not like that. For 'Man can do better than Heaven.'"[28]

Hu Shih was a Huxleian—and a confessed one at that![29] He wanted man to combat the way of natural selection. "The way to save the world is to substitute the Way of man for the way of animals."[30] It was to substitute a humane Way for a brutal way. "The struggle for survival," said Hu Shih, "is dreadfully cruel."[31] But, as Chou Tso-jen added, "although we recognize the fact of the struggle for survival in biological phenomena, we do not have to have it in the life of the human race."[32] "The gruesome drama of the struggle for survival," said Hu Shih, was "a goad to man in all his behavior," but "not even the struggle for survival can turn man into a ruthless, heartless animal. Perhaps, instead, it can give him a special sympathy for his fellow men. Perhaps it can give him a special belief in the importance of mutual aid. Perhaps it can give him a special concern for human action and effort to reduce the cruelty and waste of natural competition."[33]

Hu Shih, Liang Ch'i-ch'ao, Yen Fu, even Sun Yat-sen, all either consciously or unconsciously had come back to Huxley. They might seem to be Kropotkinites when they talked about mutual aid (even our "proto-Kropotkinist," Liang Ch'i-ch'ao, said the necessity of mutual aid was one of the great lessons of the Great War),[34] but none of them showed any Kropotkinite enthusiasm for animal life. They were not for mutual aid because the birds do it and the bees do it. They were for mutual aid in opposition to brutality, as Huxley had been all along. They still, when all was said and done, had a Confucian view of the animal kingdom.

But did that mean that Hu Shih was no longer for evolution? Of course not. It simply meant that he was for "human evolution," however such a thing as that could be. He was for "caring for the old and aiding the weak, supporting the crippled and healing the sick," not for "eliminating people because they are old, weak, crippled or sick."[35] He was for "tolerance" (*jung-jen*)—the newest word in Darwin's second Chinese day—and non-violent, rational competition.[36] He was for William James's "meliorism" (*kai-liang-chu-i*),[37] and he was for gradualism. And that is what he thought Darwin was for:

> Darwin's theory of biological evolution has given us one great lesson. It has taught us to understand that biological evolution, whether by natural change or human choice, all comes from bit-by-bit change. . . . Pragmatism gets its start from Darwinism. Therefore it can recognize only steady bit-by-bit improvement as true and reliable evolution.[38]

Pragmatism was Hu Shih's Darwinism, and pragmatism was something fairly new, not so much in its gradualism or meliorism, or its rationalism, or even in its scientific method, however now refined, for all that was roughly in the thought of Yen Fu and Liang Ch'i-ch'ao, but in its gentleness, tolerance, and compassion. And yet little of *that,* in China proper, was to survive. Hu Shih's pragmatism led from the New Culture Movement into politics, and there, in the arena of the second great struggle of Darwin's second day, it was defeated. Humane evolution was not to be the Way. In politics Hu the Fit proved Hu the Unfit.

But who then was fit?—The revolutionaries were—first the Nationalists and then the Marxists. Their fitness, of course, would eventually "grow out of the barrel of a gun,"[39] but, before things came to that, they began, especially from 1920 to 1926, to prove themselves "fitter" than the reformers in their intellectual arguments, which were very similar to those made by the original Republican Revolutionaries from 1905 to 1911. Sun Yat-sen, the "originalest" Republican Revolutionary, now preparing in Canton for his anti-warlord Northern Expedition, was the man of the hour, and he, of course, had all his old Darwinian arguments for revolution still to draw upon; but now there were also Marxists, sometimes helping him, sometimes helping themselves, and the Marxists had their own reasons for revolution.

Most of their reasons were similar to Sun Yat-sen's. The Marxists too believed in fixed evolutionary stages. They believed, as Ch'en Tu-hsiu said, that "from feudalism to republicanism and from republicanism to socialism—this is the fixed path of evolution."[40] And they believed, of course, that one had to struggle from stage to stage. They agreed with Wang Hsing-kung that "a struggle philosophy is an evolutionary philosophy."[41] And they believed that revolution was necessary for survival.

That last belief, however, an old evolutionary favorite of the Republicans, again threatened paradoxically to turn revolution against evolution. As Ch'en Tu-hsiu put it:

> If an intellectually immature and unorganized people like the Chinese, pressed harder and harder every day by outside aggression, both political and economic, do not adopt the precipitate progress of revolution, will time allow us the gradual progress of evolution?[42]

Revolutionary progress was the only way the backward could catch up. Evolutionary progress seemed a luxury for the advanced.

That conclusion did not, however, mean that the revolutionaries were ready to surrender Darwin and evolution to the reformers. Evolution and revolution, said Chou Fo-hai, only seemed contradictory: "Actually the two are like a person's two legs." Anticipating (or perhaps suggesting?) Mao Tse-tung's later metaphor, Chou Fo-hai insisted that to get anywhere one had to walk on two legs and that one leg could not do without the other. "And just so," he said, "we can say that natural evolution cannot do without man-made revolution, nor man-made revolution do without natural evolution."[43]

It was at this point that Marx put forward *his* best foot. Chou Fo-hai said, "I really respect Marx,"[44] and Ts'ai Ho-sen, the friend who perhaps converted Mao Tse-tung to Marxism,[45] gave the reason why: "As I see it, the marrow of Marxism is its *synthesis* of revolution and evolution."[45]

That synthesis, real or imagined, gave Marx tremendous power, for it seemed to mean that Marx had gone beyond Darwin, that he had become a "Super-Darwinist." Marx had supposedly proved that revolution was evolution's way for man, by discovering in class struggle the social law that corresponded to the biological law of the struggle for existence. As a Japanese, cited in *The New Youth,* reported Engels's words at Marx's funeral, "Darwin made clear individual biological evolution; Marx made clear human social evolution,"[47] or as a Chinese Marxist put it, "One discovered the hidden secrets of the biological world; the other discovered the true nature of human history."[48] Marx was *the* Social Darwinist.

Tsou Jung and Sun Yat-sen, of course, with Liang Ch'i-ch'ao's help, had synthesized revolution and evolution themselves, by declaration, which is all that Marx had done, but Marx proved a fitter Social Darwinist than they, because he was a *Western* "social scientist," in a period in which Chinese still did not quite trust themselves, because the law of class struggle did strike a raw nerve in Chinese society, and because Soviet Russia proved Marx "right," before Sun Yat-sen proved himself right—which he never did.

Long before the fitness contest between Chinese Marxists and Nationalists was ever decided, however, Soviet Russia gave both camps something far more substantial than an example. It armed each with a Leninist party, a move which very clearly helped the forces of revolution against those of reform, as well as the forces of dictatorship against those of democracy. The two Leninist parties, which together did in the liberals, were the last "new" elements in Darwin's second Chinese day. But, although they did in the Darwinian liberals, they too were supported by Darwin.

In China, Lenin barely deserved credit for the success of his party system, for it was successful not because of its newness but because of its fitness. It survived—and thrived—because it fit so well with Darwinian ideas already enunciated by Sun Yat-sen and with Confucian ideas already enunciated by Mencius, or rather with pre-Confucian ideas enunciated, if we are to believe Mencius, by the noble minister of the early Shang, I Yin! For "I Yin said," said Mencius, "'When Heaven gave birth to the people, it set the prescient to awaken the unprescient and the aware to awaken the unaware.'"[49] That was the task the leaders of the Reform Movement gave themselves as early as 1898,[50] and that was the task Sun Yat-sen gave his Revolutionaries as early as 1905, when he mixed I Yin's language and Darwin's in the dedicatory editorial of the *Min Pao:*

> In any people [*ch'ün*] there is a small group of the best minds that can spur its people on to progress, that can make the fittest forms of government fit its people, and adopt its people's progress to the world. That is the sacred duty of the prescient, and that is what our *Min pao* is set to do.[51]

I Yin had named the task and Darwin made sense of it. The fittest would awaken the less fit, or the fittest would awaken the rest of the fit and lead them against the unfit. It was the task of the Kuomintang, and that was the task of the Communist Party. And each thought its task sacred. "Without the Kuomintang," Chiang K'ai-shek would say, "there would be no China."[52] "Without the efforts of the Communist Party," Mao Tse-tung would counter, "China's independence and liberation are impossible."[53] With two prescient groups, each convinced it was the fittest one in the East, a showdown was inevitable.

Even without the threat of the other, however, each was convinced that, while the people were waking up, the fittest would have to dictate. Sun Yat-sen spoke of the necessity of "political tutelage." Ch'en Tu-hsiu spoke—more openly—of the necessity of "enlightened dictatorship." But that was not a new Way. It was the famous Way Liang Ch'i-ch'ao had taken up from Republican Revolutionary Ch'en T'ien-hua's suicide note![54] And Ch'en Tu-hsiu knew that full well, and yet he still said, "China's reformation will have to go through a period of enlightened dictatorship."[55] He said:

> I dare boldly state that, unless we instigate a policy of strict interventionism in politics and education, there will never be a day of remedy for the corruption and degeneration of our Chinese race. Our only hope, therefore, is to hope that people of conscience, knowledge, and ability from throughout the land will band together and as soon as possible create in name and fact an enlightened dictatorship, which can rescue us from below the basic level of humanity and raise us up above it.[56]

Some of Ch'en Tu-hsiu's fellow Communists criticized his use of Liang Ch'i-ch'ao's old term, but they did not criticize his concept. Chang Sung-nien, who had been reading a lot of Lenin and Trotsky, wrote from Paris:

> We should not say "enlightened dictatorship" too much. We should always formally advocate "the dictatorship of workers and peasants" or "the dictatorship of the proletariat." Although what we call "enlightened dictatorship" and "the dictatorship of the proletariat" mean the

> same thing, we still must regrettably use the monkey-keeper's technique. Otherwise we will needlessly arouse resentment, which will do us no good. It is perfectly true, of course, that most Communists today know that "the dictatorship of the proletariat" does not mean the dictatorship of the entire body of the proletarian masses. It is only the dictatorship of the minority of the vanguard, the minority of the forerunners. Even Lenin says so.[57]

Lenin believed in "enlightened dictatorship," in party dictatorship, in the dictatorship of the fit (as he saw them), and Lenin was becoming popular. In the April 22, 1925 "Lenin Issue" of *The New Youth,* Chiang Kuang-ch'ih quoted a "coolie" as saying, "Everybody knows Lenin. Before, there was Confucius, and now there is Lenin."[58] And Chiang Kuang-ch'ih himself broke into verse: "Oh, Lenin! Your noble work is like the red sun that courses through the Heavens."[59] Lenin was eclipsing Confucius and stealing light from the future Chairman. But Lenin did not enlighten a China totally in the dark. Chinese had already seen the need for what he gave them. Men as different as Ch'en T'ien-hua, Liang Ch'i-ch'ao, Sun Yat-sen, and Ch'en Tu-hsiu had already concluded that enlightened dictatorship of some sort was the fittest thing for China. Some said China was fit for nothing else, but all believed enlightened dictatorship could make China the fittest (again) the fastest. Only a few ill-fated liberals violently disagreed, but they, needless to say, were unwilling or unable violently to resist. Their own non-violent weapons, of moral and intellectual persuasion were not persuasive enough—this time, and so party dictatorship won.

First the Nationalists won and then the Communists won, and, when the Communists won, Darwin's second day was over. For the Communists had the power to decree that Marx had superseded Darwin—and they did. So finally we have the right to ask, "When all was said and done, what did Darwin do to China?" But, when all is said and done, that is still a most difficult question.

In the beginning Darwin frightened China, but also at the same time gave it hope. For Darwin made some kind of sense out of China's disastrous world plight, and even a frightening sense was better than none at all. The vision of a jungle world may have driven some to despair, but others, of a Taoist bent, who were ready to see power in knowledge of the *Tao,* in the ways of that evolutionary jungle, saw in Darwinism the way to recovery. So, even though Darwin for many delivered a final blow to cultural and racial self-confidence, he also helped start to rebuild it.

Darwin inspired, or helped inspire, a radical reappraisal of everything Chinese, and his theory, therefore, became an unprecedented weapon for change; and much of that change, it surely cannot be denied, was for the good. The abolition of foot-binding is only one example. And yet it is an aptly perplexing example, for as far as the "Darwinian" arguments against foot-binding are concerned, its abolition is proof of good coming out of nonsense. Shocked Chinese argued that a fit race must have fit feet, but a thousand years of foot-binding had not had any *genetic* effect on Chinese feet whatsoever, not even among the gentry. Moreover, Chinese did not need Darwin to tell them that foot-binding was unfitting. It was quite enough for other foreigners to tell them it was bad. For that either shamed or shocked Chinese into realizing that of course it was. And yet Darwin *was* called in to fight foot-binding. So where are we?

Darwin helped inspire a true renaissance of Chinese thought by specifically challenging (or seeming to challenge) certain favorite traditional ideas and by discrediting all ancient authority. For twenty years, all the great questions of philosophy were rethought. Individual philosophies were changed. Individual lives were changed. And, if it was a renaissance that produced nothing newly profound, it at least held great promise as the freest period of thought since the classical age. But it was cut short—by the early imposition of a neo-orthodoxy, the Thought of Mao Tse-tung.

That "imposition," of course, also owed much to Darwin. For Darwin had legitimized violent change and revolution. Surely that was one of the most momentous things Darwin did to China, even if it was the furthest thing from his own mind. Darwin offered a substitute for the traditional idea of the Mandate of Heaven, having himself "disproved" that theory by "disproving" Heaven's existence. His substitute, to be sure, was not much of a substitute, for the fittest was still the survivor, but that perhaps made it the perfect substitute. At any rate, those Chinese who were convinced that China needed rebellion were desperately in need of some legitimizing theory, for without the Mandate of Heaven rebellion for three thousand years had been one of the two cardinal sins (the other being filial impiety). It was that powerful sense of sin that Mao Tse-tung, Wu Chih-hui, Sun Yat-sen, and even Liang Ch'i-ch'ao combated so strenuously in all their Darwinian protestations that revolution was legitimate. Mao Tse-tung finally claimed that Marxism-Leninism could all be boiled down to one sentence, *tsao fan yu li*—"To rebel is justified"[60]—but that standard translation obscures the force of the *li* (reason or principle) that rebellion was now said to have. That Neo-Confucian word in new context really meant that rebellion was a natural law, and *that* lesson had been taught to Mao Tse-tung not by Marx but by Sun Yat-sen and Liang Ch'i-ch'ao, who had learned it, rightly or wrongly, from Darwin. For the *li* of revolution, they had said, lay in evolution.

Darwin justified revolution and thereby helped the cultural revolutions of Liang Ch'i-ch'ao, Hu Shih, and Mao Tse-tung (and, of course, so many others), and the political revolutions of Sun Yat-sen, Chiang K'ai-shek, and Mao Tse-tung. As things turned out, therefore, he seemed to help Mao Tse-tung the most, and indeed he did. He helped make the Marxists the fittest.

Darwin created the ideological vacuum that cried out for something like Marxism, and he established the criteria for what that something should be. The new fit and fittening ideology had to be based on the Western science of evolutionary progress. It had to identify inevitable, natural stages of human social development. It had to promise historical inevitability and yet at the same time recognize the vital importance of human action. It had to be based on struggle and yet stress mutual aid among members of the *ch'ün*. It had to provide a non-racial enemy to explain China's inner and outer troubles without damning the Chinese—and it had to give the underdog a chance.

That last stipulation was not Darwin's by any stretch of the imagination, but every Chinese Darwinist we have seen forced Darwin to give the underdog a chance. For the Chinese needed David-and-Goliath stories, despite their country's size. Darwin originally, of course, seemed only to offer a cosmic pat on the head to the top dog, or to the King of the Mountain, or to the Lord of the Jungle. If Christianity, as Nietzsche said, was a religion for the weak, Chinese should have leapt for that, not Darwinism. Or they should have retreated into Buddhism. But they chose instead a top-dog philosophy and forced it to fit the underdog, even if that took heroic illogic. They found will power the power to fitten the unfit. And therefore any fit ideology had to believe in an evolution in which will power could overcome all odds.

And the Thought of Mao Tse-tung fit perfectly. It championed the oppressed, told them that history was on their side; and that they had a way if they had a will. It provided an international enemy responsible for all China's troubles, thereby absolving the Chinese "people" from all guilt for China's weakness and concentrating all blame on a minority group at home *and* a minority group abroad, so that the Chinese people could be further reassured by immediately having "friends the world over." It provided a Western science with an evolutionary vocabulary and yet a Western science that loathed the power of the West and called it unfit. It fit all the criteria for a fit ideology that earlier Chinese Social Darwinists had set up, and it claimed to have sophisticated, philosophical, and scientific resolutions to all the theoretical contradictions that earlier Chinese Social Darwinists had stumbled upon, even though most of those resolutions had been formulated in less sophisticated language by those same early Chinese Social Darwinists before China had ever heard of Marx. The Marxists resolved the mutual struggle–mutual aid conflict, for example, the same way Yen Fu and Liang Ch'i-ch'ao had. Mutual aid was the rule within the *ch'ün,* or the people, and struggle was the rule without–the first lesson that Chinese had ever learned from Darwin. The Marxists simply had a slightly different *ch'ün*, not all the people, but workers and peasants (and those who aspired to their consciousness), who would *ch'ün* against capitalists and landlords. If the Marxists went the earlier Social Darwinists one better, it was only in their repeated use of struggle to forge *ch'ün* unity.

Marxists, I assume, would not like this analysis. They would probably say that Social Darwinists were not responsible for their victory, that capitalists, landlords, facists, and imperialists were, and, of course, they would, as always, have something—but only something. There was indeed "people power" at work at the end of the Communist Revolution, people power generated by landlord oppression, capitalist exploitation, and imperialist (at the last, Japanese) aggression. But that people power could have been tapped by many forces. (The Nationalists *could* have tapped it.) It was tapped by Marxists because there were Marxists ready to tap it. But the Marxists were intellectuals. Marxism proved itself fit intellectually long before it proved itself fit socially or politically. Marxism converted intellectuals—but intellectuals who were already converted to Darwinism. If the intellectual Marxists were the "prescient," the *hsien chih hsien chueh,* who awakened the masses, China's earlier Social Darwinists, Yen Fu, Liang Ch'i-ch'ao, Sun Yat-sen, Li Shih-tseng, Wu Chih-hui, were the "prescient" who awakened the Marxists. They fit them for Marxism.

Mao Tse-tung said in 1957 that "socialism, in the ideological struggle, now enjoys all the conditions to triumph as the fittest."[61] He was right. He would have been right had he said the same thing, as he probably did, thirty years earlier. But for him still to use Darwinian idiom as late as 1957 was fitting indeed, for Darwinian idiom was one of the conditions for the triumphant fitness of his own Sinicized socialism.

The above, I believe, can be defended as fact, or at least as seeming fact. But we cannot stop there. The question remains, "In fitting China for Marxism and the Thought of Mao Tse-tung, what did Darwin do to China?" The question must be asked. It is, indeed, *the* question. But the answer, any answer, pulls us beyond the realm of fact or even seeming fact and into the realm of exceedingly controversial opinion, into the realm of moral judgments, and the reader, if there still is one, may not wish to be dragged there. Nevertheless, here is a personal view.

If one can believe in Marxism, then Darwinian preparation for Marx, even if acknowledged, becomes preparation for a Sage, and any Social Darwinian nonsense along the way is of no account. If one cannot believe in Marx, however, any Social Darwinian nonsense along the way, if it lives on, does count. In my view Darwin did China both good and ill. He both liberated China and helped partially to re-enslave it. Darwin liberated China from many superstitions of the past, but he helped ensnare China in new superstitions of the present. Darwin's greatest service was in revealing certain "facts of life" which have to be taken into account in any sane philosophy. The trouble, however, from the beginning, has been that people have been too sure that they have known what to make of those facts of life, and so after them has arisen a host of camp-follower non-facts of life (some of them, it must be admitted, the children of Darwin himself) which have led people to jump to all sorts of unfounded (and confounded) conclusions. A true discovery has led to faulty theories and then to myths—new myths, but not necessarily any better than many of the old.

This has been a worldwide phenomenon. It is nothing to blame on the Chinese. But the Chinese have not escaped it, and some of the least fortunate "Darwinian" myths live on, it seems to me, in China today, the most important of which are the myths of the omnipotence of science, historical inevitability, "futurism," "perfectionism," "prescientism" and "it's-them-or-us-ism."

At the beginning of the twentieth century, China desperately needed science, and the scientific and technological revolution that Darwin, oddly enough, helped start in China has been wildly good for China. But, without wishing here to reopen the 1923 Chinese debate on science and metaphysics, the belief in the omnipotence of science, or even the belief that matter can know all that matters, seems patently absurd. The weakest idea is that any "social science" can make final sense out of human life, and it was Darwinism, not Marxism, that first convinced Chinese that a social science could. I have tried to point out all along how insufficient and illogical Darwinian slogans are as final answers to anything, however disturbingly useful Darwinian analogies may be (and I have, of course, used them myself, even in this conclusion), but the naturalism of Darwin and Marx together has so impressed the Chinese that few are willing to entertain even the suspicion that there is nothing in nature enough like us to explain us, that we are in fact beyond analogy.

Historical inevitability is another strange concept that owes much, in China, to Darwin. China encountered in Darwinism two seemingly contradictory theories of nature and history—of "natural history"—at the same time: the idea that we are products of nature and that our behavior and history are therefore naturally controlled; and the idea that through science we can conquer nature, even our own nature, and thus "control our destiny" (ridiculous phrase), that the created can best the creator (as absurd in natural terms as in theological). Darwin helped Chinese accept both at once. Marxists have stopped Chinese from wondering how both could be.

Kidd-like "futurism," introduced by Liang Ch'i-ch'ao, has taken China from *po chin hou ku* (denigrate the present and glorify the past) to *hou chin po ku* (glorify the present and denigrate the past) to a futurized version of *hu chin t'u yuan* (neglect the near and plan for the distant),[62] which in practice can be cruel. In practice, of course, it is never absolute, but the theory that the present forever exists for and must be sacrificed to the future reaches absurdity well before it reaches infinity.

"Perfectionism," the idea that a perfect human being or even society is in the making, is another "Darwinian" idea with no Darwinian evidence. To be sure, the neo-orthodoxy is unsure here. The latter-day Mao Tse-tung, seemed to think that "continuous revolution" was necessary just to fight some sort of original bourgeois sin, to keep each new generation from degeneration. But still the faith seems to be that someday all people will evolve into spiritual proletarians and that, when they do, the human species will be a new and infinitely better species. In the meantime, however, devout belief in perfectionism can have painful consequences, for better or for worse. If one believes in perfectionism, then the washing of men's minds becomes an act of compassion. If one does not, it looks like taking the mote from others' eyes while overlooking the beam in one's own. In 1906, Liang Ch'i-ch'ao—first even in this—almost literally called for "brain-washing." He called for *hsin-li hsi-ti* (mind-washing). Sun Yat-sen later called for *hsin-li chien-she* (psychological reconstruction).[63] Mao Tse-tung called for *ssu-hsiang kai-tsao* (thought reform). All thought to help people and help evolution along. Mao Tse-tung said that "intellectual struggle is not like other kinds of struggle: One cannot use violent methods of force, one can only use subtle methods of reason."[64] But, when one is dead set in one's convictions, and on evolution's side, and when one gets impatient, subtlety can veer towards violence.

And violence brings us to the last two myths, "prescientism" and "it's-them-or-us-ism." The notion that one can be prescient of evolution's Way has led some to feel that the prescient have special rights, if not duties, in the struggle they believe that Way requires. And so Darwin has ironically helped produce a new kind of religious know-it-all-ism, and a concomitant new kind of religious self-righteousness and religious intolerance. Were it not for Hu Shih's Darwinism, we might say that all Social Darwinism leads either to ruthless intolerance or a tolerance of ruthlessness. And Hu Shih's tolerance was not tolerated.

Mao Tse-tung in an angry moment (as late as 1964) swore that "all demons shall be annihilated."[65] He dehumanized his enemies, partly in traditional hyperbole, partly in Social Darwinian "realism." Like the Anarchists, he saw reactionaries as evolutionary throwbacks, who deserved extinction. The people's enemies were non-people, and they did not deserve to be treated as people. Even Lu Hsun had warned that one should have no qualms about "beating dogs that have fallen in the water."[66] And yet surely Mao Tse-tung usually longed to be an exorcist, not a witch-burner. It was only when he felt challenged that he fell back on it's-them-or-us-ism. For there was a contradiction in his thought. Vestigial Confucianism, a belief in the power of virtuous example and in the normal person's capability for study and self-cultivation, led him to instruct his cadres to "cure the illness and save the patient."[67] Darwinism, however, led him sometimes to accept the words if not the Way of a Taoist Sage and "not be humane."[68] Of course, in a real it's-them-or-us situation, that was not strange. And Mao Tse-tung had been in many. The only trouble was that he saw more than there were. His own experience, too much of modern Chinese history, *and* the ideas of Marx and Darwin all taught Mao Tse-tung that the world always had been, was, and would long, if not forever, be an it's-them-or-us world, and that it could not be otherwise.

The Thought of Mao Tse-tung was and remains a powerful mixture of Darwinian ironies and contradictions. It believes in a jungle world and yet rejects the law of the jungle, at least as an excuse for imperialism, even as it admits that imperialists must naturally do their worst. It humanely rejects Darwinian dog-eat-dog capitalism, but inhumanely accepts Darwinian dog-eat-dog Marxism. Moral outrage is its greatest strength, but its morality is not broad enough to extend to all men. Science is its claim to legitimacy, but pseudo-science is its greatest weakness. And Darwinian myths are at the bottom of its pseudo-science. There is much that is undeniably heroic, noble, and good in the Thought of Mao Tse-tung, and much that is not. But, good or bad, it is based on false claims, Darwinian and yet un-Darwinian, scientific and yet unscientific, humane and yet inhumane. When all is said and done, it seems to me that Huxley deserves the last word, translated back again from Yen Fu, "The fittest is not necessarily the best."[69]

We could stop here, were it not for one last nagging question: Even if all or any of the above did happen to China, was it really Darwin's doing? Darwin certainly did not accomplish the above single-handed. Did he really have a hand in it at all? If, as we have pointed out all along, much of what was called Darwinism was not really Darwinian, then what was it? Was Darwin accepted, or used and abused? And if he was abused beyond recognition, did *he* do anything?

We have seen that Darwin was Sinicized in all directions. We have seen Taoist Darwinism and Confucian Darwinism and Legalist Darwinism and even Buddhist Darwinism. And the Darwinism of many Chinese was actually a synthesis of all those Darwinisms. One can see something of all of them in the thought both of Liang Ch'i-ch'ao and of Mao Tse-tung. Taoist Darwinism supported the idea of natural law, constant change, and a Way of history. Confucian Darwinism made that Way moral and preserved a place for human action, self-cultivation, and human perfection. Legalist Darwinism supported the dictatorship of the fittest, so that human action might be kept in accord with the Way, and Buddhist Darwinism made clear the dreadful accountability in evolution of man's every action.

Chinese first saw Darwin through Taoist, Confucian, Legalist, and Buddhist eyes, and thus they first made sense of it. But, as they did so, they seemed to preserve all of the old choices of traditional Chinese philosophy. Therefore, even if there was a new synthesis, was it not really only a synthesis of Chinese thought, fitting beautifully, by coincidence, with a Marxian synthesis of Western thought?

Or is there not even a more devastating possibility? Might not all of Chinese Darwinism have been rationalization? Did not Chinese force Darwin to say what they wanted to hear and what they would have found some way to say by themselves if they had not had Darwin? Did Darwin really change men's minds, or was he changed to fit men's minds?

There *was* rationalization, and rationalization is probably at least always half irrational. And there was Sinicization and that too was probably half unconscious. Surely there was a psychological factor as well as an intellectual factor at work in Chinese Social Darwinism, whatever the difference. But which was more important? When I began this study I vainly, doubly now I realize, hoped to cast light on the nature of the influence of ideas. But we do not know enough to know how rational we are or are not. What really had more force in shaping modern Chinese "frames of mind," traditional predispositions, new visions, or subconscious desires? On that level, God only knows what Darwin did to China.

And yet Darwin was there. His name was on men's lips. His ideas in reasonably true form were somewhere in men's heads. His theory was something I am convinced all the Chinese we have looked at tried honestly to take into account. They felt they had to. They did their best. People's heads are funny things. People's philosophies are wild mixtures of reasonings, feelings, and unrecognized longings. But Charles Darwin honestly entered those mixtures in Chinese heads and made them different. So his influence was real. Chinese of course confused Darwin's ideas and were confused by them, and of course they got confused in Chinese directions, but small wonder. Every people has gotten confused. For the fact of the matter is, when all is said and done, that *no one* knows what to make of evolution.

Notes

Bibliography

Glossary

Abbreviations Used in Notes

CIP	*Ch'ing i pao*
HCN	*Hsin ch'ing-nien*
HHHP	*Hsiang hsueh hsin pao*
HMTP	*Hsin min ts'ung pao*
HP	*Hsiang pao lei tsuan*
HSC	*Hsin shih-chi*
KFCC	*Kuo fu ch'üan chi*
KFCS	*Kuo fu ch'üan shu*
MP	*Min pao*
SLHC	*Hsin-hai ko-ming ch'ien shih nien chien shih-lun hsuan chi*
SWP	*Shih-wu-pao*
TYL	Yen Fu, *T'ien-yen lun*
YF	*Yen Fu shih wen hsuan*
YPS	Liang Ch'i-ch'ao, *Yin ping shish wen chi*

Notes

PROLOGUE

1. John K. Fairbank, *Trade and Diplomacy on the China Coast* (Cambridge, Mass., 1953), p. 173.
2. See Liang Ch'i-ch'ao, *Tzu-yu shu* (Taipei, 1960), p. 23, originally published in the October 15, 1899 issue of *Ch'ing i pao. Yu sheng lieh pai* was *not* Yen Fu's translation. Liang Ch'i-ch'ao brought it into Chinese from Japan.
3. Fairbank, p. 173.
4. This information comes from the excellent paper by Wang Tzu-ch'un and Chang Ping-lun, "Ta-erh-wen hsueh-shuo tsai Chung-kuo te ch'uan-po he ying-hsiang," delivered in Peking on May 19, 1982, at the opening session of a conference on Darwin commemorating the 100th anniversary of his death. Messrs. Wang and Chang shrewdly guess that these early references probably did not strike fire precisely because they failed to mention struggle. See pp. 1–4.
5. I have not gone through the endless volumes of the *Wan kuo kung pao,* but my father-in-law did, and this is his report. See acknowledgments. For the Advisor's comments see *Yang wu yun-tung,* ed. Chung-kuo shih-hsueh hui (Shanghai, 1961), VIII, 397.
6. See Lu Hsun, *Erh hsin chi* (Hong Kong, 1964), p. 47.
7. See below, pp. 317–318.
8. Teng Ssu-yü and John K. Fairbank, *China's Response to the West* (Cambridge, Mass., 1954), p. 33.
9. Ibid., p. 55.
10. *HP,* p. 208.
11. Teng and Fairbank, p. 53 (slightly altered).

1. PROGRESS

1. Anne Robert Jacques Turgot, *Oeuvres de M. Turgot* (Paris, 1808), II, 19–20.
2. Ibid., p. 212.
3. Ibid., p. 215.
4. *The Philosophy of Kant,* ed. Carl J. Friedrich (New York, 1949), p. xliii.
5. Ibid., p. 123.
6. Ibid., p. 119.
7. Frank E. Manuel, *The Prophets of Paris* (New York, 1965), p. 62.
8. Ibid., p. 92.
9. Henri de Saint-Simon, *Social Organization, The Science of Man, and Other Writings,* ed. and tr. Felix Markham (New York, 1964), p. 4.
10. Ibid., p. 116.
11. Ibid., p. 68.
12. Elie Halévy, *The Growth of Philosophical Radicalism* (Boston, 1966), p. 471.
13. John Stuart Mill, *On Liberty* (New York, 1947), p. 86.
14. Ibid., p. 88.
15. Ibid., p. 72.
16. Alexis de Tocqueville, *Democracy in America,* ed. Philips Bradley (New York, 1945), I, 409, 414.
17. Gertrude Himmelfarb, *Victorian Minds* (New York, 1970), p. 324.
18. *Philosophy of Kant,* p. 27.
19. Ibid., p. 130.
20. de Tocqueville, I, 6.
21. Loren Eiseley, *Darwin's Century* (New York, 1958), p. 95.
22. Ibid., p. 97.
23. *Philosophy of Kant,* p. 118.
24. Eiseley, *Darwin,* p. 193.
25. Charles Darwin, *On the Origin of Species* (Cambridge, Mass., 1966), p. 489.
26. The later prevailing word for "progress" (and as often as not for "evolution" itself), *chin-hua,* does appear in K'ang Yu-wei's famous *Ta t'ung shu*, a draft of which he said he finished in 1885, but Liang Ch'i-ch'ao said the book was not put in final form until 1901–1902 (see *Nan-hai hsien-sheng shih chi* [Hong Kong, 1966], p. 16), by which time it was no coincidence, I am sure, that all of K'ang Yu-wei's evolutionary vocabulary was the same as that invented by Yen Fu or the Japanese. Also, see below, pp. 89–91.
27. Wang Ch'ü-ch'ang, *Yen Chi-tao nien-p'u* (Shanghai, 1936), pp. 7–8. Also see Benjamin Schwartz, *In Search of Wealth and Power: Yen Fu and the West* (Cambridge, Mass., 1964), p. 29.

28. K'ang Yu-wei, *K'ang Nan-hai tzu pien nien-p'u,* in Chin tai Chung-kuo shish liao ts'ung k'an, Ser. 2, No. 11 (Taipei, 1966), p. 7.
29. *YPS,* III, 6:60.
30. K'ang Yu-wei, *Nien-p'u,* p. 10.
31. Ibid.
32. *YPS,* III, 6:61.
33. K'ang Yu-wei, *Nien-p'u,* p. 11.
34. Ibid., p. 15.
35. Ibid., p. 16.
36. Manuel, p. 267.
37. *YPS,* III, 6:69.
38. K'ang Yu-wei, *Nien-p'u,* p. 14.
39. Ibid., p. 15. Actually he said he put together "Jen lei kung li" in 1887.
40. *Pu jen tsa-chih hui-pien,* ed. K'ang Yu-wei (Shanghai, 1914; Taipei, reprint, 1968), I, 368.
41. Ibid., I. 376.
42. Ibid., I, 327. See also *Meng Tzu i chu* (Hong Kong, n.d.) 3.6, p. 79.
43. Ibid., II, 824.
44. Professor Yü Ying-shih alerted me to K'ang Yu-wei's considerable unacknowledged borrowings from Liao P'ing (1852–1932). See Ch'ien Mu, *Chung-kuo san pai nien hsueh-shu shih* (Shanghai, 1937), Chapter 14, and T'ang Chih-chün, *Wu-hsu pien-fa shih lun ye* (Hong Kong, 1973), pp. 106–153. Professor Hans Bielenstein alerted me to the Bellamy case. See Richard Campbell Howard's "The Early Life and Thought of K'ang Yu-wei, 1858–1895" (PhD dissertation, Columbia University, 1972), pp. 307–308. One of the earliest references to Saint-Simon and Comte I have found is in Liang Ch'i-ch'ao's biography of K'ang Yu-wei begun in the December 21, 1901 issue of *Ch'ing i pao,* in which Liang Ch'i-ch'ao expressly claimed that K'ang Yu-wei had not known of Saint-Simon or Comte when he thought up his own utopia. See *YPS,* III, 6:73.
45. *Pu jen tsa-chih,* I, 327.
46. *YPS,* III, 6:61–62.
47. Adrian Arthur Bennett, *John Fryer, The Introduction of Western Science and Technology into Nineteenth-Century China* (Cambridge, Mass., 1967), pp. 43–44.
48. *Tso chih ch'u yen,* tr. Fu-lan-ya (John Fryer) and Ying Tsu-hsi, in *Hsi hsueh fu-ch'iang ts'ung shu,* ts'e 25 (n.p., 1896), p. 1.
49. Ibid., p. 30.
50. Ibid., pp. 2b–3.
51. Ibid., p. 3b.
52. Ibid., p. 4b.
53. Ibid., p. 5.

54. Ibid., p. 8b.
55. Ibid., p. 22.
56. Ibid., pp. 23b–24.
57. Ibid., p. 23b.
58. Ibid., p. 18b.
59. Ibid., pp. 6b–7.
60. Ibid., p. 38.
61. Bennett, p. 107.
62. *YPS,* III, 6:87.
63. K'ang Yu-wei, *Ch'un ch'iu Tung shih hsueh* (Taipei, 1969), 2:4.
64. *Ch'un ch'iu Kung-yang chuan,* in Ssu pu pei yao, "Ching pu" (Shanghai, 1927), p. 1:7b–8.
65. K'ang Yu-wei, *Ch'un ch'iu,* 4:1b–2.
66. *Pu jen tsa-chih,* II, 721–722.
67. A free translation. *Wei* should, of course, be read in the second tone, meaning "to be."
68. *Pu jen tsa-chih,* II, 723–724.
69. Laurence G. Thompson, tr., *Ta T'ung chu: The One-World Philosophy of K'ang Yu-wei* (London, 1958), p. 49.
70. *Meng Tzu,* 6.9, p. 154.
71. *Lun-yü,* in *Ssu shu hsin chie* (Tainan, 1961) 14.41, p. 231.
72. *The Philosophy of Hegel,* ed. Carl J. Friedrick (New York, 1954), p. 21.
73. Ibid., p. 16.
74. Ibid., p. xxxv.
75. K'ang Yu-wei, *Ch'un ch'iu,* 5:2b.
76. Ibid., 6A:10.
77. Ibid., 6B:46.
78. *Lun-yü,* 2.4, p. 70.
79. *Pu jen tsa-chih,* II, 731.
80. Liu P'an-sui, *Lun heng chi chie* (Shanghai, 1957), p. 384.
81. Ibid., p. 382.
82. Ibid., p. 383.
83. Ibid., p. 391.
84. K'ang Yu-wei, *Hsin hsueh wei ching k'ao* (Taipei, 1962), p. 2.
85. Teng and Fairbank, p. 138.
86. Robert Mackenzie, *The 19th Century: A History* (London, 1889), pp. 459–460.
87. *T'ai Hsi hsin shih lan-yao,* tr. Li T'i-mo-tai (Timothy Richard), (Shanghai, 1895), 23:7.

2. ENTER DARWIN

1. *Wu-hsu pien-fa,* ed. Chien Po-tsan et al. (Shanghai, 1953), II, 132.
2. Yen Fu, "Lun shih pien chih chi," in *YF,* p. 6.
3. Ibid., p. 3.
4. Ibid., p. 5.
5. Ibid., p. 3. See also Schwartz, p. 44.
6. *Meng Tzu,* 11.10, p. 265.
7. See my article, "K'ang Yu-wei and *Pao-chiao:* Confucian Reform and Reformation," *Papers on China,* 20:144–176.
8. Yen Fu, "Lun shih pien chih chi," p. 3. For Yen Fu's reliance on the *I ching (Book of Changes),* see Fung Yu-lan, *A History of Chinese Philosophy* (Princeton, 1953), 2, 460–461.
9. Yen Fu, "P'i Han" (In righteous refutation of Han Yü) in *YF,* p. 85.
10. Yen Fu, "Lun shih pien chih chi," p. 4.
11. Ibid., p. 3. See also Schwartz, p. 54.
12. Friedrich Nietzsche, *The Birth of Tragedy and the Geneology of Morals,* tr. Francis Golffing (New York, 1956), p. 155.
13. Chow Tse-tsung, *The May Fourth Movement* (Cambridge, Mass., 1960), p. 300.
14. Yen Fu, "Lun shih pien chih chi," p. 5.
15. Ibid.,
16. Ko Kung-chen, *Chung-kuo pao hsueh shih* (Taipei, 1964), p. 158.
17. Ibid.
18. Ibid.,
19. Ibid.
20. *Ko-chih* (the extension of knowledge through the investigation of things), taken originally from the first chapter of *Ta hsueh* (The great learning), in *Ssu shu hsin chie,* p. 2.
21. Yen Fu, "Yuan ch'iang," in *Yen Chi-tao shih wen ch'ao* (Taipei, 1969), p. 24.
22. Ibid., pp. 24–25.
23. *Lao Tzu tao te ching,* commentary and punctuation by Yen Fu, notes by Wang Pi (Shanghai, 1935), Chapter 42, 2:5.
24. *Li i erh fen shu.* See *Chung-kuo che hsueh shih tzu-liao hsuan-chi, Sung, Yuan, Ming chih pu* (Peking, 1962), pp. 209–210.
25. Darwin, *Origin,* p. 484.
26. Ibid., p. 488.
27. Eiseley, *Darwin,* p. 175.
28. *Hsun Tzu chien shih,* ed. Liang Ch'i-hsiung (Taipei, 1969). See Chapter 5, p. 51; Chapter 9, p. 108.
29. Richard Hofstadter, *Social Darwinism in American Thought* (Boston, 1955), p. 25.

30. Yen Fu, "Yuan ch'iang," p. 15.
31. See, for example, *HHHP,* IV, 2810.
32. Yen Fu, "Yuan ch'iang," p. 25.
33. Ibid.,
34. *Hsun Tzu,* Chapter 10, p. 119.
35. *Lao Tzu,* Chapter 68, 2:19a–b.
36. Ibid., Chapter 76, 2:22b.
37. Yen Fu, "Yuan ch'iang," in *YF,* p. 23.
38. Ibid.
39. Hofstadter, p. 21.
40. Schwartz, p. 43.
41. Yen Fu, "Yuan ch'iang," *YF,* p. 15.
42. *Hsun Tzu,* Chapter 9, p. 108.
43. Ibid., p. 109.
44. Hofstadter, p. 35.
45. Ibid., p. 49.
46. Yen Fu, "Yuan ch'iang," *YF,* p. 15.
47. Ibid., p. 17.
48. Ibid., p. 16.
49. Ibid.
50. Ibid., p. 17.
51. Ibid.
52. Schwartz, p. 59.
53. Ko Kung-chen, p. 158.
54. See, for example, *Meng Tzu,* 2.11, p. 45; *Hsun Tzu,* Chapter 11, p. 148; and *Lun-yü,* 2.1, p. 69. Much later, Mao Tse-tung would also insist that "a small state can defeat a big state; a weak state can defeat a strong state," but for him, although the ultimate might was still right, the right would have to fight.
55. *Meng Tzu,* 11.10, p. 265.
56. Yen Fu, "Yuan ch'iang," *YF,* p. 17.
57. Ibid.
58. Hofstadter, p. 85.
59. Yen Fu, "Yuan ch'iang," *YF,* p. 20.
60. Ibid., p. 22.
61. Ibid., pp. 20–21.
62. Ibid., p. 21.
63. Ibid., p. 20.
64. Yen Fu, "P'i Han," p. 88.
65. Yen Fu, "Yuan ch'iang," *YF,* p. 32.
66. Schwartz, pp. 70, 72.
67. Yen Fu, "Yuan ch'iang," *YF,* p. 21.

68. Halévy, p. 15.
69. Yen Fu, "P'i Han," p. 86. See also *Meng Tzu* 14.14, p. 328, and *Chuang Tzu tsuan chien,* ed. Ch'ien Mu (Hong Kong, 1962), Book 10, p. 76—Yen Fu attributed the saying to Lao Tzu. To sound more proverbial, I have taken the liberty of saying "enthroned" instead of something more literal like "bemarquissed." But, on second thought, it would probably be better to say, "He who steals a fief becomes ennobled."
70. Ibid.
71. Ibid.
72. Ibid., p. 87.
73. Ibid.
74. Yen Fu, "Yuan ch'iang," *YF,* p. 25.
75. Ibid., p. 26.
76. Ibid.
77. Yen Fu, "P'i Han," p. 88.
78. Yen Fu, "Yuan ch'iang," *YF,* p. 23.
79. See below, pp. 413–419.
80. Yen Fu, "Yuan ch'iang," *YF,* pp. 23–24.
81. Ibid., p. 24. For the "paper tiger" see *Mao Chu-hsi yü-lu* (Peking, 1967), p. 66.
82. Walt Kelly, *Impollutable Pogo* (New York, 1970), pp. 31, 39.
83. See above, p. 14.
84. Hofstadter, p. 45.
85. Asa Gray, *Darwiniana,* pp. 126, 70–71.
86. Schwartz, p. 198.
87. *Meng Tzu,* Chapter 7.7, p. 168.
88. Ibid., Chapter 3.2, p. 62.
89. Hofstadter, pp. 47–48.
90. Schwartz, pp. 73–80.
91. *Mao Chu-hsi yü-lu,* p. 3. The quotation is from Lin Piao's "Preface to the Second Edition" (*Tsai pan ch'ien-yen*).
92. Halévy, p. 500.

3. DARWIN AND THE CHINESE PROGRESS

1. An epithet bestowed in conversation by John K. Fairbank.
2. Ko Kung-chen, p. 162.
3. Yeh Te-hui, ed., *I chiao ts'ung pien* (Taipei, n.d.), p. 463.
4. Ting Wen-chiang, *Liang Jen-kung hsien-sheng nien-p'u ch'ang pien ch'u kao* (Taipei, 1962), I, 34.
5. Ibid.
6. Yeh Te-hui, p. 464.

7. Ibid., p. 463.
8. Ting Wen-chiang, p. 34.
9. *YPS,* I, 1:107.
10. Ibid.
11. Ibid., p. 110. The dating of *T'ien-yen lun* is still problematical. The standard "first edition" of 1898 has a preface by Wu Ju-lun dated May-June 1898, but Yen Fu's own preface in that edition is dated October 15, 1896. According to Chin Ch'ung-chi, *T'ien-yen lun* was published in installments in 1896 in the *Kuo wen hui pien* (a Tientsin magazine, published every ten days). See Chin Ch'ung-chi, "*T'ien-yen lun* ho Chung-kuo chin-tai fan K'ung ssu-ch'ao," *Ta kung pao,* December 24, 1973. But Wang Shih in his *Yen Fu chuan* (A biography of Yen Fu) claimed to have seen a "reprint" of *T'ien-yen lun* in a Shensi bookstore in 1895! See *YF,* p. 95.
12. K'ang Yu-wei, *Nien-p'u,* p. 22.
13. *Chuang Tzu,* Book 18–6, p. 143.
14. K'ang Yu-wei, *Nien-p'u,* p. 23.
15. See K'ang Yu-wei, *K'ung Tzu kai chih k'ao* (Peking, 1922).
16. See K'ang Yu-wei, *Ch'un ch'iu Tung shih hsueh.* For later works in which he did use Yen Fu's vocabulary, see *Chung-yung chu* (Taipei, 1968; original preface dated spring 1901), p. 15b, and *Meng Tzu wei* (Taipei, 1968; original preface dated December 22, 1901), 3:3.
17. K'ang Yu-wei, *Nien-p'u,* p. 23.
18. K'ang Yu-wei eventually used Yen Fu–Japanese Darwinian slogans to support his Three Age-*Ta t'ung* theories and his racial (and racist) views. ("In terms of the evolutionary principle of the survival of the fittest, the three Miao races were eliminated from our China's south by our divine race of descendants from the Yellow Emperor." See *Ta t'ung shu,* p. 117 and, for the worst of K'ang Yu-wei's racism, pp. 117–125, or Thompson, pp. 140–148.) But he also attacked those slogans and Darwin himself (and he was one of the first to do so) from a Confucian humanitarian point of view. Darwin, one of those "madly mistaken people who only half understand what they are talking about . . . created the theory of evolution . . . and led people to take competition as the greatest virtue . . . . His crime is worse than a flood." See *Ta t'ung shu*, p. 285 and, for more Darwinian and anti-Darwinian utterances, ibid., p. 237, *Men Tzu wei,* 3:3; *Chung-yung chu,* pp. 15, 39, 41–42; and Thompson, pp. 214–215, 258.
19. *YPS,* I, 1:110.
20. *SWP,* I, 7.
21. D. W. Y. Kwok's term. See his *Scientism in Chinese thought 1900–1950* (New Haven, 1965).

22. *YPS,* I, 1:1.
23. *SWP,* I, 7.
24. Ibid., p. 75.
25. Ibid., p. 271.
26. Ibid., p. 77.
27. Ibid., p. 70.
28. Ibid., p. 77.
29. Ibid.
30. *Lun-yü,* 15.18, p. 241.
31. *SWP,* I, 71. Quoted from *Meng Tzu,* 7.8, p. 170.
32. *SWP,* I, 138.
33. My italics. See the frontispiece to William Hinton, *Iron Oxen* (New York, 1970). For the original, see *Mao Chu-hsi yü lu,* p. 169.
34. *SWP,* II, 895.
35. Ibid., I, 246.
36. Ibid., II, 1039 (my italics).
37. Ibid., I, 275.
38. Ibid., I, 136.
39. Ibid., I, 137.
40. Ibid., II, 1033.
41. Ibid., I, 137.
42. Ibid., I, 621.
43. Ibid., III, 1250–1251.
44. Ibid., II, 1082–1083.
45. Ibid., I, 238.
46. Ibid., I, 361–362.
47. Ibid., I, 650. "Wolseley" is a last-minute guess at the identity of "Hua-erh-ssu-lei." Lieutenant-Colonel G. J. Wolseley fought against the Chinese in the Arrow War and wrote *Narrative of the War with China 1860.* See references to Wolseley in Jack Beeching, *The Chinese Opium Wars* (New York, 1975).
48. Ibid., II, 918–919.
49. Ibid., I, 652.
50. Ibid., II, 1242.
51. Ibid., IV, 2051. See also Wang Ch'ü-ch'ang, p. 30.
52. Ibid., III, 1527.
53. Ibid., III, 1528.
54. T. H. Huxley and Julian Huxley, *Evolution and Ethics 1893–1943* (London, 1947), p. 46.
55. Ibid., p. 82.
56. *SWP,* III, 1528.
57. Ibid.

58. Ibid.
59. Ibid.
60. Ibid.
61. Charles Darwin, *The Descent of Man* and *Selection in Relation to Sex* (New York and London, first published 1874, second edition, n.d.), "preface to second edition."
62. Ibid., p. 37.
63. Eiseley, *Darwin*, p. 216.
64. *YPS*, I, 1:109.
65. *SWP*, III, 1528.
66. Gerald Durrell, *Two in the Bush* (New York, 1971), pp. 77–78.
67. Lu Hsun, *Je feng* (Hong Kong, 1964), p. 15.
68. *YPS*, I, 1:109.
69. *SWP*, III, 1727.
70. Darwin, *Origin*, p. 75.
71. *TYL*, Part 1, p. 30.
72. Darwin, *Descent*, p. 144.
73. Ibid., p. 145.
74. Ibid., p. 152.
75. Huxley, *Evolution and Ethics*, p. 64.
76. Ibid., p. 39.
77. Darwin, *Descent*, p. 146.
78. Ibid., p. 147.
79. *TYL*, Part 1, p. 32.
80. *SWP*, III, 1727.
81. Ibid., III, 1728.
82. See James J. Y. Liu, *The Chinese Knight-errant* (Chicago, 1967).
83. *SWP*, IV, 2137.
84. Ibid., IV, 2138.
85. Ibid., IV, 2143.
86. Ibid.
87. Ibid.
88. Ibid., III, 1634.
89. Ibid., III, 1635.
90. Ibid., II, 1159.
91. Ibid., V, 2836.
92. Ting Wen-chiang, p. 40.
93. *SWP*, IV, 1936.
94. Ibid., IV, 2529. The phrase quoted in *Shih-wu pao* was *yu-che sheng-ts'un, lieh-che mieh-wang.*
95. Ibid., III, 1851.

96. Ibid., IV, 2069.
97. Ibid., IV, 2070.
98. Ibid., IV, 2071.
99. Ibid., IV, 2074–2075.
100. See, for example, *SWP,* III, 1730; ibid., VI, 3177.
101. See *SWP,* III, 1730 and ibid., VI, 3177.
102. *SWP,* VI, 3177.
103. Ibid., IV, 2071.
104. Ibid., III, 1731.
105. Ibid., III, 1733.
106. Ibid., III, 1732.
107. Ibid., III, 1731.
108. Ibid., III, 1733.
109. Ibid., III, 1730.
110. Ibid., V, 2765.
111. Ibid., V, 2768.1.
112. Ibid., V, 2768.2.
113. Ibid., V, 2768.1.
114. Ibid., V, 2768.2.
115. Ibid.
116. Ibid., V, 2769.
117. Ibid., V, 2765.
118. Ibid., V, 2768.
119. Ibid., IV, 2075.
120. Ibid., V, 2769.

4. DARWIN IN HUNAN

1. Hao Chang, *Liang Ch'i-ch'ao and Intellectual Transition in China, 1890–1907* (Cambridge, Mass., 1971), p. 124.
2. Ting Wen-chiang, p. 43.
3. *HP,* II, 455.
4. Ibid., II, 457–458.
5. *Mao Chu-hsi yü-lu.* p. 33.
6. "Ko kuo chung-lei k'ao," by "P'ing-p'i-tzu" in *HHHP,* III, 2115–2166, and I, 395–439.1, and "Wu chou feng-su i-t'ung kao," by "T'an-t'an-tzu," in *HHHP,* I, 477–523. See below, n. 8.
7. *HHHP,* III, 2009.
8. "P'ing-p'i-tzu" was T'ang Ts'ai-ch'ang, who was the main writer for the history section of the *Hsiang hsueh hsin pao.* "T'an-t'an-tzu" may also have been T'ang Ts'ai-ch'ang.

9. *HHHP,* III, 2115.
10. Ibid., III, 2115. "Moreover struggle does not have to entail murderous attacks with claw and tooth."
11. Ibid., III, 2116.
12. Ibid., I, 433.
13. Ibid., I, 434.
14. Ibid., I, 436.
15. Ibid., III, 2146–2147.
16. Ibid., I, 433. See also ibid., I, 506.
17. Ibid., III, 2120.
18. Ibid., I, 438.
19. Ibid., III, 2156.
20. Ibid., III, 2148.
21. Ibid., III, 2117.
22. Ibid., III, 2118.
23. Ibid., III, 2147.
24. Ibid., III, 2148.
25. Ibid., III, 2147–2148.
26. Ibid., III, 2154.
27. Ibid., I, 426.
28. And he only came a little closer to dealing with the problem in *The Descent of Man.* See *Descent,* p. 150.
29. *HHHP,* III, 2115.
30. *HP,* I, 160.
31. Huxley, *Evolution and Ethics.*
32. Bennet, pp. 100, 122.
33. *HP,* II, 445.
34. *HHHP,* IV, 2762.
35. *HP,* II, 446.
36. *HP,* II, 640–641.
37. *HHHP,* IV, 2585.
38. *HP,* I, 31.
39. *HHHP,* I, 478. See above, n. 8.
40. Ibid., IV, 2592.
41. *HP,* I, 31.
42. *HHHP,* I, 476.
43. *HP,* I, 19–20.
44. *HHHP,* IV, 2681.
45. Ibid., IV, 2518.
46. Ibid., II, 874.
47. *HP,* II, 650.
48. *HHHP,* II, 1264.

49. Ibid., II, 1294.
50. Ibid., I, 468.
51. *HP,* II, 450–451.
52. Ibid., II, 451.
53. *HHHP,* IV, 2699.
54. *HP,* I, 123.
55. Ibid.
56. *HP,* I, 125.
57. Ibid., I, 273.
58. Ibid., I, 279.
59. Ibid., I, 189.
60. Ibid., II, 709.
61. Ibid., II, 716–717.
62. Pusey, "K'ang Yu-wei," p. 161.
63. *HP,* I, 168.
64. Ibid., I, 181.
65. See ibid., I, 148, 340, 352.
66. Ibid., I, 181.
67. Pusey, "K'ang Yu-wei," pp. 146–147.
68. *HP,* I, 19, 22.
69. Ibid., I, 19.
70. Ibid., I, 21.
71. Ibid., I, 23.
72. Ibid., I, 23–24.
73. Yeh Te-hui, p. 419.
74. K'ang Yu-wei, *Ta t'ung shu,* pp. 114–115. The "cinnabar complexions" come from a poem in the *Shih ching.* See *Shih ching,* in James Legge, tr., *The Chinese Classics* (Taipei, 1963), IV, 197.
75. Yeh Te-hui, p. 442.
76. Ibid., p. 441.
77. Ibid., p. 235.
78. Ibid.
79. *YPS,* II, 3:19.
80. See *Meng Tzu,* 3.6, pp. 79–80, and 11.6, pp. 258–259.
81. *Lun-yü,* 7.29, p. 138.
82. Yeh Te-hui, p. 235.
83. Ibid., pp. 244, 248.
84. *Meng Tzu,* 8.19, p. 191.
85. Yeh Te-hui, p. 332.
86. Ibid., p. 245.
87. Ibid., pp. 360–361. Centaurs were living proof that the Greeks had horsed around.

88. Ibid., p. 361.
89. Ibid., p. 373.
90. Ibid., pp. 164–165.
91. Ibid., p. 166.
92. Ibid., p. 171.
93. Chang Chih-tung, in *HHHP,* III, 1757.

5. A GOOD BOOK FOR BAD TIMES

1. Hu Shih, *Ssu-shih tzu shu* (Taipei, 1959), pp. 49–50.
2. Ts'ao Chü-jen, *Lu Hsun p'ing chuan* (Hong Kong, 1961), p. 25.
3. *TYL.* See abbreviation list preceding notes.
4. Huxley, *Evolution and Ethics,* pp. 78, 82.
5. Schwartz, p. 103.
6. Ibid., pp. 100–101.
7. Ibid., p. 101.
8. Ibid., p. 111.
9. Said Lu Hsiang-shan (1139–1193): "If in our study we know the fundamentals, then all the Six Classics are my footnotes." See Wing-tsit Chan, *A Source Book in Chinese Philosophy* (Princeton, 1963), p. 580.
10. Huxley, *Evolution and Ethics,* p. 64.
11. Ibid., p. 47.
12. Ibid., p. 51.
13. Ibid.
14. *TYL,* Part 1, p. 32.
15. Huxley, *Evolution and Ethics,* p. 54.
16. *TYL,* Part 1, p. 33.
17. Ibid., p. 32.
18. Again see *Hsun Tzu,* Chapter 9, pp. 108–109.
19. Huxley, *Evolution and Ethics,* pp. 49, 59.
20. *Lun-yü,* 12.2, p. 187.
21. Huxley, *Evolution and Ethics,* p. 52.
22. Ibid., p. 36. "As a natural process . . . ," said Huxley, "evolution excludes creation and all other kinds of supernatural intervention." In *T'ien-yen lun,* Yen Fu apparently could not quite wait for Huxley to announce this "good news," and so announced it first himself, in his first commentary. See *TYL,* Part 1, pp. 3–4.
23. Huxley, *Evolution and Ethics,* p. 74.
24. Ibid., p. 79.
25. Ibid., p. 59.
26. Ibid., p. 83.
27. Ibid., p. 84.

28. Professor Schwartz's words. Schwartz, p. 104.
29. *TYL,* Part 1, p. 4.
30. Ibid., Part 1, p. 2.
31. Ibid., Part 2, p. 44.
32. Ibid.
33. Ibid., Part 1, p. 48.
34. Ibid., Part 1, p. 33.
35. Ibid., Part 1, p. 47.
36. Hsun Tzu accepted the saying, "The man in the street can become a Yü" (one of the three Sage-Kings). See *Hsun Tzu,* Chapter 23, p. 337.
37. *TYL,* Part 2, p. 48.
38. Ibid.
39. *Lun-yü,* 14.41, p. 231.
40. Huxley, *Evolution and Ethics,* p. 84. Huxley's quotation alters the order of Tennyson's lines.
41. *TYL,* Part 2, p. 51.
42. *TYL,* Wu Ju-lun's introduction, p. 2.
43. Huxley, *Evolution and Ethics,* pp. 39, 68.
44. *Lun-yü,* 18.6, p. 277.
45. *Hsun Tzu,* Chapter 9, p. 107.
46. *TYL,* Wu Ju-lun's introduction, p. 1.
47. *SWP,* III, 1727 (also *YPS,* II, 2:3); Liang Ch'i-ch'ao used the same title at least once again in the June 10, 1897 issue of the *Shih wu pao.* See *SWP,* IV, 1932 (or *YPS,* I, 1:75).
48. Huxley, *Evolution and Ethics,* p. 78.
49. Ibid.
50. *TYL,* Yen Fu's introduction (*tzu hsu*), p. 3.
51. Huxley, *Evolution and Ethics,* p. 64.
52. *TYL,* Part 1, p. 11.
53. Ibid., Part 1, pp. 11, 14.
54. *Lao Tzu,* Chapter 5, 1:3b.
55. *TYL,* Part 2, p. 14.

6. THE REVOLUTIONARY LIANG CH'I-CH'AO

1. *HMTP,* XV, 85:1.
2. *YPS,* IV, 9:42. This was in Liang Ch'i-ch'ao's extremely important article "Shih ko" (An explanation of revolution), the first installment of which was published in the December 14, 1902 issue of *Hsin min ts'ung pao.* See *HMTP,* IV, 22:1.
3. *HMTP,* V, 29:101.
4. *CIP,* I, 7–8.
5. Ibid., I, 9.

6. Hao Chang, pp. 135–136.
7. *CIP,* I, 10.
8. Ibid.
9. Ibid., I, 11.
10. Ibid.
11. Ibid.
12. *CIP,* XI, 5513. A slightly, but only slightly, free translation. Also in Tsou Jung, *Ko-ming chün* (Taipei[?], 1954), pp. 41–42.
13. Ibid., III, 1401.
14. Ibid., II, 785.
15. *HMPT,* VII, 38–39:31.
16. Ibid., II, 11:8.
17. *CIP,* III, 1334.
18. Ibid., III, 1445.
19. Schwartz, p. 142.
20. Liang Ch'i-ch'ao, *Tzu-yu shu,* Preface, p. 1.
21. *YPS,* III, 6:25–26.
22. Ibid., II, 5:46.
23. Liang Ch'i-ch'ao, *Hsin min shuo* (Taipei, 1959), pp. 44–45.
24. Ibid., p. 78.
25. *HMTP,* IV, 22:3–4. *Ko ming,* Liang Ch'i-ch'ao objected, sounded as if the only thing meant to be changed was the imperial house, that is, the dynasty.
26. *HMTP,* IV, 22:3.
27. Ibid., IV, 22:4.
28. Ibid.
29. Ibid., IV, 22:7.
30. Ibid., IV, 22:7–8.
31. Liang Ch'i-ch'ao, *Hsin min shuo,* p. 60.
32. Ibid.
33. Ibid.
34. Ibid., pp. 8–9.
35. Ibid., pp. 64–65.
36. *HMTP,* XV, 84:1.
37. Ibid., IV, 22:8.
38. *YPS,* IV, 9:9.
39. *Mao Chu-hsi yü lu,* p. 65.
40. K'ang Yu-wei, *Nien-p'u,* p. 23.
41. See Liang Ch'i-ch'ao, "Chung-kuo shih hsu lun" (An introduction to Chinese history), *YPS,* III, 6:1–12, and "Kuo-chia ssu-hsiang pien-ch'ien

i-t'ung lun'' (On variations in the evolution of the concept of nationhood), *YPS,* III, 6:12–22.

42. *YPS,* IV, 9:8.
43. Ibid., IV, 9:9.
44. Mao Tse-tung, *Mao Tse-tung hsuan-chi* (Peking, 1969), IV, 1401.
45. *Shu ching,* in James Legge, tr., *The Chinese Classics* III, 283.
46. *YPS,* IV, 9:9.
47. Ibid.
48. *The Philosophy of Kant,* p. 123.
49. *The Philosophy of Hegel,* p. 9.
50. *YPS,* IV, 9:9.
51. Ibid., IV, 9:11.
52. Ibid., IV, 9:12.
53. Ibid., IV, 9:15.
54. Ibid.
55. Ibid., IV, 9:20.
56. Ibid., IV, 9:7.
57. Ibid., IV, 9:1.
58. Ibid.
59. Ibid., IV, 9:3.
60. Ibid., IV, 9:10.
61. Ibid., IV, 9:11.
62. Ibid., III, 6:25.
63. Ibid., IV, 9:8.
64. Ibid., III, 6:1.
65. Ibid., IV, 9:7.
66. Ibid., IV, 9:2.
67. Ibid., IV, 9:4.
68. Ibid., IV, 9:6.
69. Ibid., IV, 9:4.
70. Ibid., III, 6:3.
71. Pusey, "K'ang Yu-wei,'' p. 169.
72. Darwin, *Descent,* p. 707.
73. Ibid., p. 72.
74. Ibid., p. 70.
75. Ibid., p. 156.
76. Ibid., p. 188.
77. Ibid., p. 707.
78. Ibid.
79. Ibid., p. 162.

80. To be distinguished from its homonym, *pu hsiao* (unfilial).
81. *HMTP,* X, 52:91.
82. T'an Ssu-t'ung had also attacked certain Confucian virtues in his *Jen hsueh* (A study of love). See *Sources of Chinese Tradition,* ed. Wm. Theodore de Bary et al. (New York, 1960), pp. 750–753.
83. Lu Hsun, *Fen* (Hong Kong, 1964), pp. 99–100.
84. Only four days before Chiang K'ai-shek, erstwhile director of the Whampao Military Academy, tried to annihilate the Communists in Shanghai.
85. Lu Hsun, *Erh i chi* (Hong Kong, 1964), pp. 11–12.
86. Lu Hsun, *San hsien chi* (Hong Kong, 1964), p. 4.
87. Ibid., p. 3.
88. Lu Hsun, *Lu Hsun hsiao shuo chi* (Hong Kong, 1964), p. 8.
89. Eiseley, *Darwin,* p. 52.
90. Dr. Herbert Margoles, my orthodontist of many years ago, who pronounced me an advanced human being because I was born *without* wisdom teeth.
91. Lu Hsun, *Erh i chi,* p. 11.
92. Richard Solomon, *Mao's Revolution and the Chinese Political Culture* (Berkeley, 1971), p. 474.
93. Edgar Snow, *Red Star Over China* (New York, 1961), p. 137.
94. Ibid., p. 140.
95. Ibid., p. 145.
96. Solomon, p. 251.
97. *Mao Chu-hsi yü lu,* p. 174.
98. Lu Hsun, *Erh i chi,* p. 4.
99. *HMTP,* III, 18:17.
100. Ibid., IV, 9:41.
101. Ibid., III, 18:20.
102. Darwin, *Origin,* p. 5.
103. Ibid., p. 144. See above, pp. 109–110.
104. Jerome B. Grieder, *Hu Shih and the Chinese Renaissance* (Cambridge, Mass., 1970), p. 70.
105. Ibid., pp. 227–228.
106. *Lun yü,* 1.2, p. 61.
107. *CIP,* IX, 4885–4886, "Nan-nü hun-yin tzu-yu lun," *CIP,* issue 76 (April 19, 1901).
108. *YPS,* III, 7:107.
109. Ibid.
110. Ibid., III, 7:108.
111. Ibid., III, 7:110.
112. Ibid., III, 7:109.

113. Ibid., III, 7:110. I have taken the liberty of making an eagle out of an osprey.
114. Ibid., III, 7:109.
115. Ibid., III, 7:110.
116. *Meng Tzu,* 3.6, p. 79.
117. Ibid., 7.26, p. 182.
118. *YPS,* III, 7:110.
119. Ibid., III, 7:113.
120. Joseph R. Levenson, *Confucian China and its Modern Fate* (Berkeley and Los Angeles, 1968), III, 76–82.
121. Pusey, "K'ang Yu-wei," p. 160.
122. *YPS,* III, 6:67.
123. Ibid., II, 3:55.
124. Ibid.
125. Ibid., II, 3:59.
126. *HMTP,* II, 12:61.
127. Pusey, "K'ang Yu-wei," p. 161. See also preceding discussion from p. 156 in same work.
128. *YPS,* I, 1:109.
129. Ibid., IV, 9:50.
130. Ibid., IV, 9:53.
131. Ibid., IV, 9:52.
132. Ibid.
133. Ibid.
134. Ibid., IV, 9:57.
135. Ibid.
136. Ibid., IV, 9:58.
137. Ibid., IV, 9:51.
138. Ibid., IV, 9:59.
139. Ibid., IV, 9:55.
140. Ibid., IV, 9:55–56.
141. Ibid., IV, 9:56.
142. Ibid., IV, 9:58.
143. Ibid., IV, 9:59.
144. Pusey, "K'ang Yu-wei," pp. 164, 169.
145. Liang Ch'i-ch'ao, *Hsin min shuo,* p. 15, "Lun kung te," in *HMTP,* I, 3:1 (March 10, 1902).
146. Liang Ch'i-ch'ao, *Hsin min shuo,* p. 48.
147. Ibid., p. 47.
148. See *Lun-yü,* 6.21, p. 122.
149. Chou Chen-fu, *Yen Fu ssu-hsiang shu-p'ing* (Taipei, 1964), p. 75.

150. Ibid., pp. 71–71. See also Schwartz, p. 177.
151. Ibid., p. 72.
152. Ibid., p. 74.
153. Schwartz, p. 179.
154. *YPS,* I, 1:109.
155. Ibid., III, 7:101.
156. Ibid., III, 7:102.
157. Ibid., III, 7:101.
158. Ibid.
159. Ibid., III, 7:100–101.
160. Ibid., III, 7:101.
161. Ibid., III, 7:101–102.
162. Liang Ch'i-ch'ao, *Hsin min shuo,* pp. 59–60.
163. Ibid., p. 59.
164. Ibid., p. 29. T'an Ssu-t'ung had already made similar complaints of "Confucian" distortions of the teachings of Confucius. See *Sources,* pp. 750–753.
165. Ibid.
166. *YPS,* IV, 9:55–56.
167. *HMTP,* V, 28:112.
168. *P'i K'ung* (criticize Confucius) was the term used in the *p'i Lin p'i K'ung* (criticize Lin Piao and Confucius) campaign that started in 1974 in the People's Republic of China.
169. See Loren Eiseley, *The Immense Journey* (New York, 1957).
170. *Hsun Tzu,* Chapter 21, p. 190.
171. *Mo Tzu chien ku* (Taipei, 1970), Chapter 11, p. 44.
172. Hu Shih, *Hu Shih wen hsuan* (Taipei, 1962), p. 25.
173. Liang Ch'i-ch'ao, *Hsin min shuo,* p. 15. "To hold the whip for someone" means to serve him. The phrase comes from *Lun-yü,* 7:11, p. 130: "The Master [Confucius] said, 'If wealth were something to be sought, then I would even be a whip-holding retainer if need be. But if it is not something to be sought, I shall follow what I love.'"
174. Hofstadter, p. 87.
175. Ibid., pp. 86–87.
176. Huxley, *Evolution and Ethics,* p. 64.
177. *Meng Tzu,* 11.10, p. 265.
178. Mencius's *hung shui meng shou.* See *Meng Tzu,* 6.9, p. 155.
179. Liang Ch'i-ch'ao, *Hsin min shuo,* pp. 12–13.
180. Ibid., p. 12.
181. The famous "plate of loose sand" (*i p'an san sha*) was not Sun Yat-sen's own metaphor, although he used it to best advantage. It was a foreigner's insult which he and many others picked up (see "Min-

ch'üan-chu-i, ti erh chiang" [Democracy, lecture two] in *KFCS,* p. 222). He did use the cement metaphor, but in his case, of course, the cement was to be the Three People's Principles (ibid., p. 226).

182. P. A. Kropotkin, *Mutual Aid* (London, 1919), p. 14.
183. Ibid., p. 62.
184. *Shan ch'ün.* See *Hsun Tzu,* Chapter 9, p. 109.
185. Again see *Hsun Tzu,* Chapter 9, p. 108. Kropotkin would not have appreciated this echo, for his quarrel with Huxley, after all, arose precisely from the fact that Huxley—like Hsun Tzu—refused to give adequate recognition to the "social virtues" of animals.
186. Liang Ch'i-ch'ao, *Hsin min shuo,* p. 15.
187. Schwartz, p. 100, and Huxley, *Evolution and Ethics,* p. 82.
188. Liang Ch'i-ch'ao, *Hsin min shuo,* p. 131.
189. Ibid., p. 15.
190. *Mao Tse-tung hsuan chi,* II, 499.
191. Ibid., II, 500.
192. The *locus classicus* of these well-worn slogans still eludes me. See, however, *Li-shih yen-chiu,* 4: 2nd unnumbered page (1975).
193. Huxley thought he was proscribing not prescribing evolutionary moralities, but his ethics, although turned against the parent that had produced them, were for that reason alone as "evolutionary" as any of the others.
194. Lu Hsun, "K'uang-jen jih-chi," in *Lu Hsun hsiao shuo chi,* p. 15. Lu Hsun was surely not the only one of his generation revolted by "the old morality," but his charge that it "ate people" was the charge that became the rallying cry for its opponents.
195. Liang Ch'i-ch'ao, *Hsin min shuo,* p. 14–15.
196. *Mo Tzu,* Chapter 16, pp. 72–73.
197. *YPS,* V, 13:31.
198. Ibid.
199. Ibid., p. 34.
200. Huxley, *Evolution and Ethics,* p. 82.
201. Liang Ch'i-ch'ao, *Tzu-yu shu,* p. 92.
202. *YPS,* II, 5:43.
203. Ibid., V, 13:38.
204. Liang Ch'i-ch'ao, *Tzu-yu shu,* p. 91.
205. *YPS,* II, 5:48.
206. Ibid., II, 5:49.
207. Ibid., V, 13:32.
208. Ibid., II, 5:49.
209. Ibid., V, 13:39.
210. Ibid., II, 5:49.
211. Ibid.

212. Kropotkin, *Mutual Aid,* p. 6.
213. Ibid., pp. 222–223.
214. Ibid., pp. 87–88.
215. Liang Ch'i-ch'ao, *Tzu-yu shu,* p. 92.
216. *YPS,* V, 13:36.
217. Liang Ch'i-ch'ao, *Tzu-yu shu,* p. 98.
218. *YPS,* V, 13:36.
219. Ibid., p. 28.
220. Ibid., p. 39.
221. *Shu ching,* Part II, Book II, Chapter I, 5. My translation from text given in Legge., tr., *The Chinese Classics,* III, 54.
222. Ibid., Part IV, Book III, Chapter II, 3. Again my translation from Legge, *The Chinese Classics,* III, 186–187.
223. *Chuang Tzu,* Book 4.7, p. 38.
224. Ibid., Book 4.5, p. 36.
225. Ibid., Book 1.6, p. 7.
226. *YPS,* V, 13:36.
227. The fourth noble truth was that morality, as prescribed in the Eightfold Path, was the way to stop the desire that caused the suffering that was life. See *Sources,* pp. 306–308.
228. See T'ang Hsiao-wen, "Wei-shem-me shuo Hsun Tzu shih fa-chia?" (Why do we say Hsun Tzu is a Legalist?), in *Hung ch'i* 1:50–56 (1975).
229. *Han Fei Tzu chi shih,* ed. Ch'en Ch'i-yu (Peking, 1959), Volume 2, Chapter 49, p. 1042: "Benevolence and righteousness worked in ancient times; but they do not work today."
230. *Hsun Tzu,* Chapter 12, p. 168.
231. Ibid., Chapter 15, p. 196.
232. Ibid., Chapter 19, p. 257.
233. *Lun-yü,* 12.18, pp. 195–196.
234. *Meng Tzu,* 11.18, p. 272.
235. Ibid., 1.5, p. 10.
236. Ibid., 7.7, p. 168. See also *Mo Tzu,* Chapter 26, p. 120: "Those who follow the will of Heaven . . . shall surely gain reward. Those who go against the will of Heaven . . . shall surely suffer punishment."
237. *Mo Tzu,* Chapter 16, p. 78.
238. *Meng Tzu,* 1.1, p. 1.
239. *Lun-yü,* 4.16, p. 96.
240. Ibid., 15.8, p. 238.
241. *Meng Tzu,* 3.2, p. 63.
242. Ibid., 8.8, p. 188.
243. *YPS,* V, 13:35–36.
244. *Meng Tzu,* 3.6, pp. 79–80.

245. Kropotkin, *Mutual Aid,* pp. 5–6.
246. Ibid., p. 51.
247. *HMTP,* III, 18:17.
248. *YPS,* IV, 9:45.
249. Liang Ch'i-ch'ao, *Tzu-yu shu,* p. 45.
250. *YPS,* IV, 9:46.
251. Ibid., pp. 44–45.
252. Ibid., p. 45.
253. Ibid.
254. Liang Ch'i-ch'ao, *Hsin min shuo,* p. 118.
255. Huxley, *Evolution and Ethics,* p. 82.
256. Ibid.
257. Liang Ch'i-ch'ao, *Hsin min shuo,* p. 108.
258. Ibid.
259. Ibid., p. 114.
260. "The Yen River's muddy. The Yen River's clear./My beloved has gone for a soldier./He has joined the army to resist Japan./Only good iron's for nails." From a Chinese song taught me by a Chinese friend who would probably wish to remain anonymous.
261. Liang Ch'i-ch'ao, *Hsin min shuo,* p. 113.
262. Ibid.
263. *Han Fei Tzu,* Chapter 49, p. 1042.
264. *Mao Tse-tung hsuan chi,* III, 828.
265. Ibid., II, 553.
266. Ibid., II, 460. Of course, the chivalry of Duke Hsiang of Sung was old time indeed. He had been ridiculed for 2,000 years, and his sort of chivalry was surely not to blame for China's defeats in the nineteenth century.
267. Ibid., IV, 1362.
268. Ibid., II, 553.
269. Liang Ch'i-ch'ao, *Hsin min shuo,* pp. 108–109.
270. Ibid., p. 109.
271. Ibid., pp. 109–110.
272. Liang Ch'i-ch'ao, *Tzu-yu shu,* p. 37.
273. Liang Ch'i-ch'ao, *Hsin min shuo,* p. 109.
274. Liang Ch'i-ch'ao, *Tzu-yu shu,* p. 37.
275. Tu Fu, "Ping chü hsing," "Tu Fu shih hsuan chu," ed. Li Tu, p. 4, in *Shih tz'u hsin-shang,* Volume II (Kaohsiung, 1962).
276. Liang Ch'i-ch'ao, *Hsin min shuo,* p. 10.
277. Ibid., p. 111.
278. Ibid., p. 112.
279. Ibid., p. 123.

280. Ibid., p. 23.
281. *CIP,* XI, 5693.
282. Liang Ch'i-ch'ao, *Hsin min shuo,* p. 23.
283. Ibid., p. 10.
284. Ibid., p. 96.
285. Huxley, *Evolution and Ethics,* p. 82.
286. Ibid., p. 60.
287. Liang Ch'i-ch'ao, *Hsin min shuo,* p. 96.
288. Ibid., pp. 29–31.
289. *Meng Tzu,* 3.2, p. 62.
290. Liang Ch'i-ch'ao, *Tzu-yu shu,* p. 72.
291. Liang Ch'i-ch'ao, *Hsin min shuo,* p. 115.
292. Liang Ch'i-ch'ao, *Tzu-yu shu,* p. 38.
293. Ibid.
294. Liang Ch'i-ch'ao, *Hsin min shuo,* p. 70.
295. Ibid., pp. 68–69.
296. Ibid., p. 68. The main allusion is to *Lun-yü,* 9.26, p. 162: "The Master said, 'When the year turns cold, then one knows that the pine and the cedar are the last to lose their green.'"
297. *HMTP,* X, 49:32.
298. Ibid., 50:1.
299. *Lun-yü,* 7.29, p. 138.
300. Liang Ch'i-ch'ao, *Tzu-yu shu,* p. 74.
301. Ibid.
302. Ibid., p. 75.
303. Liang Chi'-ch'ao, *Hsin min shuo,* pp. 104–105.
304. Huxley, *Evolution and Ethics,* p. 82.
305. *YPS,* IV, 9:48.
306. Liang Ch'i-ch'ao, *Hsin min shuo,* pp. 130–131.
307. Ibid., p. 122.
308. Ibid., p. 130.
309. Huxley, *Evolution and Ethics,* p. 51.
310. Ibid., p. 43.
311. Ibid., pp. 81–82. *Nota bene* that Huxley too was for "mutual aid."
312. Ibid., p. 84.
313. Ibid., p. 59.
314. *Lun-yü,* 12.1, p. 186.
315. Liang Ch'i-ch'ao, *Hsin min shuo,* pp. 50, 47.
316. *Lun-yü,* 12.1, pp. 186–187.
317. Liang Ch'i-ch'ao, *Hsin min shuo,* p. 14.
318. Ibid., p. 50.
319. *Lao Tzu,* Chapter 33 (Part 1, p. 19).

320. Liang Ch'i-ch'ao, *Hsin min shuo,* p. 49.
321. Ibid., p. 50.
322. Ibid., p. 141.
323. Wang Wei, "Kuo Hsiang Chi Ssu," in *T'ang shih san pai shou hsiang hsi* (Taipei, 1963), p. 148.
324. Liang Ch'i-ch'ao, *Hsin min shuo,* p. 141.
325. See Lu Hsun, *Wei tzu-yu shu* (Hong Kong, 1964), pp. 59, 162, and Grieder, p. 228.
326. *Mao Chu-hsi yü lu,* p. 75.
327. *Wei jen-min* (For the people) is now a slogan by itself. See *Hung ch'i,* 12:41 (1974) for one example out of thousands.
328. Hofstadter, p. 99.
329. *YPS,* V, 12:78.
330. Benjamin Kidd, *Social Evolution* (New York, 1894), pp. 36–39.
331. Ibid., p. 1.
332. Ibid., p. 315.
333. Ibid., p. 317.
334. Ibid., pp. 317, 316.
335. Ibid., p. 39.
336. Ibid., p. 52.
337. Ibid., pp. 313–314.
338. Hofstadter, p. 54.
339. Kidd, p. 314.
340. Ibid., p. 241.
341. Ibid., p. 61.
342. Ibid., p. 62.
343. Ibid., pp. 239–240.
344. Ibid., pp. 245–246.
345. Ibid., p. 293.
346. Ibid., pp. 45, 52.
347. Ibid., p. 116.
348. Ibid., p. 246.
349. Huxley, *Evolution and Ethics,* pp. 83–84.
350. Kidd, p. 14.
351. Ibid., p. 65.
352. See above, n. 310.
353. Kidd, p. 293.
354. *YPS,* V, 12:79.
355. Ibid., p. 81.
356. Ibid., p. 84.
357. Ibid., p. 86.
358. *Mutatō nomine de tē fabula narrātur* (Under a different name the story

is told of you)—a line from Horace's *Satires* (1.1, 69), quoted several times by Marx. See *Marx and Engels: Basic Writings on Politics and Philosophy,* ed. Lewis S. Feuer (New York, 1959), p. 150.

359. Kidd "introduced" Marx to China, that is, when Liang Ch'i-ch'ao mentioned Kidd's mention of Marx. See *YPS,* V, 12:79, 86.
360. *YPS,* V, 12:81.
361. *YPS,* V, 12:82.
362. *YPS,* V, 12:83.
363. Ibid.
364. Lu Hsun, *Jeh feng,* p. 44.
365. *Mao Chu-hsi yü lu,* p. 149.
366. See *YPS,* V, 12:82.
367. Kidd, pp. 290–291.
368. *YPS,* V, 12:83.
369. Ibid.
370. Ibid., V, 12:82.
371. Hu Shih, *Wen hsuan,* p. 78.
372. Ibid., p. 84.
373. *YPS,* VI, 17:3.
374. Ibid.
375. Ibid., VI, 17:1.
376. Ibid., VI, 17:9.
377. Ibid., VI, 17:8.
378. Ibid.
379. See *Tz'u hai* (Shanghai, 1948), p. 106.
380. *YPS,* VI, 17:5.
381. Ibid.
382. Ibid.
383. Ibid., VI, 17:10.
384. Ibid.
385. Ibid.
386. Ibid., VI, 17:7.
387. Ibid., VI, 17:6–7.
388. *Meng Tzu,* 8.19, p. 191.
389. *YPS,* VI, 17:9.
390. Ibid., VI, 17:4–5.
391. Ibid., VI, 17:3.
392. Darwin, *Origin,* p. 11.
393. Ibid., p. 82.
394. Ibid., p. 123.
395. Ibid., p. 43.
396. Ibid.

397. Ibid., p. 13.
398. Ibid., p. 30.
399. Ibid., p. 82.
400. Ibid., p. 131.
401. Ibid., p. 466.
402. Ibid., p. 84.
403. Darwin, *Descent,* p. 707.
404. *YPS,* VI, 17:4.
405. Bergson's book was published in 1908. Li Ta-chao lauded Bergson in April, 1915 (see Maurice Meisner, *Li Ta-chao and the Origins of Chinese Marxism* [Cambridge, Mass., 1967], p. 23), and Ch'en Tu-hsiu followed suit on the third page of the first issue of *Hsin ch'ing-nien* on September 15, 1915. Liang Ch'i-ch'ao finally acknowledged Bergson in 1919 or 1920, on page 18, of his *Ou yu hsin-ying lu chie-lu* (Taipei, 1966). If, as Jerome Grieder says, he "turned to Bergson with particular enthusiasm," however, it was not because "he discovered in the Bergsonian idea of 'creative evolution' a means of salvaging the faith in evolutionary progress that had lain close to his heart since the early days of his association with K'ang Yu-wei" (Grieder, p. 133). It was because Liang Ch'i-ch'ao found, at last, foreign confirmation of a faith in creative evolution that he had had almost from the beginning.
406. *YPS,* VI, 17:4.
407. *HMTP,* V, 27:91.
408. Even when he did show interest in improving the physical fitness of his people through the inheritable acquired characteristics that would accrue from "pre-natal education" (see above, p. 102), he was more interested in building character than physique. For one thing, as mothers would have to make themselves fit before their sons could be, it would take character to build character, which was a *ch'i ma chao ma* (ride a horse to hunt for one) paradox that did not bother Liang Ch'i-ch'ao in the least.
409. Darwin, *Descent,* p. 148.
410. Ibid., p. 699.
411. Ibid., p. 142.
412. Charles Darwin, *The Expression of the Emotions in Man and Animals* (Chicago, 1970), p. 354.
413. Ibid., p. 356.
414. Ibid., p. 355.
415. Darwin, *Descent,* p. 699.
416. My translation of a line from Mao Tse-tung's poem, "Sung wen chen," given in Wong Man, tr. and ed., *Poems of Mao Tse-tung* (Hong Kong, 1966), p. 57.

417. Ta-erh-wen (Charles Darwin), *Wu chung yuan shih,* Ma Chün-wu, tr. (Taipei, 1957), preface, p. 1.
418. *YPS,* VI, 17:3.
419. Ibid., VI, 17:7.
420. Ibid., VI, 17:9–10. K'ang Yu-wei had said the same thing as early as May 30, 1899: "When our four hundred million people realize that they cannot protect themselves, their desire to look out for themselves will turn into a desire to look out for all together. When they know that they cannot preserve their families, their desire to protect their families will turn into a desire to protect the country as a whole." See *CIP,* II, 997.
421. *The People's Comic Book,* Endymion Wilkinson, tr., (New York, 1973), p. 244.
422. Ibid., p. 247.
423. Ibid., p. 220.
424. Ibid., p. 252: My translation of "Lei Feng's" notes shown in the 177th frame.
425. Ibid. Wilkinson's translation of the text.
426. Stuart R. Schram. *The Political Thought of Mao Tse-tung* (New York, 1972), p. 304.
427. *Lun-yü,* 2.4, p. 70.
428. Teng and Fairbank, p. 25.
429. *CIP,* IV, 1931.
430. *HMTP,* I, 1:5.
431. Ibid., I, 2:31.
432. Ibid., I, 2:75.
433. Ibid., I, 1:6.
434. Ibid., I, 2:34.
435. Liang Ch'i-ch'ao, *Tzu-yu shu,* p. 24.
436. Ibid., p. 23.
437. *YPS,* II, 4:40, quoting *Meng Tzu,* 7.8, p. 170.
438. *Lun-yü,* 15.18, p. 241.
439. *CIP,* XII, 6506.
440. *HMTP,* II, 8:29.
441. Ibid., I, 5:36.
442. *CIP,* III, 1190.
443. Ibid., III, 1196.
444. *HMTP,* VI, 34:70.
445. Ibid., XIII, 71:95.
446. Liang Ch'i-ch'ao, *Tzu-yu shu,* p. 80.
447. *HMTP,* V, 25:52. For the same sentiment in verse, see Kao Hsieh (Tz'u-shih)'s poem *HMTP,* VI, 34:103. As to the question "To be or not to be a super power," Liang Ch'i-ch'ao longed for a Chinese Napoleon (see

Liang Ch'i-ch'ao, *Tzu-yu shu,* p. 58), but thirty-four years earlier, in 1867, Tseng Kuo-fan, in proto-Mao fashion had disclaimed super-power ambitions for China. See Teng and Fairbank, p. 67.

448. *HMTP,* VI, 34:70.
449. *YPS,* III, 6:39.
450. Ibid.
451. *SLHC,* Ia, 419.
452. *CIP,* VIII, 4188.
453. Ibid., VIII, 4239.
454. Ibid., V, 2147.
455. *SLHC,* Ia, 199.
456. *CIP,* XII, 6564.

7. *THE REAL REVOLUTIONARIES*

1. Sun Yat-sen did once use a novel translation of the title, *The Origin of Species, Sheng-wu ben yuan,* in a 1904 article, "Po *Pao huang pao*" (In refutation of the *Pao huang pao,* see *KFCS,* p. 372), but Yen Fu himself used two different translations, *Wu-chung t'an yuan* in "Yuan ch'iang," *YF,* p. 14, and *Wu-chung yu lai* in *T'ien-yen lun,* Part 1, p. 3.
2. *KFCC,* VI, 211.
3. Tsou Jung, p. 2.
4. *MP,* II, 12:49.
5. *KFCS,* p. 390.
6. Tsou Jung, p. 1.
7. *SLHC,* Ib, 689. The article from which this quotation comes, "Po 'Ko-ming po-i'" (A refutation of "A refutation of revolution"), *Su pao,* June 12–13, 1903, was supposedly written jointly by Chang Ping-lin, Tsou Jung, Liu Ya-tzu (Liu Ch'i-chi), and Ts'ai Yeh-min (Ts'ai Yin). See Chang Ching-lu, Li Sung-nien, "Hsin-hai ko ming shih-ch'i chung-yao pao-k'an tso-che pi ming lu," *Wen shih,* Volume 1, October, 1962 (Peking), p. 110, n. 1.
8. John K. Fairbank, Edwin O. Reischauer, Albert M. Craig, *East Asia: The Modern Transformation* (Boston, 1965), p. 634.
9. Tsou Jung, pp. 18–19.
10. Ibid., p. 49.
11. Ibid., p. 48. The *chopping* is free. A baffling literal translation would be: "I grasp a great sword ninety-nine."
12. Hu Han-min, "Shu Hou-kuan Yen shih tsui-chin cheng-chien," *MP,* I, 2:13.
13. Hu Han-min, "Min pao chih liu ta chu-i," *MP,* I, 3:7.
14. *MP,* IV, 21:102.

15. *Han shu,* 48:13a–b. I owe my discovery of this passage to Professor Yü Ying-shih's work, *Trade and Expansion in Han China* (Berkeley and Los Angeles, 1967). See his translation of an extended passage on page 11.
16. Tsou Jung, p. 26.
17. Wang Ching-wei, "Min-tsu ti kuo-min," *MP,* I, 2:22.
18. *MP,* IV, 26:31.
19. *Shu ching,* Part III, Book II, Chapter I, 3. My translation of text given in Legge, II, 153.
20. Ibid., Part V, Book I, pt. I, 9. My translation of text given in Legge, III, 297.
21. *MP,* III, 14:89.
22. Ibid., II, 12:107.
23. Chu Chih-hsin, "Lun Man-chou sui yü li hsien erh pu neng," *MP,* I, 1:31, 36.
24. *MP,* I, 5:18.
25. *Mao Chu-hsi yü lu,* p. 10.
26. The *jang pu cheng-ts'e* was a "heretical" theory of history first suggested, I think, by the historian Chien Po-tsan. He and Wu Han were attacked in 1966, if not before, for propagating it. See my *Wu Han: Attacking the Present through the Past* (Cambridge, Mass., 1969), p. 65.
27. *Ta-che hung ch'i fan hung ch'i* (waving the red flag to oppose the red flag).
28. See Schwartz, pp. 183–184 and Chou Chen-fu, p. 284.
29. *SLHC,* IIb, 657.
30. Wang Ching-wei, "Min-tsu," *MP,* I, 2:24. Also see Hu Han-min, "Shu," *MP,* I, 2:5.
31. Wang Ching-wei, "Min-tsu," *MP,* I, 2:28.
32. Ibid., I, 2:6.
33. Hu Han-min, "Shu," *MP,* I, 2:1.
34. Ibid., I, 2:2.
35. Ibid., I, 2:3.
36. Ibid., I, 2:17.
37. See Schwartz, p. 185.
38. Hu Han-min, "Shu," *MP,* I, 2:5.
39. *MP,* II, 12:122–124.
40. *SLHC,* IIb, 648.
41. Ibid., IIb, 648–649.
42. Ibid., IIb, 649.
43. *KFCC,* III, 9.
44. Ibid., III, 8–9.

45. Chang Ping-lin, *Chang shih ts'ung shu* (Taipei, 1958), II, 828. Also in *SLHC,* IIb, 656.
46. Chang Ping-lin, II, 829.
47. Ibid.
48. Ch'en T'ien-hua, "Ch'en Hsing-t'ai hsien-sheng chueh ming shu," *MP,* I, 2:4.
49. See *YPS,* V, 13:67. Bluntschli's *International Law,* of course, was already known and appreciated at the time of the Reform Movement. See above, p. 140.
50. *MP,* II, 13:30.
51. Ibid.
52. Ibid., II, 13:25.
53. Ibid., II, 13:26.
54. Ibid., II, 13:27.
55. Ibid., II, 13:31.
56. Ibid., II, 13:33.
57. *YPS,* V, 13:74, 76.
58. Ibid., V. 75.
59. *KFCS,* 480.
60. Ibid.
61. Harold Z. Schiffrin, *Sun Yat-sen and the Origins of the Chinese Revolution* (Berkeley, 1970), p. 43.
62. *KFCS,* p. 392.
63. *KFCC,* III, 215–216.
64. Ibid., III, 164.
65. *SLHC,* 1a, 545.
66. Ibid., 1a, 540.
67. de Tocqueville, I, 6.
68. *The Philosophy of Hegel,* p. 16 (my italics). Cited above, p. 38.
69. de Tocqueville, I, 6.
70. *HMTP,* IX, Liang Ch'i-ch'ao, "Hsin ta-lu yu-chi," p. 61.
71. Ibid., IV, 21:10.
72. See above, p. 72, n. 72.
73. *YPS,* VI, 17:50.
74. Ibid., VI, 17:67.
75. Ibid., VI, 17:77.
76. Ibid., VI, 17:38.
77. Ibid., VI, 17:22.
78. Ibid., VI, 17:39.
79. Ibid., VI, 17:37–38.
80. Ibid., VI, 17:30.

81. Ibid., VI, 17:34.
82. Liang Ch'i-ch'ao, *Tzu-yu shu,* p. 23.
83. Himmelfarb, p. 315.
84. *YPS,* VI, 17:20.
85. Ibid., VI, 17:36. The *legs* come from *Chuang Tzu,* Book 8, p. 68; the *rice shoots* from *Meng Tzu,* 3.2, p. 62.
86. Schwartz, p. 67.
87. *YPS,* VI, 17:21.
88. Ibid., p. 13. I have inverted Ch'en T'ien-hua's clauses to avoid ending two phrases with *dictatorship.*
89. Hao Chang, p. 252.
90. *Sources,* p. 283.
91. *KFCS,* p. 372.
92. Ibid., p. 364.
93. Ibid.
94. The Chinese idiom comes not, of course, from the Bible but from Ssu-ma Ch'ien's *Shih chi.*
95. *KFCS,* p. 364.
96. Ibid., p. 554.
97. See above, p. 73, n. 77.
98. *KFCS,* p. 364.
99. Ibid., p. 706.
100. Ibid., p. 365.
101. *KFCC,* III, 3–4.
102. Having written this paragraph, I rediscovered an article in *Hsin min ts'ung pao* by Huang Kuo-k'ang, who translated a passage from a book by a Japanese, Kumagai Gorō, who actually used such a metaphor: "The universe is like a huge ship, and the human race are people on it. If the ship sails east and the people instead go west against it, they must sink into the river. Therefore if people do not take the universe's goal as their goal, but willfully move in opposition to the universe, they will only hasten their own demise." See *HMTP,* XVII, 91:97.
103. Eiseley, *Darwin,* p. 321.
104. *KFCS,* p. 706.
105. Eiseley, *Darwin,* p. 312.
106. Ibid., p. 307.
107. *KFCC,* III, 253.
108. Thomas H. Huxley, *Man's Place in Nature* (Ann Arbor, 1961), p. 71.
109. *Lun-yü,* 7.29, p. 138.
110. *KFCC,* II, 75.
111. *Meng Tzu,* 1.7, p. 15.
112. *KFCC,* II, 78.

113. Ibid., II, 79.
114. Ibid., II, 2. See *Mao chu-hsi yü lu,* p. 172, for Mao Tse-tung's retelling of the *yü kung yi shan* (the foolish old man moved the mountain) fable.
115. Huxley, *Evolution and Ethics,* p. 44.
116. Ibid., p. 82.
117. Ibid., p. 83.
118. Ibid., p. 59.
119. Ibid., p. 60.
120. Eiseley, *Darwin,* p. 313.
121. *KFCC,* II, 73.
122. *TYL,* Part 1, p. 17.
123. Hu Han-min, "Shu," *MP,* I, 2:11.
124. *SLHC,* Ia, 326.
125. *MP,* IV, 22:1.
126. *SLHC,* Ib, 716.
127. Ibid., Ib, 715.
128. Ibid., Ib, 715–716.
129. Ibid., Ib, 716–717.
130. See above, n. 66.
131. *MP,* II, 10:83.
132. *KFCS,* p. 365.
133. *Lun-yü,* 17.19, p. 269.
134. See Hao Chang, pp. 262–271, and Martin Bernal, "The Triumph of Anarchism over Marxism, 1906–1907," in Mary Clabaugh Wright, *China in Revolution* (New Haven, 1968), pp. 97–112.
135. *YPS,* III, 6:73.
136. *HMTP,* V, 25:83.
137. See above, p. 74, n. 80.
138. *HMTP,* XVI, 86:6.
139. *KFCS,* p. 480.
140. Ibid., p. 481.
141. Ibid., p. 483.
142. Ibid., p. 481. See above, p. 73.
143. Ibid., p. 480.
144. Ibid.
145. Ibid.
146. Ibid.
147. *YPS,* VII, 20:1.
148. *KFCS,* p. 480.
149. *YPS,* III, 6:32.
150. *HMTP,* VII, 40–41:97.

151. Ibid.
152. Ibid., X, 54:6–7.
153. Ibid., XVI, 86:18.
154. Ibid., VII, 40–41:103.
155. Ibid., VII, 40–41:106.
156. Ibid., VII, 40–41:101.
157. Ibid., VII, 40–41:104.
158. Ibid., XVI, 86:20.
159. Ibid., XVI, 86:27.
160. Ibid., XVI, 86:20.
161. Ibid., XVI, 86:16–17.
162. *KFCS,* p. 487.
163. *MP,* I, 4:104.
164. *KFCS,* p. 481.
165. Ibid.
166. *HMTP,* XVI, 86:13.
167. *MP,* II, 12:104–105.
168. Ibid., II, 59.
169. *MP,* I, 5:79.
170. *KFCS,* pp. 481, 483.
171. *KFCC,* III, 49.
172. *KFCS,* p. 487.
173. *MP,* 4:102–103.
174. *KFCS,* p. 483.
175. Ibid., p. 481.
176. Ibid.
177. *MP,* II, 12:127.
178. *KFCS,* p. 487.
179. *SLHC,* IIb, 886.
180. Ibid., IIb, 889.
181. *HMTP,* XIII, 70:16.
182. Hofstadter, p. 57.
183. Ibid., p. 51.
184. *MP,* I, 4:103.
185. Mary Backus Rankin, *Early Chinese Revolutionaries: Radical Intellectuals in Shanghai and Chekiang, 1902–1911* (Cambridge, Mass., 1971), p. 107.
186. Robert A. Scalapino and George T. Yu, *The Chinese Anarchist Movement* (Berkeley, 1961), p. 33.
187. Ibid., p. 37.
188. Hu Han-min, "Min pao," *MP,* I, 3:9.

189. Scalapino and Yu, p. 5.
190. Kwok, p. 34.
191. Wu Ching-heng (Wu Chih-hui), *Wu Ching-heng hsuan-chi: Hsu-pa, yu-chi, tsa-wen* (Taipei, 1967), p. 214.
192. Ibid., p. 56.
193. Ibid., See also ibid., p. 55, for an explanation of his "novel's" title: "*Shang-hsia* refers to the universe; *ku-chin* refers to history."
194. See above, p. 151–152.
195. *HSC,* 39:154 (March 21, 1908).
196. Ibid.
197. Ibid.
198. Ibid.
199. *HSC,* 39:155.
200. Ibid.
201. Ibid.
202. Ibid., 39:156.
203. Ibid.
204. Darwin, *Descent,* p. 74.
205. Ibid., p. 111.
206. Wu Ching-heng, *Tsa-wen,* p. 165. Given Chü P'u's argument and the untutored nature of his readers, the rather crudely literal rendition "promiscuous copulation" is, I think, the fittest translation of *tsa-chiao,* but Chü P'u may well have picked up the term from Ma Chün-wu's 1903 translation of Chapter 6, "Natural Selection," of the *Origin,* in which it was used, borrowed most likely from the Japanese, to translate Darwin's words *intercrossing* and *hybridism.* Of course, Chü P'u may have borrowed the term himself from the Japanese. *Tsa-chiao* remains the modern word for hybridization—but its etymology is as stark as ever. See Wang Tzu-ch'un and Chang Bing-lun, p. 8.
207. Wu Ching-heng, *Tsa-wen,* p. 164.
208. Ibid.
209. Ibid., pp. 164–165.
210. Ibid., p. 165.
211. Ibid.
212. Ibid., pp. 165–166.
213. Ibid., p. 166.
214. Ibid., p. 167.
215. Ibid., p. 170.
216. Ibid., pp. 170–171.
217. Ibid., p. 169.
218. Ibid., pp. 168–169.

219. See above, pp. 304–305.
220. Eiseley, *Darwin,* p. 224. Mendel's work was rediscovered in 1900. Wu Chih-hui was writing in 1904.
221. Wu Ching-heng (Wu Chih-hui), *Wu Ching-heng hsuan-chi: Shang-hsia ku-chin t'an* (Taipei, 1967), I, preface, p. 2.
222. Ibid., I, preface, p. 1.
223. Wu Ching-heng, *Tsa-wen,* p. 55.
224. Wu Ching-heng, *Shang-hsia,* I, preface, p. 1.
225. Lu Hsun, *Fen,* pp. 12–13. Compare with Wu Ching-heng, *Shang-hsia,* II, 182.
226. Wu Ching-heng, *Shang-hsia,* II, 179–183.
227. Ibid., II, 184.
228. Ibid., II, 186.
229. Ibid., II, 179–180.
230. Ibid., II, 181.
231. Ibid., II, 183.
232. Ibid., II, 188.
233. Ibid., II, 170–171.
234. Huxley, *Evolution and Ethics,* p. 46. See above, p. 101.
235. Wu Ching-heng, *Shang-hsia,* II, 181.
236. *Shih chi hui-chu k'ao-cheng,* ed. Takigawa Kametaro, II, 7:60.
237. Wu Ching-heng, *Shang-hsia,* II, 190.
238. Eiseley, *Darwin,* p. 312.
239. Darwin, *Origin,* p. 208.
240. Again see above, p. 206.
241. Wu Ching-heng, *Shang-hsia,* II, 182.
242. Ibid., II, 188.
243. Ibid., II, 157.
244. P. A. Kropotkin, *Memoirs of a Revolutionist* (New York, 1970), p. 155.
245. Kropotkin, *Mutual Aid,* p. 2.
246. Ibid., p. 1.
247. Ibid., p. 2.
248. Kropotkin, *Memoirs,* p. 498.
249. Kropotkin, *Mutual Aid,* p. 3.
250. Darwin, *Descent,* p. 119.
251. Kropotkin, *Mutual Aid,* p. 52.
252. Ibid., p. 3.
253. Ibid.
254. Kropotkin, *Memoirs,* p. 498.
255. Kropotkin, *Mutual Aid,* p. 7.
256. Kropotkin, *Memoirs,* p. 498.
257. Ibid., p. 499.

258. Kropotkin, *Mutual Aid,* p. 6. Three of Kropotkin's chapters, it must be admitted, were published as articles before Huxley gave his lectures. See Ibid., p. 10.
259. See above, p. 159.
260. See Schwartz, pp. 104–107, and above, p. 164.
261. Kropotkin, *Mutual Aid,* p. 13.
262. Ibid., pp. 36–37.
263. Ibid., p. 37.
264. Ibid., p. 222.
265. Ibid., p. 18. See *Chuang Tzu,* Book 22.6, p. 177:

> Master Tung Kuo asked Chuang Tzu, "Where is this 'Way'?"
> Chuang Tzu replied, "There is nowhere that it is not."
> Master Tung Kuo said, "But you must be specific."
> Chuang Tzu said, "It is in ants."
> "What, so low as that?"
> "It is in panic grass."
> "Ah? Even lower?"
> "It is in tile and brick."
> "Can it be *that* low?"
> "It is in excrement and urine."
> Master Tung Kuo made no reply.

266. Kropotkin, *Mutual Aid,* p. 62.
267. Ibid., p. 49.
268. Ibid., p. 218.
269. Ibid., p. 60.
270. See above, p. 393, n. 263.
271. Huxley, *Evolution and Ethics,* p. 59.
272. Ibid., p. 49.
273. Ibid., p. 59.
274. Ibid., p. 49.
275. Ibid., p. 59.
276. Ibid., p. 49.
277. Kropotkin, *Mutual Aid,* opposite frontispiece: "Some Press Opinions": "*The Athenaeum:* Prince Kropotkin has written a most suggestive and stimulating study, showing . . . a most attractive and generous personality," which is an opinion I entirely share.
278. Ibid., p. 87.
279. Huxley, *Evolution and Ethics,* p. 64, and above, p. 160.
280. Kropotkin, *Mutual Aid,* p. 45.
281. Ibid., p. 37. I have changed *this is* to *there is.*
282. Ibid., pp. 222–223.

283. Ibid., p. 5.
284. Wu Ching-heng, *Shang-hsia,* I, 47.
285. Kropotkin, *Mutual Aid,* p. 13.
286. *Lun-yü,* 15.8, p. 238.
287. Huxley, *Evolution and Ethics,* p. 82. See above, p. 102.
288. Darwin, *Descent,* pp. 151–152.
289. *Lun-yü,* 9.17, p. 158.
290. Ibid., 7.30, p. 138.
291. Ibid., 2.4, p. 70.
292. Darwin, *Descent,* p. 152.
293. Ibid., pp. 126–127.
294. Ibid., p. 141.
295. Huxley, *Evolution and Ethics,* p. 78.
296. Ibid., p. 84.
297. Darwin, *Descent,* p. 142.
298. Ibid., p. 141.
299. Ibid., p. 52.
300. *Hsun Tzu,* Chapter 23, p. 330.
301. Huxley, *Evolution and Ethics,* p. 39.
302. Ibid., p. 83.
303. Kropotkin, *Mutual Aid,* p. 206.
304. *Meng Tzu,* 3.6, p. 79.
305. Kropotkin, *Mutual Aid,* p. 195.
306. Ibid., pp. 212–213.
307. *Meng Tzu,* 12.2, p. 276.
308. Kropotkin, *Mutual Aid,* p. 195.
309. Ibid., "Some Press Opinions," opposite frontispiece.
310. Ibid., p. 8.
311. Ibid., pp. 13–14.
312. Ibid., p. 8.
313. Ibid., pp. 14–15.
314. Ibid., p. 3.
315. Ibid., p. 2.
316. Ibid., p. 2–3.
317. Ibid., p. 60.
318. Ibid., p. 52. See above, p. 391.
319. Ibid., pp. 60–61.
320. Ibid., p. 61.
321. Ibid.,
322. Darwin, *Origin,* p. 87.
323. Ibid., p. 62: "I should premise that I use the term Struggle for Existence in a large and metaphorical sense. . . ."

324. Kropotkin, *Mutual Aid,* p. 54.
325. Ibid., p. 57.
326. Ibid., p. 54.
327. Ibid., p. 53.
328. G. Ledyard Stebbins, *Processes of Organic Evolution* (Englewood Cliffs, 1966), p. 65.
329. Kropotkin, *Mutual Aid,* p. 53.
330. Ibid., p. 56.
331. Ibid., p. 57.
332. Ibid.
333. Ibid., pp. 59–60.
334. Stebbins, p. 64.
335. Kropotkin, *Mutual Aid,* p. 14. See above, p. 401.
336. Ibid., p. 61.
337. Ibid., p. 2.
338. Ibid., p. 62.
339. Ibid., p. 61.
340. Ibid., p. 55.
341. *Hsun Tzu,* Chapter 1, p. 1.
342. Stebbins, p. 147.
343. Ibid.
344. Darwin, *Origin,* p. 79.
345. Kropotkin, *Mutual Aid,* p. 63.
346. *Meng Tzu,* 7.7, p. 168.
347. *Chung-yung,* Chapter 30, p. 54.
348. *Lao Tzu,* Chapter 81, Part 2, p. 24b.
349. *SLHC,* IIb, 1044.
350. *HSC,* 8:30.
351. See Ibid. 12:45 and 13:121. I am not sure that Li Shih-tseng coined the term *hu chu.* Perhaps a member of the Tokyo group did, or perhaps some Japanese.
352. Ibid., 5:17.
353. *SLHC,* IIb, 1001.
354. Ibid., IIb, 1044.
355. Ibid., IIb, 1043.
356. Ibid., IIa, 482–483.
357. Chang Ping-lin, II, 893.
358. *SLHC,* IIa, 483.
359. Chang Ping-lin, II, 893.
360. Ibid., II, 859.
361. Ibid., II, 885.
362. *SLHC,* IIa, 483–484.

363. Ibid., IIa, 484.
364. Ibid., IIa, 485.
365. Ibid., IIa 483.
366. Chang Ping-lin, II, 889.
367. Ibid., II, 893.
368. Ibid., II, 889.
369. Ibid., II, 891.
370. Ibid., II, 890.
371. Ibid., II, 887.
372. Ibid., II, 893.
373. See above, p. 398.
374. Chang Ping-lin, II, 886–888.
375. Ibid., II, 891.
376. Ibid., II, 890.
377. Ibid., II, 887–888.
378. Ibid., II, 888.
379. Ibid., II, 892.
380. Ibid.
381. Ibid., II, 888.
382. Ibid., II, 893.
383. *MP,* IV, 22:13.
384. Ibid., IV, 21.
385. *SLHC,* IIb, 1041.
386. *HSC,* 10:37.
387. *Mao Chu-hsi yü-lu,* p. 65.
388. Ibid., p. 76.
389. Wu Ching-heng, *Wu Ching-heng hsuan chi: Cheng lun* (Taipei, 1967), I, 29.
390. *SLHC,* IIb, 924.
391. Wu Ching-heng, *Cheng-lun,* I, 25. I have reversed subject and predicate.
392. Ibid., I, 9.
393. *SLHC,* IIb, 1035–1037.
394. Wu Ching-heng, *Cheng lun,* I, 9.
395. Ibid., I, 25.
396. *SLHC,* IIb, 1041.
397. *HSC,* 8:30.
398. Wu Ching-heng, *Cheng lun,* I, 79.
399. *HSC,* 1:1.
400. *Lun-yü,* 12.1. See above, p. 277.
401. Wu Ching-heng, *Cheng lun,* I, 50. For *fraternité,* Wu Chih-hui used the Confucian (albeit Mo-Tzian sounding) term *po ai* (universal love).
402. Eiseley, *Darwin,* p. 352. Eiseley brings his magnificent book towards

a close with these lines of Darwin: "If we choose to let conjecture run wild, then animals, our fellow brethren in pain, disease, suffering and famine—our slaves in the most laborious works, our companions in our amusements—they may partake of our origin in one common ancestor—we may be all melted together—noble lines except that they offer no resolution to the sorry fact that animals are our brethren, our slaves, our companions—and our meat, as many men have oft been theirs."

403. *SLHC,* IIb, 918–919.
404. Kropotkin, *Mutual Aid,* p. 58.
405. Wu Ching-heng, *Tsa-wen,* pp. 183–185.
406. Ibid., p. 186.
407. *HSC,* 64:389.
408. Ibid., 54:233–234.
409. Ibid., p. 234.
410. Ibid.
411. Ibid.
412. *Lun-yü,* 4.6, p. 93.
413. Ibid., 7.29, p. 138.
414. *Meng Tzu,* 1.7, p. 15.
415. Wu Chih-hui did once say that the Chinese people's nature, like that of all barbaric peoples, was short on spirit, but he still implied that they could have spirit if they wanted it. See *HSC,* 61:345.
416. *SLHC,* IIb, 907.
417. Ibid.
418. Ibid., IIb, 919–920.
419. Ibid., IIb, 923.
420. See *HSC,* 63:375.
421. *SLHC,* Ib, 870–871. Professor Nicholas Clifford, of Middlebury College, has suggested that Ssu-t'u-huo "may be Maria W. Miller Stewart, 1803–1879, a black woman born in Hartford, active in the Abolitionist movement; a collection of her writings was published in 1879 as *Meditations from the Pen of Mrs. Maria W. Stewart.*"
422. *HSC,* 53:216.
423. Ibid., 58:298.
424. *SLHC,* IIb, 914.
425. *Wo-men ti p'eng-yu pien t'ien-hsia,* another ubiquitous slogan, the *locus classicus* of which escapes me.
426. *SLHC,* IIb, 949.
427. See above, p. 316.
428. *SLHC,* IIb, 1022.
429. *HSC,* 39:154.
430. Ibid., 60:323.

431. *SLHC,* IIb, 1034, 1037.
432. Ibid., IIb, 1039.
433. Wu Ching-heng, *Wu Ching-heng hsuan chi: Shu hsin* (Taipei, 1967), I, 3.
434. *HSC,* 61:348.
435. Li Shih-tseng, *Ko-ming* in *Hsin shih chi ts'ung shu* collection 1, (1907), *HSC,* p. 553.
436. Wu Ching-heng, *Cheng lun,* p. 130.
437. *SLHC,* IIb, 924, 930.

WHEN ALL WAS SAID AND DONE

1. *KFCC,* III, 49.
2. *KFCS,* p. 530.
3. Ibid., pp. 544–545.
4. *KFCC,* III, 131.
5. Conversation with Professor Mou Jun-sun.
6. See above, p. 180.
7. *SLHC,* IIb, 1043.
8. *HSC,* 1:3.
9. Ibid., 2:7.
10. *YPS,* II, 3:58.
11. *SLHC,* IIb, 1050.
12. Lu Hsun, *Je feng,* p. 15.
13. Ibid., p. 34.
14. Lu Hsun, *San hsien chi,* p. 12.
15. Lu Hsun, *Erh hsin chi,* p. 24.
16. Ibid., p. 47.
17. Chou Chen-fu, p. 270. See also Schwartz, p. 235.
18. Liang Ch'i-ch'ao, *Ou yu,* p. 12.
19. Ibid., p. 9.
20. Ibid., p. 12.
21. Chou Chen-fu, p. 270.
22. Liang Ch'i-ch'ao, *Ou yu,* p. 38.
23. Grieder, p. 135.
24. See above, p. 292.
25. Hu Shih, *Wen hsuan,* p. 56. See also Liang Ch'i-ch'ao, *Ou yu,* p. 12.
26. *HCN,* III, 5:1–2.
27. Ibid., III, 5:4.
28. Ibid., III, 5:5.
29. Hu Shih, *Wen hsuan,* Introduction, p. 2.
30. *HCN,* III, 5:5.

31. Hu Shih, *Wen hsuan,* p. 4.
32. *HCN,* VI, 3:274.
33. Hu Shih, *Wen hsuan,* p. 72.
34. Liang Ch'i-ch'ao, *Ou yu,* p. 19.
35. *HCN,* III, 5:5.
36. Hu Shih, *Wen hsuan,* p. 69.
37. *HCN,* VI, 4:357.
38. Hu Shih, *Wen hsuan,* Introduction, p. 3.
39. *Mao Chu-hsi yü-lu,* p. 56.
40. *HCN,* VIIIa, 3:399.
41. Ibid., VIIb, 5:5.
42. Ibid., VIIIa, 4:510.
43. Ibid., IXa, 2:152–153. The two legs that Mao Tse-tung had in mind, of course, in his injunction *liang t'iao t'ui tsou lu* (walk on two legs), were modern and pre-modern technology.
44. *HCN,* IXa, 2:153.
45. Siao-yu, *Mao Tse-tung and I were Beggars* (New York, 1973), p. 226.
46. *HCN,* IXa, 4:555 (my italics).
47. Ibid., IXa, 1:7.
48. Ibid., XIII, 3:358.
49. *Meng Tzu,* 10.1, p. 232.
50. *HPLT,* II, 404–405.
51. Sun Wen (Sun Yat-sen), "Fa k'an tz'u," *MP,* I, 1:3.
52. *Sources,* p. 812 (quoted from *China's Destiny*).
53. Mao Tse-tung, *Hsuan chi,* III, 999.
54. See above, pp. 337–340.
55. *HCN,* IXa, 3:425.
56. Ibid., IXa, 1:140.
57. Ibid., IXb, 6:799. The "monkey-keeper" was the Chuang-tzian monkey keeper who appeased his charges, dissatisfied with getting three acorns in the morning and four at night, by giving them four in the morning and three at night. See *Chuang Tzu,* Book 2.4, pp. 14–15.
58. *HCN,* XIV, 1:145.
59. Ibid., XIV, 131.
60. See above, p. 208.
61. *Mao Tse-tung lun wen-hsueh ho i-shu* (Peking, 1964), p. 100.
62. Lu Hsun, *Ch'ie chie t'ing tsa-wen erh chi* (Hong Kong, 1964), p. 175.
63. See *HMTP,* XV, 82:21, and *KFCS,* p. 1.
64. *Mao Tse-tung lun wen-i,* p. 100.
65. *Mao Chu-hsi yü-lu,* p. 75.
66. Lu Hsun, *Fen,* p. 202.
67. *Mao Chu-hsi yü-lu,* p. 225.

68. For the Lao tzu passage, see above, p. 174. For Mao Tse-tung's "acceptance" of it, see Mao Tse-tung, *Hsuan chi,* IV, 1365: "'You people are not humane.' That is quite so. We will definitely not extend humane government to the reactionaries or to the reactionary activities of the reactionary classes." Needless to say, Lao tzu was thinking of something else.

69. *TYL,* pt. 2, p. 46.

Bibliography of Works Cited

Beeching, Jack. *The Chinese Opium Wars.* New York, Harcourt Brace Jovanovich, 1975.

Bennett, Adrian Arthur. *John Fryer: The Introduction of Western Science and Technology into Nineteenth Century China.* Cambridge, East Asian Research Center, Harvard University, 1967.

Bernal, Martin, "The Triumph of Anarchism over Marxism, 1906–1907," in Mary Clabaugh Wright, *China in Revolution, The First Phase, 1900–1913.* New Haven, Yale University Press, 1968.

Chan, Wing-tsit. *A Source Book in Chinese Philosophy.* Princeton, Princeton University Press, 1963.

Chang Chih-tung 張之洞. *Ch'üan hsueh p'ien* 勸學篇 (An exhortation to study.) Reprint, Chekiang Province, 1898.

Chang Ching-lu 張靜廬 and Li Sung-nien 李松年. "Hsin-hai ko-ming shih-ch'i chung-yao pao-k'an tso-che pi ming lu" 辛亥革命時期重要報刊作者筆名錄 (A list of pen names used by writers in important journals of the 1911 Revolution period), *Wen shih* 文史 (Literature and history), Volume 1 (October 1962).

Chang, Hao. *Liang Ch'i-ch'ao and Intellectual Transition in China, 1890–1907.* Cambridge, Harvard University Press, 1971.

Chang Ping-lin 章炳麟. *Chang shih ts'ung shu* 章氏叢書 (The writings of Chang Ping-lin). 2 vols. Taipei, 1958.

Ch'ien Mu 錢穆 *Chung-kuo san pai nien hsueh-shu shih* 中國三百年學術史 (Three hundred years of Chinese intellectual history). Shanghai, 1937.

Chin Ch'ung-chi 金冲及. "*T'ien-yen lun* ho Chung-kuo chin-tai fan K'ung ssu-ch'ao" 天演論和中國近代反孔思潮 (*T'ien-yen lun* and the modern Chinese current of anti-Confucian thought), *Ta kung pao* 大公報 (The public news; December 24, 1973).

Ch'ing i pao 清議報 (Honest criticism). Liang Ch'i-ch'ao 梁啟超 ,

Feng Ching-ju 馮鏡如 et al., eds. Photolithograph. 12 vols. Taipei, 1967.
Chou Chen-fu 周振甫. *Yen Fu ssu-hsiang shu-p'ing* 嚴復思想述評 (A critical exposition of Yen Fu's thought). Taipei, 1964.
Chow Tse-tsung. *The May Fourth Movement.* Cambridge, Harvard University Press, 1960.
Chuang Tzu tsuan chien 莊子纂箋 (Chuang Tzu with selected commentary), Ch'ien Mu 錢穆, ed. Hong Kong, 1962.
Ch'un ch'iu Kung-yang chuan 春秋公羊傳 (The Kung-yang commentary to the *Spring and Autumn Annals*). Ssu pu pei yao 四部備要 edition. Shanghai, 1927.
Chung-kuo che-hsueh shih tzu-liao hsuan-chi, Sung, Yuan, Ming chih pu 中國哲學史資料選輯宋元明之部 (Selected source materials in the history of Chinese philosophy: Sung, Yuan, and Ming). Chung-kuo k'o-hsueh yuan che-hsueh yen-chiu suo, Chung-kuo che-hsueh shih tsu 中國科學院哲學研究所中國哲學史組 (The History of Chinese Philosophy Group of the Philosophy Research Institute of the Chinese Academy of Sciences), ed. Peking, 1962.
Chung-yung 中庸 (The doctrine of the mean), in *Ssu shu hsin chie* 四書新解 (A new explanation of the four books). Chang Shou-pai 張守白, ed. Tainan, 1961.

Darwin, Charles. *On the Origin of Species.* Facsimile of 1st ed. Cambridge, Harvard University Press, 1966.
———. *The Descent of Man* and *Selection in Relation to Sex.* 2nd ed. New York and London, n.d.
———. *The Expression of the Emotions in Man and Animals.* Chicago, University of Chicago Press, 1970.
Durrell, Gerald. *Two in the Bush.* New York, Viking Press, 1971.

Eiseley, Loren. *The Immense Journey.* New York, Randon House, 1957.
———. *Darwin's Century.* New York, Doubleday Anchor, 1958.

Fairbank, John King. *Trade and Diplomacy on the China Coast.* Cambridge, Harvard University Press, 1953.
Fairbank, John K., Edwin O. Reischauer, and Albert M. Craig. *East Asia: The Modern Transformation.* Boston, Houghton Mifflin, 1965.
Fung Yu-lan. *A History of Chinese Philosophy,* Volume 2. Princeton, Princeton Universtiy Press, 1953.

Gray, Asa. *Darwiniana.* Cambridge, Harvard University Press, 1963.
Grieder, Jerome B. *Hu Shih and the Chinese Renaissance.* Cambridge, Harvard University Press, 1970.

Halévy, Elie. *The Growth of Philosophical Radicalism*. Boston, Beacon Press, 1966.

Han Fei Tzu chi shih 韓非子集釋 (Collected commentary to Han Fei Tzu). Ch'en Ch'i-yu 陳奇猷, ed. Peking, 1959.

Han shu 漢書 (The history of the [Former] Han dynasty), Volume 3. I wen yin shu kuan 藝文印書館 edition. N.p.n.d.

Himmelfarb, Gertrude. *Victorian Minds*. New York, Harper Torchbook. 1970.

Hinton, William. *Iron Oxen*. New York, Random House, 1970.

Hofstadter, Richard. *Social Darwinism in American Thought*. Rev. ed. Boston, Beacon Press, 1955.

Howard, Richard Campbell. "The Early Life and Thought of K'ang Yu-wei, 1858–1895." PhD dissertation, Columbia University, 1972.

Hsi I 郗义. "Lu Hsun ti che-hsueh kuan-tien" 魯迅的哲学观点 (Lu Hsun's philosophical point of view), *Che-hsueh yen-chiu* 哲学研究 (Philosophical research), 5:82–98 (1957).

Hsiang hsueh hsin pao 湘學新報 (The Hunan study news). Photolithograph. 4 vols. Taipei, 1966.

Hsiang pao lei tsuan 湘報類纂 (Topical selections from the Hunan news), T'an Ssu-t'ung 譚嗣同, T'ang Ts'ai-ch'ang 唐才常 et al., eds. Photolithograph. 2 vols. Taipei, 1969.

Hsin ch'ing-nien 新青年 (The new youth). Photolithograph. 14 vols. Shanghai, 1936.

Hsin-hai ko-ming ch'ien shih nien chien shih-lun hsuan-chi 辛亥革命前十年間時論選集 (A selection of topical essays written during the ten years before the 1911 Revolution). Chang Nan 張枬 and Wang Jen-chih 王忍之, eds. Volumes Ia & lb, Hong Kong, 1962; Volumes 2a & 2b, Peking, 1963.

Hsin min ts'ung pao 新民叢報 (A new people). Feng Tzu-shan 馮紫珊, ed. Photolithograph. 17 vols. Taipei, 1966.

Hsin shih-chi 新世紀 (The new century). Photolithograph. Nos. 1–77. Tokyo, 1966.

Hsun Tzu chien shih 荀子柬釋 (Hsun Tzu, with selected explanations). Liang Ch'i-hsiung 梁啟雄, ed. Taipei, 1969.

Hu Shih 胡適. *Ssu-shih tzu shu* 四十自述 (An autobiography at age forty). Taipei, 1959.

———. *Hu Shih wen hsuan* 胡適文選 (Selected works of Hu Shih). Taipei, 1962.

Hung ch'i 紅旗 (Red flag). Peking.

Huxley, Thomas H. *Man's Place in Nature*. Ann Arbor, University of Michigan Press, 1961.

Huxley, T. H., and Julian Huxley. *Evolution and Ethics, 1893–1943*. London, Pilot Press, 1947.

K'ang Yu-wei 康有為. *Ch'un ch'iu Tung shih hsueh* 春秋董氏學 (Tung Chung-shu's study of the *Spring and Autumn Annals*). Taipei, 1969.

———. *Chung-yung chu* 中庸注 (Notes on the doctrine of the mean). Taipei, 1968.

———. *Hsin hsueh wei ching k'ao* 新學偽經考 (An examination of Hsin scholarship and the forged Classics). Taipei, 1962.

———. *K'ung Tzu kai chih k'ao* 孔子改制考 (Confucius as a reformer). 2nd ed. Peking, 1922.

———. *K'ang Nan-hai tzu pien nien-p'u* 康南海自編年譜 (A self-compiled chronology of the life of K'ang Yu-wei), in Chin-tai Chung-kuo shih liao ts'ung k'an 近代中國史料叢刊 (Collected modern Chinese historical materials), 2nd series, No. 11. Taipei, 1966.

———. *Meng Tzu wei* 孟子微 (The profound subtleties of Mencius). Taipei, 1968.

———. Nan-hai hsien-sheng shih chi 南海先生詩集 (Collected poems of K'ang Yu-wei). Hand copy by Liang Ch'i-ch'ao. Photolithograph. Hong Kong, K'ang T'ung-huan, 1966.

———. *Ta t'ung shu* 大同書 (The great unity). Peking, 1956.

Kelly, Walt. *Impollutable Pogo*. New York, Simon and Schuster, 1970.

Kidd, Benjamin. *Social Evolution*. New York, MacMillan, 1894.

Ko Kung-chen 戈公振. *Chung-kuo pao-hsueh shih* 中國報學史 (A history of Chinese journalism). Taipei, 1964.

Kropotkin, P. A. *Mutual Aid*. Popular ed. London, William Heinemann, 1919.

———. *Memoirs of a Revolutionist*. New York, Grove Press, 1970.

Kuo fu ch'üan chi 國父全集 (The complete works of Sun Yat-sen). Chinese Kuomintang Archives, ed. 6 vols. Rev. ed. Taipei, 1957.

Kuo fu ch'üan shu 國父全書 (The complete writings of Sun Yat-sen). Chang Ch'i-yun 張其昀, ed. Taipei, 1960.

Kwok, D. W. Y. *Scientism in Chinese Thought, 1900–1950*. New Haven, Yale University Press, 1965.

Lao Tzu tao te ching 老子道德經 (The Tao te ching of Lao Tzu). Commentary and punctuation by Yen Fu. Notes by Wang Pi 王弼. Shanghai, 1935.

Levenson, Joseph R. *Confucian China and Its Modern Fate: A Trilogy*. Berkeley and Los Angeles, University of California Press, 1968.

Li-shih yen-chiu 历史研究 (Historical research). Peking.

Liang Ch'i-ch'ao 梁啟超. *Hsin min shuo* 新民說 (Towards a new people). Taipei, 1959.

———. *Tzu-yu shu* 自由書 (On liberty). Taipei, 1960.

———. *Yin-ping-shih wen chi* 飲冰室文集 (Collected writings from the ice-drinker's studio). 16 vols. Taipei, 1960.

———. *Ou yu hsin-ying lu chie-lu* 歐遊心影錄節錄 (Selections from a record of impressions of travel in Europe). Taipei, 1966.

Liu, James J. Y. *The Chinese Knight-errant.* Chicago, University of Chicago Press, 1967.

Liu P'an-sui 劉盼遂. *Lun heng chi chie* 論衡集解 ([Wang Ch'ung's 王充] critical essays with a compilation of explanatory notes). Shanghai, 1957.

Lu Hsun 魯迅. *Lu Hsun hsiao shuo chi* 魯迅小說集 (The collected short stories of Lu Hsun). Hong Kong, 1964.

———. *Je feng* 熱風 (A hot wind). Hong Kong, 1964.

———. *Fen* 墳 (A tomb). Hong Kong, 1964.

———. *Erh i chi* 而已集 (Nothing more). Hong Kong, 1964.

———. *San hsien chi* 三閒集 (Thrice idle). Hong Kong, 1964.

———. *Erh hsin chi* 二心集 (Traitorous thoughts). Hong Kong, 1964.

———. *Wei tzu-yu shu* 偽自由書 (On fake liberty). Hong Kong, 1964.

———. *Ch'ie chie t'ing tsa-wen erh chi* 且介亭雜文二集 (A second collection of random essays from an abode in a semi-concession). Hong Kong, 1964.

Lun-yü 論語 (The *Analects*), in *Ssu shu hsin chie.* Chang Shou-pai, ed. Tainan, 1961.

Mackenzie, Robert. *The 19th Century: A History.* London, T. Nelson and Sons, 1889.

Manuel, Frank E. *The Prophets of Paris.* New York, Harper Torchbook, 1965.

Mao Chu-hsi yü-lu 毛主席語錄 (Quotations from Chairman Mao). Peking, 1967.

Mao Tse-tung hsuan chi 毛泽东选集 (Selected works of Mao Tse-tung). 4 vols. Peking, 1969.

Mao Tse-tung lun wen-hsueh ho i-shu 毛泽东論文学和艺术 (Mao Tse-tung on art and literature). Peking, 1964.

Marx and Engels: Basic Writings on Politics and Philosophy. Lewis S. Feuer, ed. New York, Doubleday Anchor, 1959.

Meisner, Maurice. *Li Ta-chao and the Origins of Chinese Marxism.* Cambridge, Harvard University Press, 1967.

Meng Tzu i chu 孟子譯注 (Mencius, with translation and notes). *Meng Tzu i chu* team of the Lanchou University Chinese Department, ed. Hong Kong, n.d.

Mill, John Stuart. *On Liberty.* Aldurey Castell, ed. New York, Appleton-Century-Crofts, 1947.

Min pao 民報 (The people's journal). Photolithograph. 4 vols. Peking, 1957.

Mo Tzu chien ku 墨子閒詁 (An annotated Mo Tzu). Sun I-jang 孫詒讓, ed. Taipei, 1970.

Nietzsche, Friedrich. *The Birth of Tragedy and the Genealogy of Morals.* Francis Golffing, tr. New York, Doubleday, 1956.

People's Comic Book, The. Endymion Wilkinson, tr. New York, Doubleday, 1973.

Philosophy of Hegel, The. Carl J. Friedrich, ed. New York, Modern Library, 1959.

Philosophy of Kant, The. Carl J. Friedrich, ed. New York, Modern Library, 1949.

Pu jen tsa-chih hui-pien 不忍雜誌彙編 (Topical selections from The compassion magazine). K'ang Yu-wei, ed. Photolithograph of 1st ed., Shanghai, 1914. 2 vols. Taipei, 1968.

Pusey, James R. "K'ang Yu-wei and *Pao-chiao,* Confucian Reform and Reformation," *Papers on China* 20:144–176. Cambridge, East Asian Research Center, Harvard University, 1966.

———. *Wu Han: Attacking the Present through the Past.* Cambridge, East Asian Research Center, Harvard University, 1969.

Rankin, Mary Backus. *Early Chinese Revolutionaries: Radical Intellectuals in Shanghai and Chekiang, 1902–1911.* Cambridge, Harvard University Press, 1971.

Saint-Simon, Henri de. *Social Organization, The Science of Man, and Other Writings.* Felix Markham, ed. and tr. New York, Harper & Row, 1964.

Scalapino, Robert A., and George T. Yu. *The Chinese Anarchist Movement.* Berkeley, Center for Chinese Studies, Institute of International Studies, University of California, 1961.

Schiffrin, Harold Z. *Sun Yat-sen and the Origins of the Chinese Revolution.* Berkeley and Los Angeles, University of California Press, 1970.

Schram, Stuart R. *The Political Thought of Mao Tse-tung.* New York, Praeger, 1972.

Schwartz, Benjamin. *In Search of Wealth and Power: Yen Fu and the West.* Cambridge, Harvard University Press, 1964.

Shih chi hui-chu k'ao-cheng (*Shiki kaichū kōshō*) 史記會注考證 (*Shih chi,* with selected notes from various commentaries, critically assessed). Takigawa Kametarō 瀧川龜太郎, ed. N.p., I wen yin shu kuan, n.d.

Shih ching 詩經 (The book of poetry). Chinese text in James Legge, *The Chinese Classics,* Volume IV, *The She King.* Reprint, Taipei, 1963.

Shih-wu pao 時務報 (Current affairs), Wang K'ang-nien 汪康年, Liang Ch'i-ch'ao et al., eds. Photolithograph. 6 vols. Taipei, 1967.

Shu ching 書經 (The book of history). Chinese text in James Legge, *The Chinese Classics,* Volume III, *The Shoo King.* Reprint, Taipei, 1963.

Siao-yu. *Mao Tse-tung and I were Beggars.* New York, Collier Books, 1973.

Snow, Edgar. *Red Star Over China.* 1st rev. and enlarged ed. New York, Grove Press, 1961.

Solomon, Richard. *Mao's Revolution and the Chinese Political Culture.* Berkeley, University of California Press, 1971.

Sources of Chinese Tradition. Wm. Theodore de Bary et al., eds. New York, Columbia University Press, 1960.

Stebbins, G. Ledyard. *Processes of Organic Evolution.* Englewood Cliffs, Prentice Hall, 1966.

Ta hsueh 大學 (The great learning), in *Ssu shu hsin chie.* Chang Shou-pai, ed. Tainan, 1961.

Ta-erh-wen 達爾文 (Charles Darwin). *Wu chung yuan shih* 物種原始 (On the origin of species). Ma Chün-wu 馬君武, tr. 2 vols. Taipei, 1957.

T'ai Hsi hsin shish lan-yao 泰西新史攬要 (The essentials of recent Western history). Timothy Richard (Li T'i-mo-t'ai 李提摩太), tr. Shanghai, 1895. A translation of Robert Mackenzie, *The 19th Century: A History.*

T'ang Chih-chün 湯志鈞. *Wu-hsu pein-fa shih lu lun ye* 戊戌变法史論業 (Studies of the Reform Movement of 1898). Hong Kong, 1973.

Teng Ssu-yü and John K. Fairbank. *China's Response to the West.* Cambridge, Harvard University Press, 1954.

Thompson, Laurence G., tr. *Ta T'ung Shu: The One-World Philosophy of K'ang Yu-wei.* London, Allen & Unwin, 1958.

Ting Wen-chiang 丁文江. *Liang Jen-kung hsien-sheng nien-p'u ch'ang pien ch'u kao* 梁任公先生年譜長編初稿 (A first draft chronological biography of Liang Ch'i-ch'ao). 2 vols. Taipei. 1962.

Tocqueville, Alexis de. *Democracy in America.* Philips Bradley, ed. Henry Reeve, tr. 2 vols. New York, Vintage, Random House, 1945.

Ts'ao Chü-jen 曹聚仁. *Lu Hsun p'ing chuan* 魯迅評傳 (A critical biography of Lu Hsun). Hong Kong, 1961.

Tso chih ch'u yen 佐治芻言 (Humble suggestions to aid government), John Fryer (Fu-lan-ya 傅蘭雅) and Ying Tsu-hsi 應祖錫, trs. in *Hsi hsueh fu-ch'iang ts'ung shu* 西學富强叢書 (A collection of Western studies for wealth and power), Volume 25. N.p., Hung wen shu chü, 1896. A translation of "Homely Words to Aid Government," in *Chambers' Educational Course.* William and Robert Chambers, ed. Edinburgh, n.d.

Tsou Jung 鄒容. *Ko-ming chün* 革命軍 (A revolutionary army). Taiwan, Taipei (?), 1954.

Tu Fu 杜甫. "Ping chü hsing," 兵車行 (War chariot song), in "Tu Fu shih hsuan chu" 杜甫詩選注 (An annotated selection of poems by Tu Fu), in *Shih tz'u hsin-shang* 詩詞欣賞 (Poetry to be enjoyed), 2nd collection. Li Tu , ed. Kaohsiung, 1962.

Turgot, Anne Robert Jacques, Baron de l'Aulne. *Oeuvres de M. Turgot.* Volume 2. Paris, 1808.

Tz'u hai 辭海 (A sea of words). Shanghai, 1948.

Wang Ch'ü-ch'ang 王蘧常. *Yen Chi-tao nien-p'u* 嚴幾道年譜 (A chronological biography of Yen Fu). Shanghai, 1936.

Wang Tzu-ch'un 汪子春 and Chang Ping-lun 張秉倫. "Ta-erh-wen hsueh-shuo tsai Chung-kuo te ch'uan-po he ying-hsiang" 达尔文学说在中国的传播和影响 (The Introduction and influence of Darwin's theory in China). Unpublished paper delivered at conference commemorating the 100th anniversary of Darwin's death, Peking, May 19, 1982.

Wang Wei 王維. "Kuo Hsiang Chi Ssu" 過香積寺 (Passing by the Hsiang Chi temple), in *T'ang shih san pai shou hsiang hsi* 唐詩三百首祥析 (A detailed analysis of three hundred T'ang poems). Taipei, 1963.

Wong Man, tr. and ed. *Poems of Mao Tse-tung.* Hong Kong, Eastern Horizon Press, 1966.

Wu Ching-heng 吳敬恒 (Wu Chih-hui 吳稚暉). *Wu Ching-heng hsuan-chi: Cheng lun* 吳敬恒選集政論 (Selected works of Wu Ching-heng: Political essays). 2 vols. Taipei, 1967.

———. *Wu Ching-heng hsuan-chi: Hsu-pa, yu-shi, tsa-wen* 吳敬恒選集序跋遊記雜文 (Prefaces, travel reports, and random essays), Taipei, 1967.

———. *Wu Ching-heng hsuan-chi: Shang-hsia ku-chin t'an* 吳敬恒選集上下古今談 (Talks of Heaven and earth and past and present). 2 vols. Taipei, 1967.

———. *Wu Ching-heng hsuan-chi: Shu hsin* 吳敬恒選集書信 (Letters). 3 vols. Taipei, 1967.

Wu-hsu pien-fa 戊戌变法 (The Reform Movement of 1898). Chien Po-tsan 翦伯贊 et al., eds. 4 vols. Shanghai, 1953.

Yang wu yun-tung 洋務運動 (The foreign matters movement). Chung-kuo shih-hsueh hui 中國史學會 (The China Historical Association), ed. 8 vols. Shanghai, 1961.

Yeh Te-hui 葉德輝, ed. *I chiao ts'ung pien* 翼教叢編 (Collected works in defense of Confucianism), in Chin-tai Chung-kuo shih liao ts'ung k'an, Series 65, Volume 647. Taipei, n.d.

Yen Chi-tao shih wen ch'ao 嚴幾道詩文鈔 (A collection of Yen Fu's poems and essays), in Chin-tai Chung-kuo shih liao ts'ung k'an, series 42, No. 417. Taipei, 1969.

Yen Fu 嚴復. *T'ien-yen lun* 天演論 (The theory of evolution). Taipei, 1967. A translation of Thomas Henry Huxley (Ho-hsu-li 赫胥黎), *Evolution and Ethics*.

Yen Fu shih wen hsuan 嚴復詩文選 (Selected poems and essays of Yen Fu), Chou Chen-fu, ed. Peking, 1959.

Yü Ying-shih. *Trade and Expansion in Han China*. Berkeley and Los Angeles, University of California Press, 1967.

Glossary

ai-kuo hsin 愛國心

bushidō 武士道

chang-ch'eng 章程
Chang Chi 張繼
Chang Sung-nien 張崧年
Ch'ang Hsing Li 長興里
Che-chiang ch'ao 浙江潮
Chen-tan 震旦
"Ch'en Hsing-t'ai hsien-sheng chueh ming shu" 陳星台先生絕命書
Ch'en I-k'an 陳儀侃
Ch'en Pao-chen 陳寶箴
Ch'en T'ien-hua 陳天華
Ch'en Tu-hsiu 陳獨秀
cheng 爭
cheng tzu-ts'un 爭自存
Chi Po 祭伯
ch'i 氣
ch'i ma chao ma 騎馬找馬
Chia I 賈誼
Chiang Chih-yu 蔣智由
Chiang Kuang-ch'ih 蔣光赤
Chiang-su 江蘇
Ch'iang 羌
Ch'iang hsueh hui 强學會
chiao-chu 教主
Chie 羯
Chie (King) 桀
chien-chung 賤種
chien-hsiung 奸雄
chien ju 賤儒
"Chien Kung" 劍公
Chien Po-tsan 翦伯贊
ch'ien chü k'o chien 前車可鑑
"Chih-kung tien-yen lun" 治功天演論
Chih-na 支那
chin-ch'ü 進取
chin-hua 進化
chih shih 志士

Chin te hui 進德會
Ch'in Shih Huang 秦始皇
Ching-an 競盦
ching-shen 精神
ch'ing ch'u yü lan erh sheng yü lan 青出於藍而勝於藍

Ch'iu Chin 秋瑾
Chou 紂
Chou Fo-hai 周佛海
Chou Wu Wang 周武王
Chu Chih-hsin 朱執信
"Chü-fen chin-hua lun" 俱分進化論
chü-luan 據亂
Chü P'u 鞠普
ch'uan chung 傳種
ch'üan p'an hsi-hua 全盤西化
ch'üan-shih 權勢
ch'ün 群(羣)
ch'ün-hsueh 群學
chung 忠
Chung hsi hsueh-hsiao 中西學校
Chung-hua 中華
Chung-kuo hsin pao 中國新報
"Chung-kuo shih hsu lun" 中國史敘論

fa 法
"Fa k'an tz'u" 發刊詞
feng-chien 封建
Feng Kuei-fen 馮桂芬
Feng Mao-lung (Tzu-yu) 馮懋龍(自由)
fu-mu kuan 父母官
fu pi 復辟
fu shen 富紳

hai ch'ün chih ma 害群之馬
Han Wu Ti 漢武帝
Han Yü 韓愈
hao sheng chih hsin 好勝之心
hao yin hao sha chih hsin 好淫好殺之心
Hei nu hsu t'ien lu 黑奴吁天錄
Ho Chen 何震
Ho Hsiu 何休
Ho Lai-pao 何來保
ho li chi ch'ün 鶴立鷄群
hou chin po ku 厚今薄古
hou lai chü shang 後來居上
Hsi Ch'iao Shan 西樵山
hsi hsing 習性
hsia 俠
Hsia Tseng-yu 夏曾祐
hsiang-lai ju tz'u 向來如此
hsiao 孝
hsiao k'ang 小康
hsiao min-tsu-chu-i 小民族主義
hsiao wo 小我

hsien 先
hsien chih hsien chueh 先知先覺
hsin 心
hsin-li chien-she 心理建設
hsin-li hsi-ti 心理洗滌
"Hsin shih hsueh" 新史學
hsing-ko 性格
hsiu-lien 修練
hsiu shou p'ang kuan 袖手旁觀
Hsiung Hsi-ling 熊希齡
Hsiung-nu 匈奴
Hsu Chi-yü 徐繼畬
Hsu Ch'in (Chün-mien) 徐勤（君勉）
Hsu Hsi-lin 徐錫麟
Hsueh Fu-ch'eng 薛福成
hsueh-hui 學會
hsun-cheng 訓政
Hsun Hsu 荀勗
hu chin t'u yuan 忽今圖遠
hu chu 互助
Hu Han-min 胡漢民
Hua-erh-ssu-lei 華爾司雷
Huang Kuo-k'ang 黃國康
Huang Tsun-hsien 黃遵憲
hun 魂
Hung Hsiu-ch'üan 洪秀全
Hung lou meng 紅樓夢
hung shui meng shou 洪水猛獸
i 義
i chih i luan 一治一亂
i-ch'uan 遺傳
i ch'ün wei t'i, i pien wei yung 以群為體 以變為用
i hsin i te 一心一德
i i chung 遺宜種
i-lei 異類
i-li 毅力
I Nai 易鼐
i p'an san sha 一盤散沙
I Yin 伊尹
iao mo kuei kuai 妖魔鬼怪
jang-pu cheng-ts'e 讓步政策
jen (man) 人
jen (to bear, forebearance) 忍
jen (compassion) 仁
jen chih 人治
jen chih suo yü t'ien pi ts'ung chih 人之所欲天必從之
jen ch'ün 人群
Jen hsueh 仁學
jen-i tao-te 仁義道德
jen jen chün-tzu 仁人君子
Jen-lei chin-hua hsueh 人類進化學
Jen-lei kung-li 人類公理
jen li 人力
jen-li-ch'e 人力車
jen li te chin-pu 人力的進步
jen tse 人擇

jen wei 人為
jen yü 人欲
jou tao 柔道
jung-jen 容忍

kai-liang-chu-i 改良主義
kai ssu 該死
k'ai-ming chuan-chih 開明專制
K'ang Tzu 康子
Kao Hou 高垕
Kao Hsieh 高爕
Katō Hiroyuki 加藤弘之
ko-chih 格致
"Ko kuo chung-lei k'ao" 各國種類考
ko ming 革命
"Ko t'ien" 革天
k'o chi fu li wei jen 克己復禮為仁
K'o-hsueh yü jen-sheng-kuan 科學與人生觀
ku chin Chung wai 古今中外
ku i yu chih 古已有之
ku wei chin yung 古為今用
Ku Yen-wu 顧炎武
"K'uang-jen jih-chi" 狂人日記
kuei-ch'ih 傀尺
kuei-chung 貴種
Kumagai Gorō 熊谷五郎
Kung fa hui t'ung 公法會通
Kung fa tsung lun 公法總論
kung te 公德
Kung Ting-an (Tzu-chen) 龔定菴（自珍）
Kuo Cheng-chao 郭正昭
"Kuo-chia ssu-hsiang pien-ch'ien i-t'ung lun" 國家思想變遷異同論
kuo ch'ün 國群
kuo-min 國民
Kuo-min jih-jih pao 國民日日報
Kuo Sung-t'ao 郭嵩燾
Kuo ts'ui hsueh pao 國粹學報
kuo wang 國亡
Kuo wen hui pien 國聞匯編

Lao Wu 老五
le-li-chu-i 樂利主義
Lei Feng 雷鋒
li (rites, propriety) 禮
li (reason, principle) 理
li (force, effort) 力
li (profit) 利
li cheng shang yu 力爭上游
Li chi 禮記
li-chi-hsin 利己心
"Li chi yü ai t'a" 利己與愛他

Li Hung-chang 李鴻章
li i erh fen shu 理一而分殊
Li Shih-tseng 李石曾

Li Ta-chao 李大釗
li-t'a-hsin 利他心
"Li yun" 禮運
Liang Ch'i-hsun 梁啟勳
liang hsin 良心
Liang Sou-ming 梁漱溟
liang t'iao t'ui tsou lu 兩條腿走路
Liao P'ing 廖平
lieh chung 劣種
lieh teng 躐等
Lin Piao 林彪
Lin Tse-hsu 林則徐
Ling Shih 靈石
Liu Chi (Po-wen) 劉基（伯溫）
Liu Ch'i-chi (Ya-tzu) 柳棄疾（亞子）
Liu Hsien-chih 劉顯志
Liu Hsin 劉歆
Liu Pang 劉邦
Liu Shih-p'ei 劉師培
Liu Ssu-fu 劉思復
Lu Hsiang-shan 陸象山
"Lun chin-ch'ü mao-hsien" 論進取冒險
"Lun kung te" 論公德
"Lun Man-chou sui yü li hsien erh pu neng" 論滿州雖欲立憲而不能
"Lun shang wu" 論尚武
"Lun shih pien chih chi" 論世變之亟
"Lun tzu-tsun" 論自尊
"Lun tzu-yu" 論自由

Ma Chün-wu 馬君武
Ma Hui 馬悔
Mai Meng-hua 麥孟華
"Meng hsing lu" 猛醒錄
min-ch'üan 民權
min-ch'üan-chu-i 民權主義
"Min-ch'üan-chu-i ti erh chiang" 民權主義二講
min li, min chih, min te 民力民智民德
"Min pao chih liu ta chu-i" 民報之六大主義
Min sheng 民聲
min-sheng 民生
min-sheng-chu-i 民生主義
min-tsei cheng-t'i 民賊政體
min-tsu-chu-i 民族主義
"Min-tsu ti kuo-min" 民族的國民
Min to tsa-chih 民鐸雜誌
Ming T'ai Tsu 明太祖
miu lun 謬論
Mou Jun-sun 牟潤孫

Nan hsueh-hui 南學會
"Nan nü hun-yin tzu-yu lun" 男女婚姻自由論
"Nan nü tsa-chiao shuo" 男女雜交說

Nan-yang hsueh-t'ang 南洋學堂
nei luan 內亂
niu 牛
nu-li hsing 奴隸性
"Nu-ts'ai hao" 奴才好

pai-hua 白話
p'ai K'ung 排孔
P'an-hu 盤瓠
pao chiao 保教
pao chung, pao kuo, pao chiao 保種 保國 保教
Pao huang pao 保皇報
Pao kuo hui 保國會
Pao-yü 寶玉
Pi Yung-nien 畢永年
P'i Chia-yu 皮嘉祐
"P'i Han" 闢韓
p'i K'ung 批孔
P'i Lu-men 皮鹿門
pien fa 變法
"Pien-fa t'ung i" 變法通議
pien-ko 變革
p'ing-min 平民
"P'ing-p'i-tzu" 洴澼子
po ai 博愛
po chin hou ku 薄今厚古
"Po 'Ko-ming po-i'" 駁革命駁議
"Po *Pao huang pao*" 駁保皇報
p'o-huai 破壞
pu hsiao 不肖
"Pu hsiu—wo ti tsung-chiao" 不朽—我的宗教
pu jen 不忍
pu jen jen chih hsin 不忍人之心
pu-k'o-ssu-i 不可思議

san min chu-i 三民主義
san shih 三世
shan ch'ün 善群
Shan hai ching 山海經
shan yu shan pao; o yu o pao 善有善報惡有惡報
shang wu 尚武
she-hui-chu-i 社會主義
shen-ming chih chou 神明之胄
sheng p'ing 升平
sheng t'ien 勝天
"Sheng-wu pen-yuan" 生物本源
shih 始
shih-che sheng-ts'un 適者勝存
Shih-fu 師復
"Shih ko" 釋革
Shih-wu hsueh-t'ang 時務學堂
shou pen fen 守本分
"Shu Hou-kuan Yen shih tsui-chin cheng-chien" 述侯官嚴氏最近政見

Shun 舜
"Shuo ch'ün" 說群
Shuo wen 說文
Ssu-ma Ch'ien 司馬遷
ssu-hsiang kai-tsao 思想改造
ssu te 私德
Ssu-t'u-huo 斯土活
Su pao 蘇報
Su Tung-p'o 蘇東坡
Sun Wen (Yat-sen) 孫文(逸仙)
Sung Chiao-jen 宋教仁
"Sung wen shen" 送瘟神

ta-che hung ch'i fan hung ch'i 打着紅旗反紅旗
ta Han-tsu-chu-i 大漢族主義
ta min-tsu-chu-i 大民族主義
Ta t'ung hsueh-hsiao 大同學校
ta wo 大我
Tai-yü 黛玉
t'ai 台
t'ai (womb) 胎
t'ai chiao 胎教
t'ai-p'ing 太平
tan ch'iang p'i ma 單槍匹馬
T'an Ssu-t'ung 譚嗣同
"T'an-t'an-tzu" 醰醰子
T'ang Hsiao-wen 唐曉文
T'ang Ts'ai-ch'ang 唐才常
tao-te 道德
Tao-te fa-lü chin-hua chih li 道德法律進化之理
t'ao-t'ai 淘汰
Ti 狄
ti-ch'iu 地球
t'i-yung 體用
t'ien 天
T'ien an men 天安門
t'ien chih 天職
t'ien-hsia 天下
t'ien-hsiah ch'ün 天下群
t'ien-hsia wei kung 天下為公
t'ien hsing 天行
T'ien i pao 天義報
t'ien ming 天命
t'ien pu pien, tao i pu pien 天不變道亦不變
t'ien sui jen yuan 天遂人願
t'ien ti jen 天地人
t'ien tse 天擇
t'ien-yen-chia 天演家
t'ien-yen chih ming 天演之命
tou chih pu tou li 鬥智不鬥力
tsa-chiao 雜交
Tsai pan ch'ien-yen 再版前言
Tsai-t'ien 載恬
Ts'ai Ho-sen 蔡和森
Ts'ai Yin (Yeh-min) 蔡寅(冶民)
Ts'ai Yuan-p'ei 蔡元培

tsao fan yu li 造反有理
ts'e-yin chih hsin 惻隱之心
Tseng Kuo-fan 曾國藩
Tso chuan 左傳
tsun chin po ku 尊今薄古
"Tsun hsia p'ien" 尊俠篇
ts'ung ku suo yu 從古所有
tu shu 獨術
T'u Jen-shou 屠仁守
t'uan-chie 團結
t'uan-chie chiu-shih li-liang 團結就是力量
t'uan-t'i 團體
Tung Chung-shu 董仲舒
Tung Hsi wen-hua chi ch'i che-hsueh 東西文化及其哲學
t'ung-hua 同化
t'ung kuei t'ien-yen 同歸天演

tzu ch'iang 自强
tzu hsiu 自修
tzu hsu 自序
tzu li 自力
tzu li keng sheng 自力更生
tzu sheng 自勝
tzu ying 自營
tzu-yu 自由
Tz'u-shih 慈石

Wai chiao pao 外交報
wai huan 外患
Wan kuo kung fa 萬國公法
Wan kuo kung pao 萬國公報
Wan kuo shih chi 萬國史記
wan shih shih piao 萬世師表
wan wu chih ling 萬物之靈
Wang Chen-sheng 汪振聲
Wang Ching-wei 汪精衛
Wang Hsing-kung 王星拱
Wang K'ang-nien 王康年
Wang Mang 王莽
Wang Pi 王弼
Wang Shih 王栻
Wang Yang-ming 王陽明
wei-hsin 唯心
wei jen-min 為人民
"Wei-shem-me shuo Hsun Tzu shih fa-chia?" 为什么説荀子是法家
wei-shih 唯識
wei-wu 唯物
wo chien 我見
wo-men ti p'eng-yu pien t'ien-hsia 我們的朋友遍天下
wu 物
wu-cheng-fu-chu-i 無政府主義

wu ching 物競
Wu Chih-hui 吳稚暉
"Wu chou feng-su i-t'ung k'ao" 五洲風俗異同考
wu chü-luo 無聚落

wu chung-sheng 無衆生
"Wu chung t'an yuan" 物種探原
"Wu chung yu lai" 物種由來
wu fa wu t'ien 無法無天
Wu Han 吳晗
wu jen-lei 無人類
Wu Ju-lun 吳汝綸
wu shih-chie 無世界
Wu Sung 武松
wu-wei 無為
"Wu wu lun" 五無論
Wu Yueh 吳樾

Yang Tu 楊度
Yang Tzu 楊子
yang wei Chung yung 洋為中用
Yao 堯
yao mo kuei kuai 妖魔鬼怪
Yen Fu chuan 嚴復傳
yen-shih-p'i-li-ch'un 煙士披里純
yi 義
yi yi chung 遺宜種
Ying-huan chih lueh 瀛環志略
Yoshida Shōin 吉田松蔭
yu-che sheng-ts'un, lieh-che mieh-wang 優者生存劣者滅亡
yu sheng lieh pai 優勝劣敗
"Yü chih ssu sheng kuan" 余之死生觀
yü chin yü kuei cheng-tang 愈進愈歸正當
yü kung yi shan 愚公移山
"Yuan ch'iang" 原強
Yuan Shih-k'ai 袁世凱
"Yuan Tao" 原道
yun 運

Index

Harvard East Asian Monographs

1. Liang Fang-chung, *The Single-Whip Method of Taxation in China*
2. Harold C. Hinton, *The Grain Tribute System of China, 1845–1911*
3. Ellsworth C. Carlson, *The Kaiping Mines, 1877–1912*
4. Chao Kuo-chün, *Agrarian Policies of Mainland China: A Documentary Study, 1949–1956*
5. Edgar Snow, *Random Notes on Red China, 1936–1945*
6. Edwin George Beal, Jr., *The Origin of Likin, 1835–1864*
7. Chao Kuo-chün, *Economic Planning and Organization in Mainland China: A Documentary Study, 1949–1957*
8. John K. Fairbank, *Ch'ing Documents: An Introductory Syllabus*
9. Helen Yin and Yi-chang Yin, *Economic Statistics of Mainland China, 1949–1957*
10. Wolfgang Franke, *The Reform and Abolition of the Traditional Chinese Examination System*
11. Albert Feuerwerker and S. Cheng, *Chinese Communist Studies of Modern Chinese History*
12. C. John Stanley, *Late Ch'ing Finance: Hu Kuang-yung as an Innovator*
13. S. M. Meng, *The Tsungli Yamen: Its Organization and Functions*
14. Ssu-yü Teng, *Historiography of the Taiping Rebellion*
15. Chun-Jo Liu, *Controversies in Modern Chinese Intellectual History: An Analytic Bibliography of Periodical Articles, Mainly of the May Fourth and Post-May Fourth Era*
16. Edward J. M. Rhoads, *The Chinese Red Army, 1927–1963: An Annotated Bibliography*
17. Andrew J. Nathan, *A History of the China International Famine Relief Commission*
18. Frank H. H. King (ed.) and Prescott Clarke, *A Research Guide to China-Coast Newspapers, 1822–1911*
19. Ellis Joffe, *Party and Army: Professionalism and Political Control in the Chinese Officer Corps, 1949–1964*
20. Toshio G. Tsukahira, *Feudal Control in Tokugawa Japan: The Sankin Kōtai System*

21. Kwang-Ching Liu, ed., *American Missionaries in China: Papers from Harvard Seminars*
22. George Moseley, *A Sino-Soviet Cultural Frontier: The Ili Kazakh Autonomous Chou*
23. Carl F. Nathan, *Plague Prevention and Politics in Manchuria, 1910–1931*
24. Adrian Arthur Bennett, *John Fryer: The Introduction of Western Science and Technology into Nineteenth-Century China*
25. Donald J. Friedman, *The Road from Isolation: The Campaign of the American Committee for Non-Participation in Japanese Aggression, 1938–1941*
26. Edward Le Fevour, *Western Enterprise in Late Ch'ing China: A Selective Survey of Jardine, Matheson and Company's Operations, 1842–1895*
27. Charles Neuhauser, *Third World Politics: China and the Afro-Asian People's Solidarity Organization, 1957–1967*
28. Kungtu C. Sun, assisted by Ralph W. Huenemann, *The Economic Development of Manchuria in the First Half of the Twentieth Century*
29. Shahid Javed Burki, *A Study of Chinese Communes, 1965*
30. John Carter Vincent, *The Extraterritorial System in China: Final Phase*
31. Madeleine Chi, *China Diplomacy, 1914–1918*
32. Clifton Jackson Phillips, *Protestant America and the Pagan World: The First Half Century of the American Board of Commissioners for Foreign Missions, 1810–1860*
33. James Pusey, *Wu Han: Attacking the Present through the Past*
34. Ying-wan Cheng, *Postal Communication in China and Its Modernization, 1860–1896*
35. Tuvia Blumenthal, *Saving in Postwar Japan*
36. Peter Frost, *The Bakumatsu Currency Crisis*
37. Stephen C. Lockwood, *Augustine Heard and Company, 1858–1862*
38. Robert R. Campbell, *James Duncan Campbell: A Memoir by His Son*
39. Jerome Alan Cohen, ed., *The Dynamics of China's Foreign Relations*
40. V. V. Vishnyakova-Akimova, *Two Years in Revolutionary China, 1925–1927*, tr. Steven I. Levine
41. Meron Medzini, *French Policy in Japan during the Closing Years of the Tokugawa Regime*
42. *The Cultural Revolution in the Provinces*
43. Sidney A. Forsythe, *An American Missionary Community in China, 1895–1905*
44. Benjamin I. Schwartz, ed., *Reflections on the May Fourth Movement: A Symposium*
45. Ching Young Choe, *The Rule of the Taewŏn'gun, 1864–1873: Restoration in Yi Korea*

46. W. P. J. Hall, *A Bibliographical Guide to Japanese Research on the Chinese Economy, 1958–1970*
47. Jack J. Gerson, *Horatio Nelson Lay and Sino-British Relations, 1854–1864*
48. Paul Richard Bohr, *Famine and the Missionary: Timothy Richard as Relief Administrator and Advocate of National Reform*
49. Endymion Wilkinson, *The History of Imperial China: A Research Guide*
50. Britten Dean, *China and Great Britain: The Diplomacy of Commerical Relations, 1860–1864*
51. Ellsworth C. Carlson, *The Foochow Missionaries, 1847–1880*
52. Yeh-chien Wang, *An Estimate of the Land-Tax Collection in China, 1753 and 1908*
53. Richard M. Pfeffer, *Understanding Business Contracts in China, 1949–1963*
54. Han-sheng Chuan and Richard Kraus, *Mid-Ch'ing Rice Markets and Trade, An Essay in Price History*
55. Ranbir Vohra, *Lao She and the Chinese Revolution*
56. Liang-lin Hsiao, *China's Foreign Trade Statistics, 1864–1949*
57. Lee-hsia Hsu Ting, *Government Control of the Press in Modern China, 1900–1949*
58. Edward W. Wagner, *The Literati Purges: Political Conflict in Early Yi Korea*
59. Joungwon A. Kim, *Divided Korea: The Politics of Development, 1945–1972*
60. Noriko Kamachi, John K. Fairbank, and Chūzō Ichiko, *Japanese Studies of Modern China Since 1953: A Bibliographical Guide to Historical and Social-Science Research on the Nineteenth and Twentieth Centuries, Supplementary Volume for 1953–1969*
61. Donald A. Gibbs and Yun-chen Li, *A Bibliography of Studies and Translations of Modern Chinese Literature, 1918–1942*
62. Robert H. Silin, *Leadership and Values: The Organization of Large-Scale Taiwanese Enterprises*
63. David Pong, *A Critical Guide to the Kwangtung Provincial Archives Deposited at the Public Record Office of London*
64. Fred W. Drake, *China Charts the World: Hsu Chi-yü and His Geography of 1848*
65. William A. Brown and Urgunge Onon, translators and annotators, *History of the Mongolian People's Republic*
66. Edward L. Farmer, *Early Ming Government: The Evolution of Dual Capitals*
67. Ralph C. Croizier, *Koxinga and Chinese Nationalism: History, Myth, and the Hero*
68. William J. Tyler, tr., *The Psychological World of Natsumi Sōseki*, by Doi Takeo

69. Eric Widmer, *The Russian Ecclesiastical Mission in Peking during the Eighteenth Century*
70. Charlton M. Lewis, *Prologue to the Chinese Revolution: The Transformation of Ideas and Institutions in Hunan Province, 1891–1907*
71. Preston Torbert, *The Ch'ing Imperial Household Department: A Study of its Organization and Principal Functions, 1662–1796*
72. Paul A. Cohen and John E. Schrecker, eds., *Reform in Nineteenth-Century China*
73. Jon Sigurdson, *Rural Industrialization in China*
74. Kang Chao, *The Development of Cotton Textile Production in China*
75. Valentin Rabe, *The Home Base of American China Missions, 1880–1920*
76. Sarasin Viraphol, *Tribute and Profit: Sino-Siamese Trade, 1652–1853*
77. Ch'i-ch'ing Hsiao, *The Military Establishment of the Yuan Dynasty*
78. Meishi Tsai, *Contemporary Chinese Novels and Short Stories, 1949–1974: An Annotated Bibliography*
79. Wellington K. K. Chan, *Merchants, Mandarins, and Modern Enterprise in Late Ch'ing China*
80. Endymion Wilkinson, *Landlord and Labor in Late Imperial China: Case Studies from Shandong by Jing Su and Luo Lun*
81. Barry Keenan, *The Dewey Experiment in China: Educational Reform and Political Power in the Early Republic*
82. George A. Hayden, *Crime and Punishment in Medieval Chinese Drama: Three Judge Pao Plays*
83. Sang-Chul Suh, *Growth and Structural Changes in the Korean Economy, 1910–1940*
84. J. W. Dower, *Empire and Aftermath: Yoshida Shigeru and the Japanese Experience, 1878–1954*
85. Martin Collcutt, *Five Mountains: The Rinzai Zen Monastic Institution in Medieval Japan*

STUDIES IN THE MODERNIZATION OF THE REPUBLIC OF KOREA: 1945–1975

86. Kwang Suk Kim and Michael Roemer, *Growth and Structural Transformation*
87. Anne O. Krueger, *The Developmental Role of the Foreign Sector and Aid*
88. Edwin S. Mills and Byung-Nak Song, *Urbanization and Urban Problems*
89. Sung Hwan Ban, Pal Yong Moon, and Dwight H. Perkins, *Rural Development*

90. Noel F. McGinn, Donald R. Snodgrass, Yung Bong Kim, Shin-Bok Kim, and Quee-Young Kim, *Education and Development in Korea*

91. Leroy P. Jones and Il SaKong, *Government, Business and Entrepreneurship in Economic Development: The Korean Case*

92. Edward S. Mason, Dwight H. Perkins, Kwang Suk Kim, David C. Cole, Mahn Je Kim, et al., *The Economic and Social Modernization of the Republic of Korea*

93. Robert Repetto, Tai Hwan Kwon, Son-Ung Kim, Dae Young Kim, John E. Sloboda, and Peter J. Donaldson, *Economic Development, Population Policy, and Demographic Transition in the Republic of Korea*

106. David C. Cole and Yung Chul Park, *Financial Development in Korea, 1945–1978*

94. Parks M. Coble, *The Shanghai Capitalists and the Nationalist Government, 1927–1937*

95. Noriko Kamachi, *Reform in China: Huang Tsun-hsien and the Japanese Model*

96. Richard Wich, *Sino-Soviet Crisis Politics: A Study of Political Change and Communication*

97. Lillian M. Li, *China's Silk Trade: Traditional Industry in the Modern World, 1842–1937*

98. R. David Arkush, *Fei Xiaotong and Sociology in Revolutionary China*

99. Kenneth Alan Grossberg, *Japan's Renaissance: The Politics of the Muromachi Bakufu*

100. James Reeve Pusey, *China and Charles Darwin*

101. Hoyt Cleveland Tillman, *Utilitarian Confucianism: Ch'en Liang's Challenge to Chu Hsi*

102. Thomas A. Stanley, *Ōsugi Sakae, Anarchist in Taishō Japan: The Creativity of the Ego*

103. Jonathan K. Ocko, *Bureaucratic Reform in Provincial China: Ting Jih-ch'ang in Restoration Kiangsu, 1867–1870*

104. James Reed, *The Missionary Mind and American East Asia Policy, 1911–1915*

105. Neil L. Waters, *Japan's Local Pragmatists: The Transition from Bakumatsu to Meiji in the Kawasaki Region*